Psychoneuroendocrinology

Clarissa S. Holmes
Editor

PSYCHONEURO-
ENDOCRINOLOGY

Brain, Behavior, and Hormonal Interactions

With 25 Illustrations

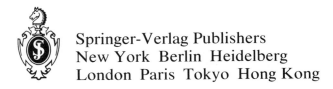
Springer-Verlag Publishers
New York Berlin Heidelberg
London Paris Tokyo Hong Kong

Clarissa S. Holmes
Departments of
Psychology and Pediatrics
Tulane University
New Orleans, Louisiana 70118
USA

Library of Congress Cataloging-in-Publication Data
Psychoneuroendocrinology: brain, behavior, and hormonal interactions
    /edited by Clarissa S. Holmes.
        p.   cm.
      ISBN 0-387-97112-2 (alk. paper)
      1. Endocrine glands—Diseases—Psychological aspects.   2. Genetic
    disorders—Endocrine aspects.   3. Psychoneuroendocrinology.
    4. Growth disorders—Psychological aspects.   I. Holmes, Clarissa S.
      [DNLM: 1. Behavior—drug effects.   2. Growth Disorders—
    psychology.   3. Growth Substances—deficiency.
    4. Neuroendocrinology.   5. Psychophysiology.   6. Sex Chromosome
    Abnormalities—psychology.   7. Sex Hormones—physiology.   8. Thyroid
    Diseases—psychology.   WL 102 P97443]
    RC649.P79   1989
    152—dc20
    DNLM/DLC   89-11544

Printed on acid-free paper.

Typeset by TCSystems, Inc., Shippensburg, Pennsylvania.
Printed and bound by Edwards Brothers, Inc., Ann Arbor, Michigan.
Printed in the United States of America.

9 8 7 6 5 4 3 2 1

ISBN 0-387-97112-2 Springer-Verlag New York Berlin Heidelberg
ISBN 3-540-97112-2 Springer-Verlag Berlin Heidelberg New York

Acknowledgments

I wish to acknowledge the helpful suggestions of Drs. Joanne Rovet and Eva Tsalikian in recruiting contributors for this volume. I am thankful for the wisdom and support of my husband Dr. William E. Cooper, for the inspiration of my daughter Ashley, and for the secretarial assistance of Kathy Funk, who provided help that was often above and beyond the call of duty. The Robert E. Flowerree Fund is gratefully acknowledged for its assistance with the manuscript preparation. Finally, I am indebted to the outstanding researchers who agreed to serve as contributors to this book, for without their hard work and prompt response to deadlines this book would not exist.

CLARISSA S. HOLMES

Contents

Contributors

CHARLES ANNECILLO, PH.D.
Johns Hopkins University, Baltimore, MD 21202, USA

GILBERT P. AUGUST, PH.D.
Endocrinology and Metabolism, Children's Hospital, Washington, DC 20010, USA

BRUCE G. BENDER, PH.D.
National Jewish Center, Denver, CO 80206, USA

DANIEL B. BERCH, PH.D.
CCDD/Psychology, Cincinnati, OH 45229, USA

SHERI A. BERENBAUM, PH.D.
Department of Psychology, University of Health Science, Chicago Medical School, North Chicago, IL 60064, USA

J. BROOKS-GUNN, PH.D.
Educational Testing Service, Princeton, NJ 08541, USA

RICHARD R. CLOPPER, PH.D.
Children's Hospital, State University of New York at Buffalo, Buffalo, NY 14222, USA

HEATHER J. DEAN, M.D.
Department of Pediatrics, University of Manitoba, Winnipeg, Manitoba R3E 0Z2, Canada

THOMAS P. FOLEY, M.D.
Division of Endocrinology, Children's Hospital of Pittsburgh, Pittsburgh, PA 15213-3417, USA

WAYNE H. GREEN, M.D.
110 Bleecker Street, New York, NY 10012, USA

CLARISSA S. HOLMES, PH.D.
Departments of Psychology and Pediatrics, Tulane University,
New Orleans, LA 70118, USA

JENNIFER A. KARLSSON, PH.D.
Department of Psychiatry, Dean Medical Center, Madison, WI 53715,
USA

CECILIA LOBATO
Johns Hopkins University and Hospital, Baltimore, MD 21205, USA

DUNCAN J. MACCRIMMON, M.D.
Department of Research, Hamilton Psychiatry Hospital, Hamilton,
Ontario L8N 1Y4, Canada

ELIZABETH MCCAULEY, PH.D.
Department of Psychiatry and Behavioral Sciences, University of Washington, Seattle, WA 98105, USA

JOHN MONEY, PH.D.
Johns Hopkins University and Hospital, Baltimore, MD 21205, USA

ROBERT L. PAIKOFF
Educational Testing Service, Princeton, NJ 98541, USA

MARY H. PUCK, M.S.
National Jewish Center, Denver, CO 80206, USA

PATRICIA A. RIESER, FNP-C
Department of Pediatrics, University of North Carolina at Chapel Hill,
Chapel Hill, NC 27599, USA

ARTHUR ROBINSON, M.D.
National Jewish Center, Denver, CO 80206, USA

JUDITH LEVINE ROSS, M.D.
Hahnemann University, Philadelphia, PA 19102, USA

JOANNE F. ROVET, PH.D.
Hospital for Sick Children, Toronto, Ontario M5G 1X8, Canada

JAMES A. SALBENBLATT, M.D.
National Jewish Center, Denver, CO 80206, USA

PATRICIA T. SIEGEL, PH.D.
Children's Hospital of Michigan, Wayne State University, Detroit, MI
48201, USA

LOUIS E. UNDERWOOD, M.D.
Department of Pediatrics, University of North Carolina at Chapel Hill,
Chapel Hill, NC 27599, USA

JEAN E. WALLACE, PH.D.
Department of Psychology, St. Joseph's Hospital, Hamilton, Ontario L8N
1Y4, Canada

DEBORAH L. YOUNG-HYMAN, PH.D.
Division of Pediatric Endocrinology, University of Maryland, Baltimore,
MD 21230, USA

1
Understanding the Role of Hormones in Brain and Behavioral Functioning

CLARISSA S. HOLMES

The interaction between *psyche* and *soma* has long interested philosophers and scientists. From the early postulation of bile *humors,* which were thought to influence mood states, to phrenology as a crude explanation of the strictures of anatomical structure on psychological expression, representative schemsa have been sought to better understand the mysteries of mind/body dualism. As is true of most science, contemporary theories and technologies shape empirical questions and methods. This shaping occurs not only in the identification and formulation of research queries, but in the techniques available to assess and quantify them. The psychological study of endocrine disorders mirrors this trend.

Within psychology, behavioral medicine and health psychology have only recently, been formalized as areas of specialization, within the last 10 to 20 years (Blanchard, 1982; Agras, 1982). Behavioral medicine, with its emphasis on disease-related psychological status, will be the focus of this discussion, rather than health psychology with its orientation toward disease prevention through health engendering life-style decisions. As psychologists have become more involved in medically related research, further subspecialization in research and clinical practice has occurred together with various medical subspecialties, such as endocrinology. In addition, concurrent study of cognitive sequelae of neurologic and other medical diseases has progressed substantially during this time interval via the field of neuropsychology.

Both behavioral medicine and neuropsychology are distinctive in their theoretical approach, compared to earlier medically related research, through adoption of more rigorous experimental design and the availability of better standardized test instruments. Early medically related psychological research was more descriptive in nature and often relied upon structured or unstructured interview as the primary means of assessing patient status. Research questions and data interpretation were often

* The author wishes to thank Dr. James Hansen for his helpful comments on an earlier draft of this paper.

steeped in the prevalent psychodynamic orientation of the day. Frequently, single or multiple case studies prevailed, with variables intuitively defined and measured, and representative of the primarily clinical emphasis in the early literature. Similarly, early studies were often lacking statistical analyses, tending to rely on readers' judgment or popular consensus as to what constituted a significant characteristic or a significant change in treatment, disposition, or functioning.

The psychological study of endocrine diseases is representative of the evolution which has occurred in philosophical orientation within psychology over the last 20 years. The early psychological and psychiatric literature focused primarily upon patients' psychic *re*actions to somatic events/diseases; increasingly, it is the *inter*action of somatic and psychologic status that is the focus of evaluation. The shift has occurred from explaining psychological responses to disease solely in terms of mediating psychic mechanisms to study of contributing or mediating physiologic mechanisms, specifically, hormonal or other biochemical influences on brain and behavior functioning. This shift in research orientation would not have been possible without the concomitant development of supportive technology. At a molecular level, advances have occurred in the description and measurement of neurotransmitters and other hormones to enable a more comprehensive understanding of endocrinology diseases themselves. Within the field of neurology and neuropsychology, advances in neuroimaging techniques, such as PET, SPECT, CT, and MRI scanning, now allow more detailed exploration of brain structure and function. As these noninvasive techniques become more widely available and less expensive, new opportunities will be present to increase our understanding of the interplay between morphologic and dynamic factors that affect the brain and behavior.

The term *psychoneuroendocrinology* reflects a growing understanding of the complexity and interdependence of psychological factors, neurologic status, and hormonal and genetic predisposition—all of which can interact and mutually influence one another. The term also reflects the greater interdisciplinary emphasis that has occurred as information in psychology/psychiatry, neurology, and endocrinology becomes increasingly complex such that practitioners and scientists must rely on colleagues in related disciplines to better understand the nuances and full manifestations of a disease. Further, the *psyche* prefix is indicative of the applied research that will be the focus of this text. Although relevant comparative and biochemical research literatures may be discussed, the purpose of the book is to review the determinants and patterns of patient functioning in association with endocrinology disease. It is well known that some emotional or psychic states affect hormonal functioning, most notably stress, however, the scope of such discussion will be limited to its relevance to the endocrine disorders under review.

The chapters in this book reflect the transition that has occurred from more intuitive study to greater empiricism. Frequently, chapters are divided into discrete sections to discuss each rather distinct era of inquiry. This approach, when used, can provide a valuable historical overview of the early clinical formulations about each disease. Even though many of the earlier research philosophies and techniques may seem so simplistic as to mitigate against their inclusion, early research hypotheses were often generated from astute observation of clinical findings and relationships. In addition to shaping later empirical questions, a review of historical antecedents provides a yardstick by which to measure the progress of more current studies, even though much is yet to be learned. As is true of any refinement of knowledge, the juxtaposition of the two approaches of study reveals that some of the early postulations about patient attributes and disease consequences have been confirmed, while other suppositions have been discarded. Although the generally subjective assessment methods used in the early studies may not have provided an optimal data base, it is interesting to note which clinical impressions were able to withstand greater empirical rigor and which were not.

The book at its inception was intended to provide a succinct introduction to psychoneuroendocrinology research for practitioners and scientists who might be relatively unfamiliar with the area. However, it quickly became apparent that the sophistication of the information could not be readily reduced without vast oversimplification and loss of substance. Therefore, although clarity to an uninitiated reader is still of paramount importance, rather than significantly dilute the content of chapters for specialists, instead an effort has been made to explain novel or difficult concepts in sufficient depth to provide a contextual framework for psychologists to decode unfamiliar endocrine concepts and vice versa. Several format decisions aided in this process. First, each endocrine disease is introduced and discussed by a leading medical researcher who has special expertise in the disease presented. Included in the medical overviews are reviews of the incidence and etiology of a disease, its major clinical features and underlying biochemical mechanisms, and finally, factors which enter into different treatment decisions, and available options. Depending on how well studied the psychological phenomenon are in association with a disease, one or more chapters follow which describe both major emotional and psychosocial features of a disorder, as well as neuropsychological and cognitive disease sequelae. Each psychological chapter was written by a leading expert who has conducted relevant research about the disease factors they describe. The psychology chapters have overview summaries at the end of each major topic section, which provide a succinct synopsis of the current state of knowledge under review. Each chapter generally concludes with a discussion of future directions for additional inquiry and study. These insights into developing research trends are particularly authoritative, given the professional stature of the authors and the intimate knowledge they have of their topics.

Growth Hormone Deficiency: Blueprint of Psychoneuroendocrinology Study

Growth failure is one of the more common pediatric endocrinology problems, following diabetes mellitus which is discussed elsewhere (Holmes, in press). Perhaps because it occurs relatively often and is so readily visible, the causes of growth failure have been studied more extensively than any other endocrine abnormality reported in this book. Growth hormone deficiency, though relatively uncommon as a cause of short stature, provides an example, in microcosm, of medically related research trends over the last 20 to 30 years, of the frequent intertwining of societal expectations and psychological adjustment, of important new information from neuropsychological studies, and most dramatically, of the impact that technological advances can exert on stimulating exciting gains in knowledge and treatment.

Culturally defined desirable physical appearance, and the psychological factors associated with significant deviation from normative expectation, both wield influence upon medical management of short stature. Medical treatment of isolated growth hormone deficiency is in itself relatively routine and consists primarily of multiple injections of replacement hormone either by pediatric patients themselves or by their parents (see Chapter 2). Few, if any, acute medical crises occur with this disorder. Further, once a reasonable height has been attained, short stature is no longer a physically handicapping condition (e.g., sufficient height to drive an automobile, to access light switches, etc.). Thus, when human growth hormone was in very limited supply during the 1970s, once a height of 5'4" to 5'7" was attained in males, who are more commonly referred for treatment, therapy with growth hormone from the National Pituitary Agency was stopped. Recent follow-up studies of adults who were recipients of growth hormone during that period indicate that many of these patients who are still short, but otherwise "normal," display evidence of psychological and social distress as adults in the form of lower rates of dating, marriage, and more controversially, relative underemployment. See Chapter 6 for a report of the demographic outcome of these patients. Part of this poor social outcome may be attributed to frequently co-occurring, but time-limited, sexual immaturity associated with some types of growth failure (see Chapter 5), or it could be a logical progression from childhood behavior patterns which may include social withdrawal and behavioral inhibition (see Chapter 4).

Most intriguingly, poor occupational outcome and perhaps poor social outcome may be related to hitherto unstudied cognitive factors (see Chapter 3). There is evidence of an increased incidence of learning disabilities associated with growth hormone deficiency, characterized by visual/motor and attentional problems (Siegel & Hopwood, 1986). Within the neuropsychological literature, visual/motor deficits have been related to social imperception, secondary to difficulty evaluating nonverbal gestures and cues (Rourke, 1985). Thus, social inhibition may result in part from relative difficulty interpreting ambiguous social situations. The negative social reinforcement that a short child experiences may worsen across time into adulthood, when social nuances become increasingly subtle and more difficult to detect. There is also preliminary evidence that growth hormone replacement in deficient adults may have a mild salutary effect on attention (Almqvist et al., 1986). Additional evidence suggests that the degree of attentional improvement correlates with ambient blood levels of replacement growth hormone (Smith et al., 1985).

While psychoneuroendocrinology studies of growth hormone deficiency have benefited from greater methodological and statistical rigor, better standardized and normed test instruments, and advances in theory development to aid in data interpretation, one glaring weakness remains in available studies. The relatively low incidence, in absolute terms, of growth hormone deficiency means that no one university center has access to a statistically adequate number of patients at any one time, who are within a developmentally homogeneous age group. Now that initial survey studies have identified relevant areas of research inquiry, larger patient groups must be studied to achieve a deeper understanding of the interactions of age, gender, and other factors which may relate to psychological status. For example, one might expect that psychological adjustment for a 5-year-old female who looks 3 years old could be quite different from an 18-year-old male who looks 12 or 13 years old. Although this supposition is well grounded in the developmental literature, it has been very difficult to test empirically because of few subjects and subsequent limited statistical power to examine second-order interactions of relevant demographic variables. Thus, the latest trend in the study of short stature, and of many of the other endocrine disorders reported in this book, is toward multicenter collaborative studies.

The exciting development of recombinant DNA technology, and the subsequent production of unlimited quantities of biosynthetic growth hormone, have made it possible to investigate aspects of growth hormone deficiency that were limited in the past by the scarcity of human growth hormone, which was laboriously harvested from the pituitaries of cadavers. The abundance of biosynthetic growth hormone has had a direct quantitative and qualitative impact upon research conducted during the last 5 years. First and foremost, it has removed the necessity of exploring growth hormone effects in only severely deficient patients, who have almost a total absence of growth hormone. When supplies were limited, only these most deficient patients could be justified as recipients of the hormone either for research or clinical applications. In the past, it was thought that the amount of growth hormone produced by an individual was either adequate or deficient. It is now apparent, though, that it is more accurately conceptualized as ranging from severely deficient to more subtle degrees of relative growth hormone insufficiency. The increased supply of growth hormone has made it possible to treat other groups of short children, such as girls with Turner syndrome and those with normal variant short stature who previously did not qualify for its use because of stringent treatment criteria borne out of the hormone shortage. The beneficial response of growth hormone recipients in these groups has contributed greatly to an understanding of the role that growth hormone plays in mediating growth and the limitations that exist in predicting who will respond to therapy.

Interdisciplinary Study, Hormonal Interactions, and the Psychosocial Environment

The role of interdisciplinary research is further demonstrated in a variety of endocrine diseases reviewed in this book. Its impact is real, as evidenced by intellectual studies conducted with patients having congenital hypothyroidism. In the past, this disorder was not detected until several months after birth when physical signs of hypothyroidism first occurred. However, by that time, the developing brain had been deprived and irreparably harmed, by the lack of thyroid hormone necessary for normal neuronal growth and differentiation. Based upon these early findings of gross mental retardation, neonatal screening programs have been almost universally instituted in Western countries. Further, intellectual outcomes associated with various thyroid-replacement regimens were a determining factor in the decision to increase thyroxine-replacement dosages to minimize intellectual damage. See Chapter 17. Recent neuropsychological studies of congenital hypothyroidism reveal that the etiology of thyroid dysfunction relates to different profiles of cognitive strengths and weaknesses. Chapter 17 provides further discussion of the types of congenital hypothyroidism and their neuropsychological correlates.

Since individuals do not operate in a vacuum but within a social context, the social milieu is another factor that can rather uniquely alter endocrine functioning. The clearest and perhaps most fascinating case of this alchemy is the disorder of psychosocial dwarfism. As detailed in Chapters 7 and 8, rather conclusive evidence is available that an impaired interaction between child and caregiver can adversely affect the secretion of neurotransmitters and growth hormone, which may result in growth failure and generally reversible intellectual impairments. Dramatic linear growth and gains in intelligence test scores (up to 30 IQ points) have been documented in as little as 3 to 6 months after a child has been removed from an emotionally aversive environment. However, another important variable appears to intervene in predicting intellectual outcome; specifically, there may be critical developmental stages or time intervals which, if passed in a psychically detrimental environment, may diminish the intellectual rebound that typically occurs in older children. See Chapter 8 for a discussion of the neuropsychological/intellectual sequelae of aversive environments and rebound effects upon removal from such an environment. Substantial controversy has centered around the precise etiology of impaired growth in this disorder, which often is its most noticeable feature. Because of the atypical eating behaviors that frequently accompany psychosocial dwarfism poor or withheld nutrition was initially thought by many to be the primary agent of arrested growth rather than an impaired caregiver relationship. See Chapter 7 for a review of theoretical aspects of this often perplexing and misunderstood condition, in addition to a comprehensive review of evidence from the relevant comparative literature implicating the impaired caregiver relationship as the primary etiologic agent.

The impact of the social environment and of prevailing social attitudes upon disease management is demonstrated in another, more traditional, fashion. Hyperthyroidism is a condition that may mimic the classical symptoms of "histrionic" behavior, including palpitations of the heart, sweating, and feelings of faintness and disorientation. See the medical overview for a complete description of the physiologic symptoms (Chapter 16). However, because the disorder predominantly affects women, and often older women, for many years it was thought to have a purely psychogenic cause. See Chapter 18 for a review of the early psychosocial history of hyperthyroidism and for a discussion of newer investigations into the behavioral correlates and neuropsychological sequelae, their severity, and the time course of these impairments.

The role of socialization is probably no more controversial than when examining the influence of sex hormones upon behavior and cognition. Puberty, in particular, has a multitude of popular folklore that accompanies this key time of transition in the life span. The medical mechanisms and hormonal changes associated with precocious, normal, and delayed puberty are reviewed in the medical overview (Chapter 12); cognitive correlates are described in Chapter 13 and behavioral factors are evaluated in Chapter 14. As the chapters reveal, many problems are present in adequately staging levels of pubertal development. This difficulty can cloud research when a child exhibits inconsistencies on different medical parameters. Unlike earlier work in the 1960s and before, research in this area has significantly advanced through the measurement and quantification of hormonal levels, with attempts to correlate them with brain and behavioral functioning. Further discussion of delayed puberty also can be found in the medical overview of growth disorders (Chapter 2), since the most prevalent cause of short stature in males is also accompanied by delayed puberty. Psychological correlates of delayed puberty are described in Chapters 3, 4, and 5.

Physiologic features of atypical sexual development are considered in the medical overviews pertaining to disorders of the sex chromosomes

(Chapter 9) and of the sex hormones (Chapter 12). Specifically, two of the better-known disorders of the sex chromosomes, that of Turner syndrome (XO karyotype) and Klinefelter syndrome (XXY karyotype), and their variants are described in some detail. In the neuropsychological study of gender differences in brain development and functioning, these two disorders play a key role in differentiating the relative influence of the X and Y chromosomes. Distinct neuropsychological profiles are associated with each, and in their more severe expression, these disorders are associated with specific learning disabilities (see Chapter 10). The controversial XYY or "super male" karyotype is reviewed, particularly its prevalence rate among males in penal populations and its behavioral expression in this and other, less skewed, samples. In addition, behavioral characteristics, and in some cases their neuropsychological underpinnings, of children and adults with Turner syndrome and Klinefelter syndrome are described in Chapter 11.

Finally, the effect of atypical sex hormone exposure in utero is discussed in Chapter 15; specifically, the presence of excess androgen upon developing fetuses. While this male sex hormone has little apparent effect upon males, it has, not surprisingly, a detectable impact upon female fetuses. Since social issues surrounding sex roles, gender identity, and especially, sexual orientation, are particularly sensitive to prevailing mores and theories, it is not unexpected that little systematic research in this area has been done. Such work is similarly hampered by less-developed empirical assessment techniques in studying sexual behaviors, with much of the early work done in the 1950s by the Kinsey Institute. Nevertheless, working within these limits, Chapter 15 describes new work that incorporates refined play observation techniques of affected young girls' behavior, among other innovations. Included is an insightful interpretation of "male sexual behavior" in that greater rates of homosexuality may not necessarily be found in these viralized females, but androgen effects may be expressed more subtlely in the form of increased sexual desire or energy. Also, neuropsychological profiles are described and reviewed in detail, since congenital adrenal hyperplasia provides unique information about the relative contribution of sex hormones in influencing traditionally accepted male and female differences in spatial and verbal abilities, respectively.

In Perspective

It is the goal of this book to compile into a single source a review of the current state of knowledge of both cognitive and psychosocial features of major endocrine disorders, exclusive of diabetes. When possible, psychological functioning is correlated with hormonal levels or degree of physical stigmata, and if known, mechanisms of action are discussed. Previously this information has been limited to several paragraphs, or an occasional chapter, within an endocrinology textbook, or it has been scattered throughout periodicals in the medical, psychological, neuropsychological, psychiatric, and behavioral medicine literature. It is hoped that this compilation will serve a need in aiding empirically based clinical practice and in stimulating additional research query. There are still many questions to be answered.

References

Agras, W.S. (1982). Behavioral medicine in the 1980's: Nonrandom connections. *Journal of Consulting and Clinical Psychology, 50,* 797–803.

Almqvist, O., Thoren, M., Saaf, M., & Eriksson, O. (1986). Effects of growth hormone substitution on mental performance in adults with growth hormone deficiency: A pilot study. *Psychoneuroendocrinology, 11,* 347–352.

Blanchard, E.B. (1982). Behavioral medicine: Past, present, and future. *Journal of Consulting and Clinical Psychology, 50,* 795–796.

Holmes, C.S. (Ed.). (in press). *Neuropsychological and behavioral aspects of diabetes.* New York: Springer-Verlag, Inc.

Rourke, B.P (1985) (Ed.), *Neuropsychology of learning disabilities: essentials of subtype analysis.* New York: Guilford Press.

Siegel, P.T., & Hopwood, N.J. (1986). The relationship of academic achievement and the intellectual functioning and affective conditions of hypopituitary children. In B. Stabler & L.E. Underwood (Eds.), *Slow grows the child: Psychosocial aspects of growth delay* (pp. 57–71). Hillsdale, New Jersey: Lawrence Erlbaum Associates, Inc.

Smith, M.O., Shaywitz, S.E., Shaywitz, B.A., Gertner, J.M., Raskin, L.A., & Gelwan, E.M. (1985). Exogenous growth hormone levels predict attentional performance: A preliminary report. *Journal of Developmental and Behavioral Pediatrics, 6,* 273–278.

2
Disorders of Growth and Short Stature: Medical Overview

Patricia A. Rieser, FNP-C. and
Louis E. Underwood

Linear growth, which reflects a child's general health and well-being, is influenced by a host of organic and psychological factors. Assessment of growth at regular intervals, therefore, is one of the best methods available for detecting disease. The purposes of this chapter are to describe the process of normal growth, to outline the elements of a growth evaluation, and to provide an overview of some of the causes of growth retardation.

Normal Growth and Growth Charts

Linear growth is rapid during the first year of life, with infants growing 18 to 25 cm by their first birthday. Growth slows to a rate of 10 to 13 cm during the second year, then continues at a steady rate of 5 to 6 cm/year until puberty. The pubertal growth spurt begins at 10 to 11 years of age in girls and 12 to 13 years in boys, lasts about 2 years, and accompanies the development of secondary sexual characteristics. Peak growth velocity during puberty is 6 to 11 cm/year in girls and 7 to 13 cm/year in boys. Linear growth ceases when sexual development is almost complete and the epiphyses of the bones fuse. The average age of epiphyseal fusion is 16 years in girls and 18 years in boys.

The cornerstone of evaluation of linear growth is a growth chart on which the child's correct height is plotted at regular intervals. A growth chart permits comparison of the child's height with the heights of many normal children of the same sex and age and facilitates assessment of the child's growth over time. Whatever the rate of growth in utero, most children enter their genetically predetermined growth channel by 2 years of age and maintain growth in this channel until puberty. Sustained deviation from the established growth channel is cause for concern regardless of the child's height and should prompt a referral to a pediatrician or pediatric endocrinologist for further evaluation.

Although not every child whose height is more than 2 standard deviations (SD) below the mean for sex and age (or less than third centile) has a growth problem, this degree of short stature usually is cause for concern. For example, a child with short parents who is growing at a normal rate slightly below the third centile line on the growth chart may not require any medical action beyond continued observation. On the other hand, a thorough evaluation should be initiated in any child whose growth rate is subnormal (regardless of height), whose height is more than 3 SD below the mean for age, or whose height is shorter than expected given parental heights (more than one SD below midparent height).

Evaluation of Growth Disorders

Accurate measurement of height and calculation of interval growth velocity are the first steps in the evaluation of a child suspected of having a growth disorder. Ongoing growth failure requires persistence in the search for an etiology until the cause is discovered. Review of previous growth records often is helpful in determining the child's growth pattern and the time of onset of the problem. A detailed history (including family history) and careful physical examination always are obtained. The history should focus on the child's gestation and birth, general health, past occurrence of severe illness, subtle symptoms of chronic illness, dietary intake, and emotional well-being. Screening studies to rule out many of the possible causes of growth failure are obtained if the child's growth rate is slow or if the child's height is more than 2 SD below the mean. These studies typically include analysis of urine and blood samples (to look for evidence of systemic disease) and x-rays of the hand and wrist (to assess skeletal maturity and growth potential) and of the skull (to examine the sella turcica, the bony pouch surrounding the pituitary gland). A karyotype is obtained in girls with short stature to ascertain if an abnormality of the X chromosome exists (Turner syndrome).

A full assessment of pituitary function is appropriate if the child is growing at a subnormal rate and other causes of growth failure have been excluded by the initial evaluation. Pituitary testing, usually performed and interpreted by a pediatric endocrinologist, may include tests of thyroid, adrenal, gonadotropin, and growth hormone secretion. These may be performed in an outpatient setting or during a brief hospitalization.

Differential Diagnosis of Growth Disorders

Many diseases and conditions may result in growth failure and short stature; these are summarized in Table 2.1. Although the list of possible causes of growth failure is long and encompasses a wide variety of disorders, the task of diagnosis is simplified by noting several general principles: (1) With the exception of some girls with Turner syndrome, patients

TABLE 2.1. Overview of growth disorders.

  I. Intrinsic defects of growing tissues
     A. Skeletal dysplasias—abnormal skeletal proportions; achondroplasia,
        hypochondroplasia, chondrodystrophies
     B. Autosomal abnormalities—characteristic physical stigmata, often mental
        retardation; Down's syndrome
     C. Abnormalities of the X chromosome—absence of or deletion from one of the sex
        chromosomes; Turner syndrome
     D. Dysmorphic or primordial dwarfism—intrauterine growth retardation, physical
        abnormalities resulting from genetic defects or in utero environmental insults;
        Prader-Willi, Russell-Silver, Noonan's syndromes
 II. Abnormalities in the environment of growing tissues
     A. Nutritional insufficiency—malnutrition, starvation
     B. Gastrointestinal disease—malabsorption, chronic inflammatory bowel disease,
        celiac disease
     C. Renal disease—chronic renal insufficiency, renal tubular acidosis, renal
        osteodystrophy
     D. Cardiac disease—congenital heart defects
     E. Diabetes mellitus in poor control
     F. Vitamin D-resistant rickets and other metabolic disorders
III. Endocrine abnormalities
     A. Thyroid hormone deficiency—congenital or acquired hypothyroidism
     B. Glucocorticoid excess—adrenal tumors, Cushing's disease, pharmacological
        therapy (steroid overdosage)
     C. Growth hormone deficiency—classic hypopituitarism, constitutional growth
        delay, psychosocial dwarfism, syndromes of growth hormone resistance

with intrinsic defects of the growing tissues usually can be identified by careful physical exam, because these patients exhibit a variety of somatic abnormalities. (2) Patients with abnormalities in the environment of growing tissues almost always are underweight for height and appear thin and undernourished. (3) Patients with endocrine abnormalities, particularly those patients with growth hormone deficiency or glucocorticoid excess, are overweight for height.

The remainder of this discussion will focus on the three growth disorders that form the basis for the chapters that follow—growth hormone deficiency, constitutional growth delay, and psychosocial dwarfism.

Growth Hormone Deficiency

Growth hormone (GH) is a peptide secreted into the bloodstream in a pulsatile manner by the pituitary gland in response to stimulation by hypothalamic growth hormone releasing hormone (GHRH). GH stimulates cellular proliferation, increases synthesis of new protein in most cells, lowers the rate of carbohydrate use, and increases the use of stored fat for energy. It plays a major role in postnatal linear growth, working in harmony with several other hormones.

Growth hormone deficiency (GHD; hypopituitarism) causes slow growth (often 2 to 4 cm/year) and severe, proportional short stature. Severe GHD occurs in about 1 in 4,000 children, and partial GH deficiency probably occurs with greater frequency. Children with GHD have normal body proportions and excess body fat, particularly on the trunk. The face may appear small and immature; this, in combination with general chubbiness, may create a cherubic appearance. Boys with congenital hypopituitarism may exhibit micropenis. Prolonged neonatal hyperbilirubinemia, especially if accompanied by hypoglycemia, suggests congenital hypopituitarism. GHD may be associated with defects in midline development of the face (such as cleft lip or palate) and forebrain (such as septo-optic dysplasia, a condition involving optic nerve hypoplasia, blindness, and absent septum pellucidum).

The most common form of GHD (idiopathic hypopituitarism) often occurs in children who have suffered prenatal or perinatal insults. Idiopathic GHD is thought to result from damage to the hypothalamus that impairs GHRH secretion. Acquired GHD may result from trauma, infection, tumors in the pituitary region (craniopharyngioma is the most common), histiocytosis, or therapeutic cranial irradiation.

The diagnosis of GHD is based on documentation of a subnormal growth rate and biochemical evidence of inadequate GH secretion. The latter is demonstrated by subnormal serum GH responses to stimulation with pharmacological agents that cause GH release in normal children. Most physicians agree that a peak GH response of less than 10 ng/dl after stimulation with at least two pharmacologic agents suggests impaired GH secretion. Measurement of GH in blood samples obtained every 20 minutes for 6 to 24 hours provides an indication of the amount of GH the child secretes under physiological conditions. The plasma concentration of somatomedin-C (SM-C)/IGF-I, a substance produced in response to GH, usually is low, which reflects little tissue exposure to GH. Unfortunately, the tests for GHD are not foolproof, and diagnosis is made difficult by the absence of a clear discrimination between subnormal and adequate GH secretion. Children with GHD may have deficiencies of other pituitary hormones as well.

Until 1985, GHD was treated with GH extracted from cadaver pituitaries. Because the supply of GH was limited, only severely affected children were treated, and treatment often was discontinued after the child reached a height in the low normal range. Biosynthetic GH, produced by using recombinant-DNA technology, became available in 1985, and although it is expensive ($5,000 to $40,000 per year), supply is unlimited. GH usually has been given by intramuscular injection three times a week, but recent studies show that subcutaneous injections are equally effective and better tolerated by children. There is mounting evidence that daily injections stimulate faster growth than the same weekly dose given three times a week.

GH-deficient children often exhibit a vigorous growth response when GH treatment is initiated, growing as much as 8 to 12 cm during the first year. This supranormal growth rate declines somewhat over time, although increasing the dose or frequency of administration may be effective in ameliorating the diminishing response. Children who are diagnosed promptly and respond well to GH treatment should reach normal adult stature. However, children with prolonged growth failure, extreme short stature, or advanced bone age at the time of diagnosis may never "catch up" completely and may remain short as adults.

The side effects of GH in the treatment of GHD are few. Some children develop antibodies to the GH preparation, but these rarely interfere with the action of the drug. Hypothyroidism may occur during GH treatment; the mechanism of this phenomenon is unknown. Several former recipients of pituitary GH have died from a rare, degenerative neurological disease called Creutzfeldt-Jakob disease, presumably from pituitary GH contaminated with the infectious agent responsible for the disease. Leukemia has been reported in several GH-treated patients, but it is not known whether GH causes the slight increase in risk of contracting this disease.

Constitutional Growth Delay

Constitutional growth delay (CGD) is the term used to describe children who are shorter than expected given their midparent height, who grow at a normal rate throughout most of childhood, who have a delayed bone age (usually 1 to 4 years behind chronological age), and who are late entering puberty. This condition is more common in boys than in girls; reliable estimates of prevalence do not exist, but these children account for a large number of referrals to growth clinics. Children with CGD often have a period of slow growth during infancy or early childhood, resulting in a height below the third centile. Normal growth resumes and height parallels the third centile line throughout childhood. Puberty is delayed by 2 or more years, but it is accompanied by a normal pubertal growth spurt. Adult height usually is consistent with genetic heritage. See Chapter 12 for additional discussion of delayed puberty.

The etiology of CGD is not known, but there is often a family history of delayed growth and puberty in male relatives. The diagnosis of CGD should be made only after other causes of short stature and delayed adolescence have been ruled out. GH responses to provocative stimuli may be normal or marginal, and some investigators believe that physiological secretion of GH is impaired in many children with CGD who respond normally to standard provocative tests. SM-C/IGF-I may be in the low normal range.

When treatment of CGD is advisable for social or psychological reasons, androgens are prescribed to older boys to stimulate sexual development and initiate a growth spurt. The only disadvantage of this treatment is that androgens may cause rapid advancement of bone age, so that the net result of treatment may be a slight decrease in adult height. Although the use of GH in patients with CGD has not been studied extensively, 50% to 70% of boys treated with GH respond with a significant increase in growth rate. It is not known whether this response is sustained over time, whether compensatory deceleration in growth rate occurs after treatment is discontinued, or whether adult height is increased. The short- and long-term psychological effects of this condition and its treatment are not well described.

The Spectrum of GH-Related Disorders

The question of whether short children who do not have unequivocal GHD might respond favorably to GH treatment was not addressed prior to the advent of biosynthetic GH because of the limited supply of pituitary GH. It is clear that a spectrum of disorders related to GH secretion exists, with complete GHD on one end and normal GH secretion on the other. For example, some children with CGD have almost normal GH secretion, while others have severe impairment of physiological GH secretion, presumably of hypothalamic etiology. The term "neurosecretory dysfunction" is sometimes used to describe the latter population. Children who have received cranial irradiation also may exhibit impaired GH secretion under physiological conditions.

A basic problem confronting pediatric endocrinologists is how to identify children who will benefit from GH therapy. Some physicians advocate a 6-month trial of GH as the ultimate test of GH status, assuming that children who need it will grow faster than those who do not. The financial and emotional investment required for a trial of GH often makes this approach unsatisfactory. GH therapy is not innocuous; there are physical and psychological risks that may outweigh potential benefits.

Several factors should be considered before and during a trial of GH in a child who is not clearly GH deficient. It is important to explain both what is known and what is not known about GH therapy; the use of GH in children who do not have clear-cut GH deficiency must be considered experimental. Treatment, if initiated, should be accompanied by regular, systematic assessment of physical and psychological responses, which is best accomplished in a controlled research trial.

Pyschosocial Dwarfism

In 1967, Powell et al. described a group of children experiencing severe psychosocial stress and growth retardation similar to that observed in hypopituitary children. These researchers hypothesized that the environmental stress produced GHD and growth failure. Most investigators agree that malnutrition does not play a major role in classic psychosocial dwarfism (PSD), and some believe that a spectrum exists, with maternal deprivation (lack of parenting, poor feeding practices, inadequate caloric intake) on one end and classic PSD (suppressed pituitary function) on the other.

In addition to growth retardation and delayed skeletal maturation, children with PSD exhibit bizarre behavior related to acquisition of food. Polydipsia, polyphagia, ingestion of contaminated or discarded food, and gorging and vomiting are reported. The child with PSD may be withdrawn, irritable, apathetic, and accident prone. Self-injury and pain agnosia may be observed. Parental pathology varies, as does the occurrence of physical abuse.

Endocrinological findings at presentation vary, although most cases exhibit subnormal GH responses to provocative stimuli immediately after hospitalization. SM-C/IGF-I is depressed in most cases. Secretion of adrenocorticotropic hormone (ACTH) also is impaired in many of these children. The abnormalities in hormone secretion and responses to stimulation revert to normal after the child's removal from the hostile environment, whereas return to the original environment most often results in rapid deceleration of growth and return of pathological behavior.

References

Green, W.H., Campbell, M., & David, R. (1984). Psychosocial dwarfism: A critical review of the evidence. *Journal of the American Academy of Child Psychiatry, 23,* 39–48.

Powell, G.F., Brasel, J.A., & Blizzard, R.M. (1967). Emotional deprivation and growth retardation simulating idiopathic hypopituitarism. I. Clinical evaluation of the syndrome. II. Endocrinologic evaluation of the syndrome. *New England Journal of Medicine, 276,* 1271–1283.

Prader, A. (1975). Delayed adolescence. *Clinics in Endocrinology and Metabolism, 4,* 143–155.

Underwood, L.E. (1984). Report on the conference on uses and abuses of biosynthetic growth hormone. *New England Journal of Medicine, 311,* 606–608.

Underwood, L.E., & Van Wyk, J.J. (1985). Normal and aberrant growth. In D.W. Foster & J.D. Wilson (Eds.), *Williams textbook of endocrinology* (pp. 155–205) (7th ed.). Philadelphia: W.B. Saunders Co.

3
Intellectual and Academic Functioning in Children with Growth Delay

Patricia T. Siegel

The relationship between intelligence and academic achievement of children with significant growth delay and subsequent short stature has been investigated in the United States and abroad for almost three decades. Investigators have generally agreed that a substantial minority of growth-delayed children have problems learning in school although theories explaining academic underachievement are conflictive. Specifically, the high incidence of academic failure has been explained in the literature by three conflicting theories: the cognitive underfunctioning theory, the low ability theory and the cognitive deficit theory (Siegel, 1982).

Prior to 1980, four studies report findings supportive of the cognitive underfunctioning theory. According to this theory, there is a discrepancy between average ability and academic achievement. Poor school performance is considered secondary to environmental and psychosocial factors including poor parenting and low self-esteem resulting from the impact of short stature (Pollitt & Money, 1964; Drash & Money, 1968; Kusalic et al., 1972; Steinhausen & Stahnke, 1976). During the same period, three studies provide empirical support for the low ability theory which views low achievement as commensurate with an intellectual compromise secondary to endocrine pathology (Frankel & Laron, 1968; Obuchowski, 1970; Spector et al., 1978). Investigations that assessed underachievement in terms of the cognitive deficit theory, which postulates that learning problems may be due to specific cognitive, attentional, and visual-spatial deficits, were not conducted until after 1980 (Abbott et al., 1982; Siegel & Hopwood, 1986; Bedway, 1988; Ryan et al., 1988). Reports of a high incidence of birth trauma and developmental delays reported among short-statured children (Craft et al., 1980; Kusalic & Fortin, 1975; Meyer-Bahlburg et al., 1978) prompted this new area of research because the correlation between birth/neonatal trauma, developmental delays, and learning problems is well recognized (Firestone & Prabhu, 1983; Hallahan et al., 1985; Kinsbourne, 1973).

Historical Overview

The lack of agreement in early clinical investigations about the relationship between cognitive and academic functioning among short-statured children can be traced to specific methodological flaws familiar to early research efforts involving newly described clinical populations. Specifically, early studies were limited by small sample sizes with wide age ranges (Pollitt & Money, 1964; Kusalic et al., 1972; Drash & Money, 1968), variable diagnostic criteria used in Europe compared to those in the United States, (Obuchowski et al., 1970; Frankel & Laron, 1968), subjective psychological measurements (Dorner, 1973; Kusalic & Fortin, 1975), and absence of control groups (Meyer-Bahlburg et al., 1978; Pollitt & Money, 1964; Rotnem et al., 1977). The most serious design flaws, however, were diagnostic heterogeneity and failure to account for socioeconomic status (SES) of the samples studied.

There are many causes of short stature including genetic, environmental, and endocrine factors, metabolic status, and home environment (Stabler, 1987). Therefore, it is reasonable to assume there may be cognitive and behavioral distinctions between children with short stature secondary to endocrine disease (i.e., growth hormone deficiency) from those whose short stature can not be attributed to endocrine pathology (i.e., constitutional delay).

Growth hormone deficiency is an endocrinological disease caused by a deficiency of growth hormone secretion secondary to either hypothalamic dysfunction or pituitary gland malfunction. Children with this disorder have delayed growth due to isolated growth hormone deficiency if growth hormone is the only hormone not secreted by the pituitary, or to multiple hormone deficiencies when more than one of the pituitary hormones are deficient. When all of the pituitary hormones are deficient the condition is known as panhypopituitarism. The majority of cases of growth hormone deficiency are unexplained or idiopathic, while approximately 25% are organic and due to craniopharyngioma, a hypothalamic tumor (Frasier, 1987).

Short stature that is secondary to constitutional delay has no known pathologic cause for growth delay. Children with constitutional delay can be characterized as having normal birth weight, growth failure between 6 months and three years, subsequent normal growth velocity, height below the fifth percentile throughout childhood, delayed onset of adolescence but potential for attaining normal adult height (Richman et al., 1986).

It is interesting to speculate that if any cognitive, psychosocial, and academic differences actually exist between children with growth hormone deficiency and constitutional delay groups, these differences may have been obscured by the collective analyses of early research efforts.

Thus, while both groups were reported to have average intelligence (Money et al., 1967; Money et al., 1967; Money, 1968; Drash & Money, 1968; Drash et al., 1968), the diagnostic heterogeneity of the groups studied preclude definitive conclusions about intelligence for either subtype. In fact, IQ differences have been noted in the few investigations that did account for diagnostic classification. Specifically, short-statured children with multiple pituitary hormone deficiencies and craniopharyngioma have been reported to have lower intellectual functioning compared to either isolated growth hormone deficient or constitutionally delayed children (Pollitt & Money, 1964; Frankel & Laron, 1968; Kusalic & Fortin, 1975; Meyer-Bahlburg et al., 1978). In addition, specific cognitive deficits in the visual-spatial domain have been reported in at least two early studies among children with endocrine pathology (Frankel & Laron, 1968) and constitutional delay (Steinhausen & Stahnke, 1976).

Failure to report the socioeconomic status (SES) of sampled children is another serious limitation of early studies (Pollitt & Money, 1964; Money et al., 1967; Kusalic et al., 1972). Intelligence data reported without SES information may be spurious because of the well-recognized correlation between the two (Rees & Palmer, 1970; Tyler, 1965; Weinberg et al., 1974). Findings from the few studies that did control for SES were equivocal; Drash and Money (1968), Spector et al. (1979), and Frankel and Laron (1968) noted that the level of intellectual functioning in their samples was lower than the elevated SES levels should have predicted. Conversely, Meyer-Bahlburg et al. (1978) found that the intelligence levels of their sample of growth hormone deficient children were consistent with that expected by their measured SES.

In summary, the first two decades of research involving children with significant short stature had a myriad of methodological flaws that precluded definitive conclusions regarding their intellectual functioning. Important observations were reported, however, which served as initial building blocks for continued research. Thus, while it was generally accepted that intellectual retardation did not accompany short stature, the importance of taking a more careful look at the cognitive and behavioral profiles of clearly identified diagnostic subtypes was established. It was also to be determined if growth-delayed youngsters had low ability when SES was controlled and whether the high incidence of academic underachievement was secondary to low ability, specific cognitive deficits, or varying diagnostic classifications. Finally, the role of psychosocial factors among these children required further clarification. Specifically, the question remained whether self-esteem was a primary causative factor in underachievement or whether low self-esteem was secondarily reactive to poor school performance that resulted from either below-average potential or specific cognitive deficits. Indeed, the relationship between a child's school performance and self-esteem is not likely unilateral.

Review of Current Research

Studies conducted in the 1980s (Table 3.1) are characterized by several methodological improvements including utilization of standarized psychometric measures with proven reliability and validity, more careful delineation of diagnostic classifications as well as SES information. However, the majority of these studies failed to look for specific learning disabilities or cognitive atypicalities in their samples of short-statured children and continued to use intelligence and achievement measures as screening instruments for study inclusion.

The following literature review will include contemporary studies investigating underachievement and its possible cognitive correlates among growth-delayed children with either idiopathic growth hormone deficiency and those with constitutional delay. Studies that focused on the psychosocial correlates of underachievement between these two diagnostic classifications are reviewed elsewhere in this volume and, therefore, only briefly addressed in this review.

Cognitive and Academic Achievement Functioning

A high incidence of academic underachievement and grade retention ranging from 17% to 50% is substantiated in recent studies of children with growth hormone deficiency (GHD) and constitutional delay (CD).

Holmes, Hayford, and Thompson (1982), in a study investigating school and behavioral adjustment among 56 short-statured children including 18 with GHD, 25 with CD, and 13 with Turner's syndrome (TS), reported a 25% grade-retention rate due to immaturity and small size. Teachers and parents had different views regarding behavior problems as measured by the Problem Behavior Checklist and the Achenbach Social Competence Scale. Teacher ratings discriminated among children's behavior by age and sex; adolescent girls were rated as the most immature and emotionally inhibited. In contrast, parents discriminated according to diagnostic classification; all children, except young CDs (<10 years) were rated as having significant school adjustment problems. In a follow-up study including 47 of the original sample, Holmes, Thompson, and Hayford (1984) investigated variables, including academic achievement, which might account for the high incidence of grade retention among the three subtypes of short-statured children. There was a high rate of retention among the group as a whole although comparisons between the etiology specific subtypes revealed that more GHD (31%) failed than either the CD (19%) or TS (20%) groups. Achievement data obtained from composite scores of the Iowa Test of Basic Skills revealed that retained children continued to achieve approximately six months below current grade placement despite prior retention. Comparisons between retained children and those not retained revealed important cognitive and behavioral differences; intelligence (pro-rated from four WISC-R subtests) and achievement standard scores were significantly lower for the retained group. Similarly, retained children had more behavioral and emotional problems at school compared to the short-statured children never retained. The authors speculated that lower ability may account for both the academic underachievement and the high incidence of behavioral problems but cautioned that no definitive, causative relationship could be concluded regarding retention, underachievement, and behavioral problems because the information used in their study was obtained after the retention had occurred. Despite this limitation, these data provide preliminary evidence that important distinctions may exist, in terms of intellectual and psychosocial functioning, not only between short-statured children of different diagnostic etiologies, but also between those who fail in school and those who do not. Unfortunately, the investigators failed to administer the entire WISC-R, precluding comprehensive examination of the cognitive profiles between the retained and not-retained groups.

Three studies characterized by the use of homogeneous groups of idiopathic GHD children document a high rate of retention among this diagnostic-specific group (Abbott et al., 1982; Lewis et al., 1986; Ryan et al., 1988).

Abbott et al. (1982) assessed the cognitive and emoitional functioning of 11 documented GHD children before and after one year of growth hormone replacement therapy. Intelligence as measured by the WISC-R fell in the low average range at pre and post testings. The authors noted that because more subjects in their sample were from low SES, the mean IQs would be expected to fall somewhat below average. Significant visual-spatial deficits were noted on both the Bender Gestalt and Beery Test of visual-motor integration and described as significantly below IQ potential. Mean achievement scores, obtained from the Wide Range Achievement Test (WRAT), fell in the low average range and, therefore, commensurate with IQ. Yet, 45% of the GHD children had failed a grade despite their low-average performance on the standardized test. The authors explained that illness, rather than achievement or psychosocial factors, accounted for retention in a majority of the children according to parent report. Therefore, it is plausible that retention was more a reflection of school policy rather than underachievement. Another possible interpretation is that individual achievement problems were obscured by the five children the authors described as earning ''above average'' scores on the achievement test, thereby inflating the group mean. Emotional functioning for the GHD group was described as significantly immature, with many of the children experiencing difficulty coping with age-appropriate developmental tasks. Peer and family relationships, in contrast, were reported as positive. The authors concluded that achievement did not appear directly related to the condition of GHD but as compatible with intellectual potential. Unfortunately, they failed to compare the cognitive profiles between

TABLE 3.1. Summary of studies completed in the 1980s.

Authors-year	N / Age range	Diagnostic classification IGHD*	MHD**	CD***	Other	Controls	Rate of retention	Psychometric measurements	Behavioral	Problems identified / Cognitive Intelligence	Visual spatial	Achievement
Holmes et al., 1982	56 / 6-12	16	2	25	13 Turner Syndrome (TS)	—	25%	WISC-R[1]: 2 Subtests Child Behavior Checklist Achenbach Social Competence Scale	Social-emotional inhibition; poor school adjustment	Pro-rated to screen for average ability	—	—
Holmes et al., 1984	47 / 6-16	14	2	21	10 (TS)	—	23%	WISC-R: 4 Subtests Iowa Test of Basic Skills Problem Behavior Checklist	Retained children had lower IQ, achievement scores, and social-emotional problems than children not retained.			
Abbott et al., 1982	11 / 4-18	4	4	—	3 (Organic)	—	45%	WISC-R Bender Gestalt Test of Visual-Motor Integration Draw-A-Person Visual Motor Integration Test	Emotional immaturity; Poor coping skills	Low Average	Below Ave Bender and VMI	
Gordon et al., 1982 Gordon et al., 1984	47 / 6-12	—	—	24		23 →	→ 21% vs → 4%	WISC-R; Bender; Peabody Individualized Achievement Test; Bender Gestalt Test; Piers-Harris Self-concept Test	Low self-esteem; Somatic complaints; Social withdrawal; emotional immaturity; Poor coping skills	No differences between CD's and controls in IQ (Average), achievement (Average) or visual-spatial skills (Average) scores.		

Study	n / age	Groups	Tests	Psychosocial findings	Cognitive findings
Lewis et al., 1986	$\frac{85}{6\text{-}20}$	45: IGHD & MHD are not reported ——→ 43%; 40 ——→ 0	Child Behavior Checklist; Peabody Picture Vocabulary Test (PPVT); Piers-Harris Self-concept Test; Child Behavior Checklist; Children's Social Desirability Scale	Low self-esteem; Anxious; Less socially competent	PPVT for GHD group significantly lower than controls.
Ryan et al., 1988	36	12: IGHD & MHD not reported ——→ 50%; 12 ——→ 0; Diabetic 12 ——→ 0	WISC-R: 4 Subtests; WRAT-R²: Spelling, Reading, Math; WMS³: Visual Reproductions; Symbol-Digit Learning Test; Money Road Map Test; Trail Making Test; Grooved Pegboard Test; Verbal Fluence Test; Piers-Harris Self-concept	No difference in self-esteem	The GHD group earned significantly lower scores on 8 of the 16 cognitive/neuropsychological tests administered.

[1]WISC-R: Wechsler Intelligence for Children—Revised
[2]WRAT-R: Wide Range Achievement Test—Revised
[3]WMS: Wechsler Memory Scale
*IGHD: Isolated Growth Hormone Deficiency
**MHD: Multiple Hormone Deficiency
***CD: Constitutional Delay

those who failed and those who did not in terms of intellectual level and visual-spatial deficits. Although the findings of this study must be interpreted with caution because the sample was small and perhaps not representative of the GHD population in terms of SES and degree of illness (several children had significant medical problems in addition to GHD) and lack of a normal control group, two important findings are noted. First, significant visual-spatial deficits as measured by two highly respected instruments were described in an etiology-specific group of short stature. Second, emotional immaturity was described for the first time as commensurate with IQ, or secondarily reactive to either illness or poor school achievement rather than as a specific, linear consequence of short stature or ineffective parenting.

Lewis et al. (1986) examined the status of 45 GHD children compared to 40 normal-statured children matched for age, sex, and SES within the classroom via peer ratings (Positive/Negative Peer Nomination Inventory, Pupil Evaluation Inventory), ratings of behavioral adjustment from teachers and both parents (Child Behavior Checklist) and measures of the children's self-concepts (Piers-Harris, Children's Social Desirability Scale). Learning potential was assessed by the Peabody Picture Vocabulary Test (PPVT), sometimes used as a screening for verbal ability. The investigators found that the GHD group earned significantly lower mean scores on the PPVT than the control group and that 43% of the GHD children had been retained at least one time, reportedly due to either academic problems or emotional immaturity. Teachers described the GHD children as earning lower grades, being less motivated, learning less, and being less happy than the controls but did not rate the GHD children as experiencing more behavior problems. Mothers and fathers rated their GHD children as less socially competent on the Child Behavior Checklist, while fathers only rated their older GHD children (>12 years) as more internalizing (withdrawn and anxious) than the younger GHD children. The GHD children perceived themselves as having lower self-esteem and more severe behavioral and interpersonal problems than the control children. The researchers suggest that reported low self-esteem may be a function of social skill deficits which make it difficult for some GHD children to maneuver within their peer group. It is equally reasonable to speculate, however, that low self-esteem may be secondarily related to chronic achievement problems which in turn may have developed from reported lower verbal potential. Failure to comprehensively assess the intelligence of the sample, however, precludes any definitive statement about the relationship between achievement and cognitive and psychosocial factors.

In a recent study investigating the neuropsychological profiles of 12 school-aged GHD children, Ryan et al. (1988) found that a full 50% had failed a grade compared to no grade failure in either the diabetic or normal control groups matched for age, sex, and SES. The groups were compared on four WISC-R subtests, all segments of the WRAT-R, the Visual Re-productions subtest from the Wechsler Memory Scale, the Symbol-Digit Learning Test, the Money Road Map Test, the Trail Making Test, the Grooved Pegboard Test, the Verbal Fluency Test, and the Piers-Harris Self-Concept Scale. The GHD group differed from the control groups on 8 of the 16 psychological tests with the greatest between-group differences on the three measures of academic achievement. Although the authors stated that the cognitive profile observed among the GHD group was not consistent with diffuse brain dysfunction, they did note significant deficits in several specialized cognitive domains including those requiring complex visuoconstructional skills, orientation in space, long-term memory, and attention span. Despite the differences observed across several cognitive processes, there were no between-group differences on the self-concept scale. Thus, while the investigators avoided drawing definitive conclusions regarding the relationship between neuropsychological deficits and underachievement, they did rule out low self-esteem as a contributing factor.

Only one group of investigators compared a group of constitutionally delayed (CD) children and normal controls (Gordon et al., 1982; Gordon et al., 1984; Richman et al., 1986). In a series of studies designed to assess personality, intellectual, visual-motor, and academic performance, this group of researchers administered an extensive battery of psychological tests to 24 children with CD and compared them to 23 normal controls matched for age, sex, and SES. No significant differences were found in mean IQ, achievement, or visual-spatial skills as measured by the WISC-R, Peabody Individualized Achievement Test, and Bender Gestalt Test, respectively. However, 21% of the CDs had failed a grade compared to 4% of the matched controls. Teacher and parent ratings were similar to those reported by the Holmes et al. studies (1982, 1984); short-statured children were rated by their parents as being more socially withdrawn and as having more somatic complaints than the normal controls, while teachers did not differentiate among the groups and did not rate the CD children as having significant behavioral problems. Parents, in turn, were not reported as overprotective, but rather as having less expectations for their short-statured children. However, the children with CD rated themselves as having low self-esteem compared to controls on the Piers-Harris Self-Concept Scale. The authors concluded that while CD children do not appear to exhibit severe psychopathology, a general behavioral profile of social withdrawal and aloofness emerges. Unfortunately, the investigators failed to directly address the high incidence of grade retention despite average ability and commensurate performance on the standardized achievement test. They imply, however, that difficulty in treating CD children in an age-appropriate manner inhibits them from developing a sense of competence and this lack of confidence, in turn, interferes with overall functioning.

Summary of Current Research

A high incidence of academic underachievement and rate of grade retention has been soundly substantiated in growth hormone deficient children and children with constitutional delay in several recent investigations. These studies were also found to provide additional evidence in support of each of the three previously described theories explaining underachievement. Specifically, low ability, specific learning disabilities, and low self-esteem were reported as possible contributing factors in children with short stature secondary to growth hormone deficiency, while psychosocial factors including low self-esteem and reduced parental expectations were found to be related to underachievement in children with constitutional delay. Thus, these data tend to support the notion that cognitive and behavioral distinctions exist between at least two etiology-specific types of short stature.

The primary limitation of these recent investigations was generally that of omission rather than commission. For example, except for the Gordon et al. studies (1982, 1984, 1986) intelligence was either not fully assessed or if tested was not comprehensively examined in terms of verbal-performance differences, factor score differences, or any specific strengths or weaknesses. Only two studies (Abbott et al., 1982; Gordon et al., 1982) measured visual-spatial skills and only the former reported performance relative to potential. Finally, none of the studies used a systematic method of identifying underachievers relative to potential or compared achievers and underachievers on cognitive measures.

The Michigan Studies

Overview

Beginning in the early eighties, several researchers from Children's Hospital of Michigan in Detroit and C.S. Mott Children's Hospital in Ann Arbor (Siegel et al., 1988) have collaborated to investigate the cognitive, behavioral, and academic functioning of a group of school-aged children who met the criteria for short stature secondary to idiopathic growth hormone deficiency and were being treated with growth hormone replacement therapy at one of the two hospitals. Specific research interests developed when clinical experience suggested that the cognitive and psychosocial profiles of GHD children known to the investigators were markedly discrepant from descriptions reported in the literature. Specifically, the high incidence of birth trauma and developmental delays coupled with the psychometric profiles of GHD children assessed in out-patient clinics due to learning and/or behavioral problems suggested that underachievement may be secondary to either significant learning disabilities or subaverage

potential. In addition, they presented as quiet, low-energy children, but
not socially withdrawn or markedly immature as typically described. Further, their mothers, while appropriately concerned, did not present as
overprotective. In fact, they were observed to provide their children with
effective rebuttals to teasing by peers and to clarify their child's age to
individuals who failed to treat the child in an age-appropriate manner.
These clinical observations prompted a series of longitudinal investgations
that challeneged several basic assumptions regarding these children including average ability, low self-esteem, and pejorative parenting. Specific
initial questions addressed included, "What does the cognitive profile of
idiopathic growth hormone deficient children look like in terms of significant WISC-R Verbal-Performance Scale differences, and specific visual-
spatial, attention, and auditory processing deficits?" and "Do GHD children have low self-esteem and mothers who are overprotective?" Finally,
we were interested in determining whether cognitive and behavioral profiles were stable over time and examining the fluid versus static components of cognitive and academic skills.

The first study (Siegel, 1982) was descriptive in nature with the specific
aim of comprehensively assessing the cognitive and affective profiles of a
group of short-statured children known at the two children's hospitals.
The subjects were 42 idiopathic GHD children, 27 with isolated growth
hormone deficiency (IGHD), and 14 with multiple hormone deficiencies
(MHD), between the ages of 6 and 16 from primarily middle-class families
as measured by the Hollingshead Two Factor Social Scale Index (Table 3.2).

Intelligence, measured by the WISC-R, fell at the low end of the average
range (M = 93.4, SD = 16.7) for the group as a whole, with a significantly
high incidence (38%) of verbal-performance differences (=/>15 points) as
well as a significantly higher mean verbal-performance difference score
(13) compared to the normative standardized sample. Additional clarification of cognitive strengths and weaknesses was noted when WISC-R

TABLE 3.2. Demographics for the 42 idiopathic growth
hormone deficient children Hollingshead two-factor index for
social position.

	Low	Low Ave	Ave	High Ave	High
Number of children	4	14	13	9	2
Age measures	Range		X	SD	
Chronological age	65–166		122	24	
Bone age	35–123		87	25	
Diagnosed	6 mos–130		66	29	
Treatment initiated	25–130		70	24	
Duration of treatment	0–86		35	33	
School grade	1–10		5	15	

factor scores were examined; on the triad of subtests (arithmetic, digit span, and coding) known as the freedom from distractibility factor (FFD), the GHD group earned a significantly lower mean score than on either the verbal comprehensive factor or the perceptual organization factor. Poor performance on the FFD factor suggests difficulties in attention, concentration, sequencing, and short-term memory (Kaufman, 1979). Also, 38% of the GHD group earned scores below the sixteenth percentile on the Bender Gestalt Visual Motor Test despite their overall average IQ scores, suggesting that a significant number of the group had specific visual-spatial deficits. Significantly lower math achievement was found in the GHD sample compared to the standardization sample; reading achievement was not significantly lower. In contrast, this group of GHD children did not rate themselves as having lower self-esteem on the Piers-Harris Self-Concept Scale compared to children in the normative sample (Table 3.3). Also, their mothers did not endorse items reflective of being more overprotective than mothers in the standardization sample on the Parent Attitude Research Instrument.

When the 27 IGHD and 14 MHD children were compared, clinically important and significant differences were also noted. First, the MHD group was found to be more physically vulnerable. More MHD children experienced developmental delays in walking (36%) and talking (42%), and also had a higher incidence of clinical seizures (36%) compared to the IGHD group (18%). This is not an unexpected finding and may reflect the increased probability of seizure activity in children with MHD secondary to hypoglycemia (Hopwood et al., 1975). Second, important differences were also found when cognitive profiles were compared. The MHD group showed almost no subtest scatter or variability with all WISC-R scaled and factor scores falling below 90. The IGHD group, in contrast, had cognitive profiles characterized by overall average intelligence but significant variability, that is, a higher incidence of verbal-performance differences and a freedom from distractibility factor score significantly lower than the verbal comprehension factor. The IGHD and MHD groups were not different in terms of self-concept scores, although mothers of MHD

TABLE 3.3. Comparisons of growth hormone deficient and normative sample means for cognitive, affective, and achievement measures.

| Measure | GHD Sample | | | Normative | | | |
	Range	X̄	SD	X̄	SD	T	P
Verbal IQ	52–127	93.9	18.3	100	15	−2.14	<.05
Performance IQ	49–123	94	16.3	100	15	−2.37	<.05
V-P difference	0–42	13	9.3	9.7	7.6	2.27	<.05
Incidence of V-P							
Differences ≥15		38%		25%			
General self-concept	43–77	60.2	11.4	51.8	13.8	4.59	<.05
Reading	52–131	96.1	16.9	100	15	− .34	NS
Math	55–126	84.7	13.9	100	15	−7.17	<.001

TABLE 3.4. Comparisons of cognitive and affective measures for the IGHD and MHD children.

	Isolated (N = 27)	Multiple (N = 15)	P
WISC-R Measure			
Verbal IQ	100	86	.06
Performance IQ	96	87	NS
Verbal comprehension factor	9.5	7.5	.07
Perceptual organization factor	10	9.0	NS
Freedom from distractibility factor	8.4	7.0	NS
V-P differences	13	7	<.05
Incidence of V-P differences ≥15 points	48%	15%	<.05
Piers-Harris			
General self-concept	62	64	NS
Parent attitude			
Overpossessiveness	1.6	1.9	0.6

children were more overprotective than mothers of IGHD children, perhaps reflecting the physiologic vulnerability noted in the former (Table 3.4).

In summary, results of this initial investigation of GHD children support overall low average ability with significant deficits in visual-spatial, sequencing, attentional, and short-term memory skills for the GHD group as a whole. Important distinctions between IGHD and MHD children were also found that substantiate earlier reports that MHD children have lower ability than IGHD children. The most important finding of the study, however, was the preliminary evidence suggesting that some IGHD children may have cognitive profiles similar to those described as learning disabled.

Michigan Three-Year Follow-Up Study

The purpose of the three-year follow-up study was to determine the extent of achievement problems in the original sample of 42 GHD children, to examine the cognitive and psychosocial profiles of GHD children with and without achievement problems, and also to determine how achievement problems could be explained relative to the low ability, cognitive deficit, or cognitive underfunctioning theories previously described (Siegel & Hopwood, 1986).

Significant achievement problems, defined as standard scores more than one standard deviation below average on either the reading or math segments of the Wide Range Achievement Test (WRAT) or the reading comprehension segment of the Peabody Individualized Achievement Test (PIAT) were found in 22 (52%) of the GHD children. More than one third had failed at least one grade in school and significantly more underachievers had been retained than those without learning problems.

There were no demographic differences between underachievers and achievers in terms of sex, SES, diagnostic classification of IGHD/MHD, chronologic or bone age, age diagnosed, number of months treated, or milestone attainment. However, underachievers had significantly different scores from achievers on all cognitive measures including WISC-R IQ scores, verbal-performance difference scores, and Bender percentiles, while mean self-concept and maternal attitude scores were not significantly different (Table 3.5).

Underachievement was examined relative to the three theories said to explain underachievement:

1. Cognitive deficit theory—at least one WISC-R scale score falls within the average range (90-110); a V-P difference >15 points and/or a significant visual-motor integration deficit (Bender %ile <16th).
2. Low ability theory—both WISC-R scale scores fall below the average range (<90).
3. Cognitive underfunctioning—low self-concept theory—both WISC-R scale scores fall within the average range; there is neither a significant V-P difference nor a significant visual-motor deficit.

The cognitive profiles of 9 (40%) underachievers (5 IGHD and 4 MHD) were best explained by the low ability theory: four had WISC-R Full Scale IQs <90; two <80; and three <70. The cognitive deficit theory best explained the profiles of 8 (36%) underachievers (7 IGHD and 1 MHD). All children in this group had at least average intelligence (>90) but also had at least one identifiable cognitive deficit (i.e., WISC-R verbal-performance differences of at least 15 points or Bender Gestalt scores below the sixteenth percentile). Thus, 76% of the GHD children with significant underachievement had at least one cognitive atypicality (low ability or a specific cognitive deficit). The cognitive underfunctioning theory explained the cognitive profiles of 5 (24%) underachievers (4 IGHD, 1 MHD). All of these children had at least average ability and no apparent cognitive deficits. Interestingly, they did not have lower self-esteem scores and their mothers were not more protective than mothers of the other undera-

TABLE 3.5. Mean comparisons of achievers and underachievers for cognitive and affective measures.

	Achievers (N = 20)		Underachievers (N = 22)		
	Mean	SD	Mean	SD	P
Verbal IQ	102.6	11.6	86.7	19.9	<.01
Performance IQ	100.3	13.1	89.2	17.2	<.05
V-P differences	9.6	7.2	15.4	10.4	<.05
Bender percentiles	45.3	29.0	22.9	26.9	<.01
General self-concept	60.5	11.7	59.3	10.9	.73
Maternal overpossessiveness	5.8	1.9	5.1	1.7	.21

chievers. The factors precluding achievement among these 6 were not apparent, but may have included motivational, stress, anxiety, or other factors not measured in this study.

In summary, results of the second Michigan study suggest that cognitive factors play an important role in the school performance of a significant majority of GHD children with underachievement while learning problems among some children remain poorly understood.

Michigan Seven-Year Follow-Up Study

The third Michigan study was a seven-year longitudinal follow-up of 28 of the 42 GHD children in the original study; 15 had IGHD and 12 had MHD (Siegel et al., 1988). The purpose of this study was to determine if cognitive, behavioral, and academic profiles persisted in the GHD group and to assess behavioral as well as cognitive indicators of attention deficits. Particular interest developed in the possible presence of attention deficits because similarities between children with GHD and attention deficit hyperactivity disorder (ADHD) were noted including a high incidence of birth/neonatal trauma and delays in milestone attainment, specific cognitive deficits in terms of verbal-performance differences and low freedom from distratibility factor scores, and a behavioral description of immaturity. Because immaturity is a well-known associated feature of children with ADHD (Barkley, 1981) it was of interest to determine if another behavioral characteristic of ADHD (hyperactivity) could be empirically demonstrated in GHD children. Therefore, in addition to the tests administered in the original study, two other behavioral measures often used to assess inattention/hyperactivity (The Child Behavior Checklist and the Conners Parent Rating Scale) were added to the test battery.

Results suggest cognitive stability for the 27 idiopathic GHD children. Overall intelligence continued to fall within the average range and the triad of subtests comprising the freedom from distractibility factor remained significantly lower than either the verbal comprehension or perceptual organization factors. Bender performance at posttesting fell in the average range and commensurate with potential, Figure 3.1. The difference in Bender performance may indicate a genuine improvement in visual-spatial skills or could also mean that prior findings reflected a maturational delay, rather than a deficit, in the development of visual-spatial skills.

Cognitive profiles between the IGHD and MHD groups also persisted. The MHD group continued to have lower intelligence scores on all WISC-R scaled and factor scores and commensurately lower achievement scores than those with IGHD (Figure 3.2). Self-concept improved for the total GHD group in terms of popularity and physical attraction but there were no differences between the diagnostic classifications (Table 3.6). Improvement in self-esteem was interpreted to reflect general satisfaction with the increase in linear stature from growth hormone replacement therapy.

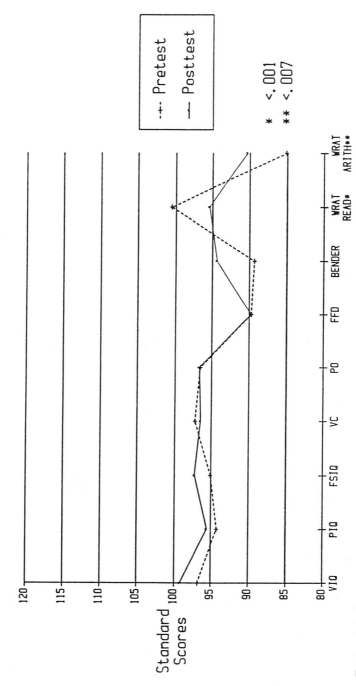

FIGURE 3.1. Pretest (1980–1981) versus posttest (1987–1988) cognitive assessment of patients with GHD (N = 27). Cognitive scales: Wechsler IQs, factor scores, Bender scores, and WRAT scores.

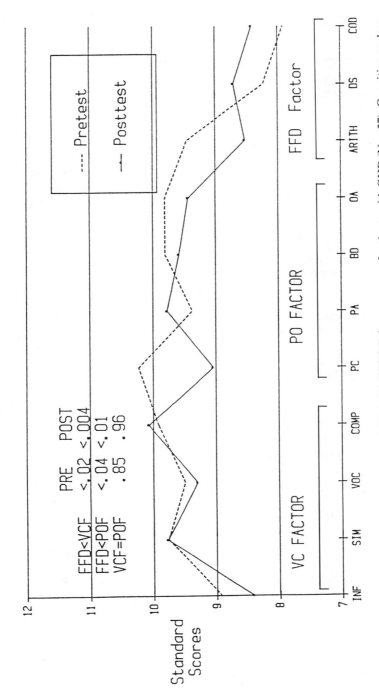

FIGURE 3.2. Pretest (1980–1981) versus posttest (1987–1988) WISC-R factor scores of patients with GHD (N = 27). Cognitive scales: Wechsler subscale scores.

TABLE 3.6. Comparisons of cognitive and self-concept measures of 15 IGHD and 12 MHD children at seven-year follow-up.

	IGHD		MHD		
	X̄	SD	X̄	SD	P
WISC-R					
Verbal IQ	105	10.5	92	14.2	.01*
Performance IQ	99	15.5	90	10.6	.11
Full Scale IQ	103	12.4	91	12.6	.02*
Verbal comprehension factor	104	11.0	87	14.5	.004*
Perceptual organization factor	100	15.9	94	11.6	.07
Freedom from distractibility factor	95	11.9	83	9.6	.01*
Bender Gestalt	98	15.9	90	20.6	.25
WRAT-R					
Reading sight word recognition	102	15.3	88	10.2	.01*
Arithmetic	98	12.8	82	15.8	.007*
Piers-Harris					
General self-concept	64.3	7.6	60.9	11.0	.39

Academic achievement problems also persisted over time with over a third underachieving at pretesting and posttesting. Underachievement was defined for each child individually and was derived in terms of each child's potential. Specifically, a child was identified as an "underachiever" if he/she had at least one standard score on one of the WRAT-R subtests that was at least one standard deviation (15 points) lower than his/her verbal comprehension factor score. This procedure operationalizes under-achievement relative to a child's learning potential and is known as a "significant discrepancy quotient." Using this procedure, 74% of those underachieving at pretest continued to underachieve at posttest. The most striking difference between underachievers and achievers continued to be in their cognitive profiles: underachievers had significantly lower WISC-R FFD scores compared to either verbal comprehension or perceptual orga-nization factor scores (Table 3.7). Those without achievement problems, in contrast, had consistently average scores across all WISC-R factors. In addition, more underachievers than achievers had clinically elevated scores on the CBCL hyperactivity and immaturity scales, providing pre-lininary evidence of behavioral characteristics associated with attentional problems.

Summary of Michigan Studies

Findings of the Michigan stuides are in general agreement with prior reports that idiopathic GHD children in general have low average ability, that those with MHD are more vulnerable to intellectual compromise than

TABLE 3.7. Comparisons of achievers and underachievers at seven-year follow-up.

	Achievers	Nonachievers	
	N = 17 (63%)	N = 10 (37%)	
	Mean	Mean	P
AGE	18.3	16.7	NS
SES	36.5	45.	NS
IGHD	9	6	NS
MHD	8	4	
WISC-R:			
VCF	93.3	103.4	.06
POF	93.4	102.2	NS
FFD	90.4	88.6	NS
FFD/VCF	90.4/92.4 NS	88.6/103.4	.001*
FFD POF	90.4/93.4 NS	88.6/102.2	.003*
BENDER	94.9	93.4	NS
PIERS HARRIS:			
ATTRACTIVENESS	8.1	9.7	NS
POPULARITY	8.8	10.6	NS
TOTAL			
SELF-CONCEPT	62.0	65.2	NS
CBCL: (=/>65T)			
HYPERACTIVITY	6 (35%)	5 (50%)	
IMMATURITY	5 (31%)	4 (40%)	

those with IGHD, and that a significant minority experience academic underachievement. Self-esteem was not found to be affected and mothers were not overprotective. Cognitive factors were found to play a prominent role in academic achievement. Finally, these studies provide preliminary evidence that some GHD underachievers have cognitive and behavioral profiles that are similar to those of children with specific learning disabilities and associated attention deficits.

We are currently conducting a comparative study of 30 recently diagnosed idiopathic GHD children, 30 of their unaffected siblings, and 30 normal controls matched for age, sex, and SES to further investigate the cognitive and behavioral profiles of GHD children (Siegel et al., 1988). Preliminary findings (Bedway, 1988) suggest that the cognitive and behavioral profiles of the GHD children differ from their siblings in several important ways. First, the mean freedom from distractibility factor (FFD) score was ten points lower than the FFD score for the siblings, a finding that is both clinically and statistically significant. Second, the GHD group demonstrated low average visual-motor skills on the Developmental Test of Visual Motor Integration compared to average skills among the sibling group, a finding that was highly significant. Third, parents rated their GHD significantly higher than their unaffected children on the hyperactivity scale of the Child Behavior Checklist. Finally, more GHD children compared to their siblings were achieving in school at a level significantly discrepant from their intellectual potential. These findings appear to be consistent with our earlier reports suggesting that cognitive and attentional deficits play an important role in the academic underachievement of GHD children.

Directions for the Future

The high incidence of academic achievement problems among growth hormone deficient and constitutionally delayed children has been well established by several studies completed during this decade. A clearer picture of growth hormone deficient children in terms of their cognitive and behavioral functioning is beginning to emerge perhaps as a consequence of recent research efforts that have altered the focus of investigations from primarily psychosocial factors to cognitive factors. The cognitive profiles of constitutionally delayed children, however, remain poorly understood because they have not as yet been comprehensively examined. The greatest caveat would be to dogmatically adhere to the assumption that it is primarily the negative impact of short stature that causes achievement problems in this group and thereby fail to examine other possibilities including cognitive and behavioral factors or family interactional patterns.

The most efficient method of further delineating the cognitive and behavioral profiles of children with short stature would be best accomplished by a national or regional collaborative study that included several groups of etiology-specific subtypes of children with short stature and compare them to matched normal controls. A study of this magnitude would enable investigators to examine possible confounding variables including age and gender which to date have only infrequently been considered because of the constraints of small sample sizes. Also, agreement could be reached regarding tests used allowing groups to be compared on the same measures. Perhaps more importantly, rules operationalizing critical variables could be clearly defined and utilized. For example, it may be beneficial for researchers to consistently use one method when identifying academic underachievement, that is, significant discrepancy quotient (SDQ) as described previously. In summary, there are rich opportunities for future research efforts in several areas involving children with significant growth delay. The potential for a significant increase in the use of growth hormone replacement therapy because of recently developed synthetic products promises an array of research challenges relative to treatment effects. The increased availability of growth hormone also increases the likelihood that groups other than growth hormone deficient children will be treated. Indeed, clinical studies suggest that children with constitutional delay and Turner's syndrome may benefit from growth hormone replacement therapy (Rosenfeld, 1987). The use of stimulant medication to attenuate significant attentional deficits, should they be confirmed, is yet another area to be explored. Finally, the literature to date is glaringly void of studies investigating family interactional patterns in terms of the differences that cohesiveness, adaptability, and triangulation may have on the overall functioning of these physically and psychologically vulnerable children.

References

Abbott, D., Rotnem, D., Genel, M., & Cohen, D.J. (1982). Cognitive and emotional functioning in hypopituitary short-statured children. *Schizophrenia Bulletin, 8*(2), 310–319.

Bahlburgh-Meyer, H., Feinman, J., MacGillivary, M., & Acedo, T. (1978). Growth hormone deficiency, brain development, and intelligence. *American Journal of Diseases in Childhood, 132,* 565–572.

Barkley, R.A. (1981). *Hyperactive children: A handbook for diagnosis and treatment.* New York: Guilford Press.

Bedway, M. (1988). *A comparative assessment of behavioral and cognitive variables among growth hormone deficient children and their siblings.* Unpublished doctoral dissertation, University of Pittsburgh, Pittsburgh, Pennsylvania.

Craft, H.W., Underwood, L.E., & VanWyk, J.J. (1980). High incidence of perinatal insult in children with idiopathic hypopituitarism. *Journal of Pediatrics, 96*(3), 397–402.

Dorner, S. (1973). Short, taught, and vulnerable. *Special Education, 62,* 12–16.

Drash, P.W., Greenberg, N., & Money, J. (1968). Intelligence and personality in four syndromes of dwarfism. In D.B. Cheek (Ed.), *Human growth: Body composition, cell growth, energy, and intelligence.* Philadelphia: Lea & Baberger.

Drash, P.W., & Money, J. (1968). Statural and intellectual growth in congenital heart disease, in growth hormone deficiency, and in sibling controls. In D.B. Cheek (Ed.), *Human growth: Body composition, cell growth, energy, and intelligence.* Philadelphia: Lea & Baberger.

Firestone, P., & Prabhu, A.N. (1983). Minor physical anomalies and obstetrical complications: Their relationship to hyperactive, psychoneurotic, and normal children and their families. *Journal of Abnormal Child Psychology, 11*(2), 207–216.

Frankel, J., & Laron, Z. (1968). Psychological aspects of pituitary insufficiency in children and adolescents with special reference to growth hormone. *Israel Journal of Medical Science, 4,* 953–961.

Frasier, S.D. (1987). Growth hormone deficiency. In R. Hintz & R. Rosenfeld (Eds.), *Growth abnormalities.* New York: Churchill Livingstone.

Gordon, M., Crouthamel, D., Post, E.M., & Richman, R.A. (1982). Psychosocial aspects of constitutional short stature: Social competence, behavior problems, self-esteem and family functioning. *Journal of Pediatrics, 101,* 477–480.

Gordon, M., Post, E.M., Crouthamel, C., & Richman, R.A. (1984). Do children with constitutional delay really have more learning problems? *Journal of Learning Disabilities, 17,* 291–293.

Hallahan, D.P., Kauffman, J.M., & Lloyd, J.W. (1985). *Introduction to learning disabilities.* Englewood Cliffs, NJ: Prentice-Hall.

Holmes, C.S., Hayford, J.T., & Thompson, R.G. (1982). Parents and teachers' differing views of short children's behavior. *Child: Care, Health, and Development, 8,* 327–336.

Holmes, C.S., Thompson, R.G., & Hayford, J.T. (1984). Factors related to grade retention in children with short statures. *Child: Care, Health, and Development, 10,* 199–210.

Hopwood, N.J., Forsman, P.J., Kenny, F.M., & Drash, A. (1975). Hypoglycemia in hypopituitary children. *American Journal of Diseases in Childhood, 129,* 918–926.

Kaufman, A.S. (1979). *Intelligence testing with the WISC-R.* New York: Wiley.

Kinsbourne, M. (1973). Minimal brain dysfunction as a neurodevelopmental lag. *Annals of the New York Academy of Sciences, 205,* 268–273.

Kusalic, M., & Fortin, C. (1975). Somatopsychic relationships: Growth hormone treatment in hypopituitary dwarfs. *Canadian Psychiatric Association Journal, 20,* 325–331.

Kusalic, M., Fortin, C., & Gauthier, Y. (1972). Psychodynamic aspects of dwarfism—response to growth hormone treatment. *Canadian Psychiatric Association Journal, 17,* 29–34.

Lewis, M.S., Johnson, S.B., Silverstein, J., & Knuth, P.A. (1986). *The adjustment of growth hormone deficient children: Parent, teacher, peer, and self-perceptions.* Paper presented at the annual meeting of the American Psychological Association, Washington, D.C.

Meyer-Bahlburg, H.F.L., Feinman, J.A., MacGillivary, M.H., & Aceto, T. (1978). Growth hormone deficiency, brain development, and intelligence. *American Journal of Diseases in Childhood, 132,* 565–572.

Money, J. (1968). Psychological aspects of endocrine and genetic disease in children. In L.I. Gardner (Ed.), *Endocrine and genetic diseases of childhood.* Philadelphia: W.B. Saunders Co.

Money, J., Cohen, S., Lewis, V., & Drash, D. (1967). Human figure drawings as index of body image in dwarfism. In D.B. Cheek (Ed.), *Human growth: Body composition, cell growth, energy and intelligence.* Philadelphia: Lea & Baberger.

Money, J., Drash, P.W., & Lewis, V. (1967). Dwarfism and hypopituitarism: Statural retardation without mental retardation. *American Journal of Mental Deficiencies, 72,* 122–126.

Obuchowski, K., Zienkiewicz, H., & Graczykowska-Koczorowska, A. (1970). Psychological studies in pituitary dwarfism. *Polish Medical Journal, 9,* 1229–1235.

Pollitt, E., & Money, J. (1964). Studies in the psychology of dwarfism, I. Intelligence quotient and school achievement. *Journal of Pediatrics, 64,* 415–421.

Rees, A.H., & Palmer, F.H. (1970). Factors related to changes in mental test performance. *Developmental Psychology Monographs, 3*(2, part 2), 1–57.

Richman, R., Gordon, M., Tegtmeyer, P., Crouthamel, C., & Post, E.M. (1986). Academic and emotional difficulties associated with constitutional short stature. In B. Stabler & L. Underwood (Eds.), *Slow grows the child.* Hillsdale, NJ: Lawrence Erlbaum Associates.

Rosenfeld, R.G. (1987). Treatment of non-growth hormone deficient short stature. In R. Hintz & R. Rosenfeld (Eds.), Growth abnormalities. New York: Churchill Livingstone.

Rotnem, D., Ginel, M., Hintz, R., & Cohen, D.L. (1977). Personality development in children with growth hormone deficiency. *Journal of Child Psychiatry, 16,* 411–425.

Ryan, C., Johnston, S., Lee, P., & Foley, T. (1988). *Cognitive deficits in short-statured adolescents with growth hormone deficiency.* Poster Presentation at the American Psychological Association Annual Convention, August 12–16, Atlanta, Georgia.

Siegel, P., Koepke, T., Bedway, M., Postellon, D., Hopwood, N., & Clark, K. (1988). *Growth hormone deficient children: A seven year follow-up of intellectual, behavioral and academic functioning.* Poster Presentation at the American Psychological Association Annual Convention, August 12–16, Atlanta, Georgia.

Siegel, P.T. (1982). *Academic achievement, intellectual functioning, and affective conditions of hypopituitary children.* Unpublished doctoral dissertation, University of Michigan, Ann Arbor.

Siegel, P.T., & Hopwood, N.J. (1986). The relationship of academic achievement and the intellectual functioning and affective conditions of hypopituitary children. In B. Stabler & L. Underwood (Eds.), *Slow grows the child.* Hillsdale, NJ: Lawrence Erlbaum Associates.

Spector, S., Faigenbaum, D., Sturman, M., & Hoffman, W. (1979). *Psychological correlates of short stature in children and adolescents.* Paper presented at the American Association of Psychiatric Services for Children. Chicago.

Stabler, B. (1987). Psychological effects of growth delay. In R. Hintz & R. Rosenfeld (Eds.), *Growth abnormalities.* New York: Churchill Livingstone.

Steinhausen, H., & Stahnke, N. (1976). Psychoendocrinological studies in dwarfed children and adolescents. *Archives of Diseases in Childhood, 51,* 778–783.

Tyler, L.E. (1965). *The Psychology of Human Differences* (3rd ed). New York: Appleton-Century-Crofts.

Weinberg, W.A., Dietz, S.G., Penish, E.C., & McAlister, W.H. (1974). Intelligence, reading achievement, physical size, and social class. *Journal of Pediatrics, 85*(4), 482–489.

4
Psychosocial Functioning and Social Competence in Growth Hormone Deficient, Constitutionally Delayed and Familial Short-Stature Children and Adolescents

Deborah L. Young-Hyman

Psychosocial Issues Associated with Short Stature: Framing the Problems

At the time of a child's birth, an assessment of the baby's physical condition is made (the Apgar score), and the weight and height are recorded. Later, these measurements are often published on a birth announcement. On each successive visit to the doctor, parental attention is focused upon the child's growth pattern. The child's growth becomes a focus of concern, a measure of physical well-being, and often a measure of the adequacy of parental care. It is unavoidable that height/stature becomes a value-laden physical characteristic.

Height has been positively correlated with perceived academic status (Wilson, 1968), starting salaries (Deck, 1968), and promotions and raises (Schumacher, 1982). Schumacher also found that people in high-ranking positions were taller than average and that tall individuals are perceived as possessing the characteristics of successful people. Before our current (1988) election, 80% of our elected presidents were taller than their opponent (Feldman, 1975). Our language also uses height to connote positive or negative personality characteristics or social traits. (Feldman, 1975). When we are proud, we "sit tall in the saddle" or "stand tall." Conversely, when we wish to indicate that a person is not smart, we may say he or she is "short on brains." When cheated, we comment that we have been "short changed." Not only does the concept of shortness frequently have negative connotations, it is often used in a gender-specific manner. Expressions that incorporate the word short in a positive context are frequently associated with femininity: "short and sweet," "good things come in small packages." The masculine stereotype in our culture dictates large size and mass. The feminine stereotype emphasizes petiteness and small mass. It would seem out of context to evaluate the literature on the effects of short stature upon the psychosocial functioning and social competence of short children without taking these predetermined biases into account. As has been noted before, short stature is both a physical and social condition (Stabler, 1979).

In addition to considering social biases when evaluating the psychosocial functioning and social competence of short-stature children, the criterion by which we judge adjustment or competence should be carefully considered. Various authors have proposed that a distinction be made between psychomaturation and physical maturation (Drash, 1969; Meyer-Bahlburg, 1985). The distinction between psychological and physical growth allows for the possibility that a growth-delayed child may form a personal identity and coping mechanisms appropriate to their size (Rotnem, 1986) and that behavior may vary depending upon the context in which a child is judged. It has been suggested that short-stature children undergo a process of "juvenilization," that is, they are assumed to be younger than their chronological age based on their physical appearance (Clopper et al., 1986). Physical maturation tends to dictate what is developmentally expected of a child, and therefore the discrepancy between height age and chronological age may represent a major psychosocial risk factor (Rotnem, 1986) or a handicapping condition (Young-Hyman, 1986) for growth-delayed children. Based on this model, developmental expectations of the child may be lowered and become age inappropriate. Resultant behavior and personality may be reciprocally shaped by the child's size and resultant lowered expectations, rather than the child's actual psychosocial competence (Rotnem, 1986).

Social competence can be defined as a person's ability to successfully accomplish developmentally appropriate tasks (Achenbach, 1979; Young-Hyman, 1986). If what is expected of short children is behavior commensurate with their height age (Rotnem et al., 1977; Rotnem, 1986), then the source of maladaptation or incompetence might well lie outside of the child. The child may be capable of mastering age-appropriate tasks, but not be expected to perform according to these standards. Thus, the concept of juvenilization applies to not only the physical appearance of short-stature children, but to developmental expectations as well.

Evaluation of growth-delayed children with psychometric measures that are based upon chronologic age norms or developmentally based behavior might produce results that reflect juvenilization. Results of various studies have shown these children to be functioning like younger children, that is, they exhibit immature behavior (Holmes et al., 1986; Rotnem et al., 1977; Holmes et al., 1982a). In a number of the studies, parental assessment of social competence has been ascertained using the Child Behavior

Checklist (Achenbach, 1979). This instrument asks parents to describe their child's social activities, social functioning, school progress, behavioral characteristics, etc. The child's reported behavior is then compared to age mates. Using this instrument, we may be measuring parental expectations of the child's competence and the child's resultant adaptation, that is, juvenilization rather than the child's ability to function age appropriately. Another example of the age versus size paradox concerns physical competence as rated by the Piers-Harris Self-Concept Scale (Piers, 1969). Various items ask children to rate their desirability as sports team members. Size, especially in males, determines ability to compete successfully in many team sports. If the short-stature child responds by answering that he is chosen last for teams, this loads the scale towards lowered self-esteem. If the short-stature child responds that he is popular, well liked by peers, and never picked last for sports teams, the current psychological literature suggests that the short-stature child may be responding according to what the child considers to be socially desirable (Lewis et al., unpublished manuscript). In the case of the Piers-Harris scale, regardless of how the child responds, we can find justification for labeling the child as maladapted based upon age-related criterion which may not be appropriate for the short-stature population.

Psychological Issues Related to Diagnosis and Treatment

Another set of issues contributing to the psychosocial functioning of short-stature children are factors associated with the diagnosis of growth delay and prognosis for future growth. The diagnostic work-up for growth hormone deficiency is typically a time of heightened anxiety for the family. For many families the initiation of diagnostic endocrine tests is the culmination of a long period of watchfulness and concern, and may be met with ambivalent feelings on the part of the parents and child. In one study parents reported that the most frequent problems related to their child's short stature affecting the family were the strain of the diagnostic evaluation and jealously of siblings created by the special attention given to the patient (Mitchell et al., 1986).

Parents who have decided to seek treatment for their child are explicitly acknowledging that there is something wrong or unacceptable about the child (Stabler, 1988). They may view treatment as the opportunity to "normalize" their child's stature and are frequently unable to accurately predict their child's projected adult height. Both parents and child may also be unrealistic about the effectiveness of intervention, that is, treatment with growth hormone (Grew et al., 1983). Increased depression in the child after the initiation of treatment has been reported by some authors (Rotnem et al., 1979). Alternatively, parents may disclaim the effects of short stature on their child's psychosocial adjustment (Richman et al., 1986) and patients have been known to refuse treatment with growth hormone (Drash, 1969). This refusal has been attributed to the patient's desire not to have to go through a major personality readjustment (Drash, 1969; Meyer-Bahlburg, 1985).

An alternative explanation may be that parents and children may experience short stature differently depending upon the etiology of the growth delay. Growth hormone deficiency may be viewed as a "medical problem" whereas constitutional delay or familial short stature may be seen as a "normal variant" growth pattern (Richman et al., 1986). Direct assessment of parents' and children's attitudes towards diagnosis and prognosis for adult height, and the relationship between this knowledge and acceptance of and psychological response to treatment has not been systematically assessed to date. In any case, the accuracy with which the diagnosis and prognosis for adult height is communicated by the medical profession to the family will influence the child's and family's adjustment to and acceptance of treatment (Grew et al., 1983).

Early Literature on Psychosocial Adjustment

Interest in the psychosocial adjustment of growth-delayed children dates back to the 1950s. These early studies focused on the effects of physical maturation on behavior and personality. Most authors approached the assessment of these children and their families using psychiatric interview techniques and traditional methods of testing psychological adjustment such as the TAT (Murray, 1943), Rorschach (Exner, 1974), and drawing techniques.

In one such study, Mussen and Jones (1957) looked at early- and late-maturing 17-year-old adolescent boys. The purpose of the study was "to investigate the relationship between maturational status and certain important, covert aspects of personality during late adolescence" (p. 244). The authors hypothesized that the sociopsychological environment that late-maturing adolescent boys are subjected to might be significantly different from that of early maturers, causing them to form differing patterns of overt social behavior and covert personality characteristics. The authors choose to assess self-conceptions, underlying motivations, and basic interpersonal attitudes. Early maturers were defined as those boys whose physical development had been "consistently advanced" and late maturers were those whose development was "consistently retarded." They found that late maturers revealed more negative self-concepts and feelings of inadequacy, were more insightful and more dependent. Early maturers were more confident, aggressive, and independent. Based upon late maturers perception of parental rejection, the authors concluded that parents of late-maturing boys were probably transmitting that their physically immature sons were unacceptable in some way and incapable of independence. They suggest that the attitudes of the parents resulted in the boys being more touchy, rebellious, impulsive, and self-indulgent; characteristics indicative of poor adjustment and psychological immaturity.

A series of studies by Lewis et al. (1973, 1977) also looked at the psychological consequences of delayed puberty, with and without statural delay, that is, adolescents who were constitutionally delayed. These authors considered a predominance of their patients to have intrapsychic or behavioral pathology regardless of stature and whether or not they experienced spontaneous puberty by age 15. Dating and heterosexual behavior was considered to be inadequate. Kusalic and Fortin (1975) studied psychological adjustment in the hypopituitary dwarf population after treatment with growth hormone replacement. They found that an overall worsening of psychopathology took place after treatment; specifically, increased regression, depression, and a lessening of peer contact and achievement were noted. These authors also reported a high degree of pathology in the families of their subjects who treated their children with ambivalence and overprotection. They concluded that hypopituitary dwarfs are a vulnerable population at high risk for psychosocial problems.

These early studies are marred by significant methodologic flaws and biases. Of central concern is that the children and families studied were almost uniformly noted to have a high incidence of psychopathology, independent of the child's pubertal or statural delay. In addition, the methods used to assess the variables chosen tended to be anecdotal and subjective. Ratings of scoring categories for such tests as the TAT and Rorschach, which also provided the basis for psychiatric diagnostic criterion, have been found to be difficult to replicate (Rapapport et al., 1974) and conclusions were often drawn from nonstandardized questionnaires, interviews, and chart review. Lastly, the subjects studied during this period of time, before the mid 1970s, often represented extreme cases of growth delay or those families who presented as most distressed by their child's growth problems, creating a sample selection bias. Normal stature or comparison control groups were rarely used. These factors probably contributed to a weighing of findings in the direction of significant pathology.

It should be noted however, that the themes which emerged from this early work have set the direction for the study of these children for the past ten to fifteen years. Following the model that was set forth earlier in the chapter, the current literature will be reviewed making distinctions between intrapsychic adjustment and personality style, overt behavioral manifestations of adjustment and patterns of interpersonal coping. Literature that deals with family dynamics, including parental assessment of the child and parent-child interactions will be handled separately, as will studies that measure social competence.

Current Research (1975 to the Present)

The self-esteem/self-concept of children with hypopituitarism, constitutional delay, and familial short stature has received considerable attention. In contrast to early reports that short-stature children suffered from uniformly lowered self-esteem, recent studies have found that these children discriminate among different aspects of their self-worth, and that not all facets of their self-esteem are low. Looking at short-stature children and adolescents as a group, excluding Turner's and Noonan's syndrome, these children describe themselves as unhappy and unpopular (Rotnem et al., 1977; Apter et al., 1981; Gordon et al., 1982) but as having high, even elevated overall self-esteem (Young-Hyman, 1986; Lewis et al., unpublished manuscript). This "overinflated" (when compared to test norms) self-image in the Young-Hyman (1986) study was based on the children's evaluation that they were intellectually competant, academically able, and possessed good social skills. In this study it is noteworthy that the population had a mean socioeconomic status well above the average of the general population. The parents of these children clearly valued verbal skills and educational achievement, providing confirmation for the children's self-worth based on their above-average intelligence and academic achievement. The mean SES for the study population is not provided in the Lewis article, however, the children's mean verbally based intelligence was in the average range. It is reported that the higher the growth hormone deficient children's IQs, the more socially outgoing they were reported to be by peers. Thus, IQ in this group may have also mediated self-esteem.

Two studies that assessed the effects of treatment on self-image also cited high overall self-esteem in their study groups. The study by Rosenfeld et al. (1982) followed adolescents with constitutional delay of growth over a one-year course of treatment with androgen. All short-stature adolescents gained in self-esteem, regardless of treatment or nontreatment status. All boys in this study did grow, with or without treatment. The relationship between height and higher self-esteem was also found in the study by Young-Hyman (1986). However, physical appearance is rated lower than other aspects of self-worth and co-varies with how socially desirable these children rate themselves to be (Rosenfeld et al, 1982; Gordon et al, 1982; Young-Hyman, 1986; Lewis et al., unpublished manuscript). As indiccated by the findings of Holmes et al. (1985), degree of height deficit may be a factor in the psychological adjustment of short-statured individuals. Younger, less deficient (in height) children fare better than their older, more height-impaired counterparts.

A study by Mitchell et al. (1986) assessed adult growth hormone deficient patients who had received treatment with growth hormone during childhood. These patients also rated themselves as having elevated self-satisfaction, personal worth, and sociability but professed significantly lower evaluation of their physical self. Mention was made that most of these subjects did not reach average adult height over the course of treatment. These studies suggest that although height may interfere with popularity and happiness, especially during the adolescent years, short-stature adults eventually experience adequate self-esteem based upon the internal strengths and coping mechanisms they develop, regardless of ultimate adult height.

Devaluation of self-worth based on physical status would suggest that short-stature children might be prime candidates for a tendency to somaticize or become preoccupied with body-image. Using the Missouri Children's Picture Series (MCPS), Drotar et al. (1980) did not find differences in somatization between 16 hypopituitary dwarfs aged 8 through 16 years and healthy controls. This finding was replicated by Holmes et al. (1982a) who did not find elevations of the somatization scale of the MCPS in her growth hormone deficient and constitutionally delayed population. Gordon et al. (1982) found his preadolescent constitutionality delayed subjects to rate their physical appearance as being within normal limits on Piers-Harris Self-Concept Scale (Piers, 1969).

Evaluation of short-stature youngsters' ability to assess their body size compared to peers has likewise shown that a high percentage are quite accurate in their appraisal of their own size (Grew et al., 1983; Young-Hyman, 1986; Lewis et al., unpublished manuscript). Assessment with drawing techniques has suggested, however, that "size-confusion" regarding a mature body type may exist (Rotnem et al., 1977) and as suggested previously, internal conflict and unhappiness about physical adequacy may coexist with average or elevated self-esteem (Richman et al., 1986).

Based upon early findings that short-stature children were more dependent and less aggressive, recent studies have sought both confirmatory information and an explanation for these characteristics by assessing locus control as well as behavioral manifestations of personality traits such as social skills, social isolation and withdrawal, immaturity, conformity, and expression of aggression. Richman et al. (1986) reported that his study group of constitutionally delayed children and adolescents did not report higher levels of helplessness or hopelessness (than controls) on the Nowicki-Strickland Locus of Control Scale (Nowicki & Strickland, 1983). Anxiety, as measured by a subscale of the Piers-Harris Self-Concept Scale has been found to be comparable to the normative population (Richman et al., 1986).

Drotar et al. (1980), in a study that measured reactions to frustration, found that his hypopituitary population perceived more obstacles to be overcome than solutions. He concluded that his study population had poorer, less mature and adaptive problem-solving skills, which represented "a withdrawal from the adaptive expression of aggression" (p. 63). In a related study, Stabler et al. (1980) tested short-stature boys' abilities to use social problem-solving skills under normative and competitive circumstances. Although subjects scored significantly lower than normal stature controls on the picture arrangement subtest of the WISC-R (Wechsler, 1974), these authors also tested the subjects' ability to respond to a competitive challenge on the same task. Under competitive test conditions, both groups significantly increased their scores. The authors suggest that short-stature males have the ability to respond aggressively, but as was suggested by Drotar et al. (1980), withdraw from the expression of assertive or aggressive behavior. The results of the Stabler study should be viewed with caution. Although the PA subtest is well accepted as a measure of social judgment (Glasser & Zimmerman, 1976) the content of the stories is not primarily aggressive, and the competitive challenge presented to the boys does not necessarily reflect assertive or aggressive behavior. Short-stature children have also been shown to have equal levels of aggression as measured by a paper and pencil scale of aggressiveness (Steinhausen & Stahnke, 1977). As suggested earlier, fantasy assertiveness and behavioral manifestations of assertiveness may be discrepant in this population. In order to test these hypotheses, studies that ask subjects and independent (nonparental or nonteacher) observers to rate the subjects' social interactions (i.e., ratings of withdrawal, aggression and maturity) in response to a scripted scenario are suggested.

Withdrawal and immaturity in social interactions has been consistently ascribed to this population regardless of diagnostic entity (Gordon et al. 1982; Holmes et al., 1982a, 1982b). Studies to date have tended either to look at a single diagnostic category (Rosenfeld et al., 1982; Abbott et al., 1982; Lewis et al., unpublished manuscript; Gordon et al., 1982) when assessing psychosocial adjustment, or combine diagnoses and present group results (Young-Hyman, 1986; Stabler, et al., 1980). An exception to this is a study by Steinhausen and Stahnke (1977) that compared the adjustment of three groups of short children and adolescents: those with isolated growth hormone deficiency, those with multiple pituitary hormone deficiencies, and children without endocrine disease. These authors found "that growth hormone deficiency had no impact on the psychological variables" (p. 263). A later study by Holmes et al. (1982b) also looked at etiology-related behavioral characteristics. These authors compared two diagnostic groups: constitutional delay of growth versus growth hormone deficiency. No etiology-specific behavior patterns were found between groups, thus confirming the earlier work of Steinhausen and Stahnke.

It is interesting to note that this behavior, withdrawal and immaturity, is reported by parents and peers of short-stature youngsters, despite the children's assertion of their own social competence. This may be due to the children's self-perception that since they attend group and social activities in the same number as normal-stature controls (Gordon et al., 1982), their behavior is adequate. As has been suggested before, denial may be at work in order to maintain adequate self-esteem. Two longitudinal follow-up studies of the social behavior of a cohort of short-stature patients that was conducted by Holmes et al. (1985, 1988) used the Achenbach scale to assess social skills. These authors found that short males experienced a significant decline in social functioning over time. Holmes found that parental report of the decline in children's participation in clubs and in socializing with peers occurred sooner (at the 3-year follow-up) for the growth hormone deficiency in males, but that a similar and significant decline occurred for the constitutionally delayed males later (at the 6-year follow-up). The authors speculate that the differential between diagnostic categories may have been attributable to height, that is, males with growth hormone deficiency were significantly shorter. The GHD and CD females assessed in these studies grew substantially and improved their social functioning. This is in contrast to the TS girls who were less socially active and were slightly shorter. These findings underscore the notion that height deficiency per se rather than diagnostic category may impact upon social functioning.

Interaction with peers has also been of interest in light of a high incidence of peer teasing (Mitchell et al., 1986) as well as the early reports of inadequate social behavior (Lewis et al., 1973, 1977). In a recent study, Lewis et al. (unpublished manuscript) used peer evaluation techniques to determine the popularity and likability of their short-stature population. The authors found that these children and adolescents (age range 5–20) were no more or less popular than classmates, and were noted as equally likable. In the same study, peers were asked to rate the attractiveness of the short-stature sample in comparison to normal-stature controls. No differences were found between the ratings of the two groups. These findings, in conjunction with somewhat inconsistent reports of withdrawn and immature behavior on the part of short-stature children, suggest that contrary to original reports of extreme social isolation, it appears that at least up through early adolescence short-stature children adopt a personality style that allows them to blend into their childhood social network. Lack of assertive and aggressive behavior, especially by males, may be a realistic personality adaptation that minimizes their psychological and social discomfort. This adaptive strategy was suggested by the findings in one study that assessed friendship patterns among growth-delayed children between the ages of 8 and 16 of varying diagnoses. This group of children reported primarily long-term stable relationships with individual friends rather than a preference for "hanging out" with groups of peers. This friendship pattern gave them the opportunity to use their highly developed social sensitivity and verbal skills (Young-Hyman, 1986). Reciprocity of friendships was also noted to contribute to good psychological adjustment by Abbott et al. (1982).

A mixed picture of developmental mastery emerges from the literature. Assessment of participation in age-appropriate sports, hobbies, and household activities has not been found to differ from normative samples for a group of short-statured subjects with varying diagnosis studied by Holmes et al. (1986), however, ratings of school performance and social competence were rated lower during early adolescence when compared to late adolescence. Young-Hyman (1986) also found an age-related decrease in participation in competitive team sports and organized group acitivities as growth-delayed children entered adolescence. Problems with age-appropriate developmental tasks of adolescence were found by Clopper et al. (1986) and Dean et al. (1986). In particular, lower rates of dating and sexual behavior have been documented in the hypopituitary population. These authors speculate that immaturity of physical appearance and poor masculinization in males may contribute to withdrawal from heterosexual activities. Rates of academic achievement and job acquisition were not found to be different from the general population (Mitchell et al., 1986).

Although short-stature children's mastery of their physical environment has not been quantified and compared to developmental norms, obvious limitations due to size have been well documented (Meyer-Bahlburg, 1985). Problems ranging from difficulty finding age-appropriate clothing, to difficulty boarding mass transit vehicles have been cited. Short-stature children further report that difficulty with mastery of their physical environment increases their dependence upon normal-stature peers and adults, thus impinging on other aspects of social competence and adaptation (Meyer-Bahlburg, 1985).

The Family Context

The effect of having a short-stature child upon family functioning has received very little systematic study, nor have the relationships between marital discord, financial burden, and the presence of a short-stature child in the home. Patterns of parent-child interaction and parental responses to diagosis and treatment have received some attention, as have parental attitudes toward the short-stature child. Parents of short-stature children have been reported to overprotect (Rotnem et al., 1977), have difficulty and be inconsistent with limit setting (Gordon et al., 1982, Rotnem et al., 1986) and exhibit a tendency to juvenilize (Clopper et al., 1986) their children. A contributory factor in these findings may be that parents of short-stature children who are not short themselves tend to view their children as less competent than those parents who are short (Young-Hyman, 1986). It is important to note that in two of the studies cited (Rotnem et al., 1977; Clopper et all, 1986), conclusions were reached based upon clinical interviews and observations, rather than standardized evaluation procedures. Standardized evaluation of parent-child interaction is essential in understanding the psychosocial sequalae of short stature.

Poor communication and less effective communication, as measured by the Family Functioning Index, has been documented by Gordon et al.

(1982) who also used the Maryland Parent Attitude Survey (Pumroy, 1966) to assess levels of discipline, indulgence, protectiveness, and rejection. A consistent but incidental finding has been that when parents and children are asked to rate the short-stature child's social competence, self-esteem, and other aspects of interpersonal functioning, parents rate their children as more socially maladjusted and less competent than the children rate themselves, regardless of the etiology of short stature (Young-Hyman, 1986; Lewis et al., unpublished manuscript). Parents also experience their child's short stature as having a greater impact upon school functioning and peer relationships than do the short-stature individuals (Mitchell et al., 1986). These findings are not surprising given the fact that "normal" children uniformly rate themselves to have less problems than their parents (Achenbach et al., 1987; Phares et al., 1989).

A picture emerges of parents who perceive their short-stature children as vulnerable and in need of protection. Increased parental anxiety has also been reported in response to the process of diagnosis and treatment (Mitchell et al., 1986) suggesting a heightened sensitivity to their child's condition. In contrast, a supportive familial environment in which parents encourage age-appropriate behavior has been shown to be effective in promoting the short-statured child's competence (Rotnem et al., 1977; Young-Hyman, 1986). With the exception of the Gordon et al. (1982) study cited previously, the conlusions reached by these authors have been based on anecdotal information provided by the parents or the child. Although subjective, contrary to early reports of subtle parental rejection and ostracism (Kusalic & Fortin, 1975; Mussen & Jones, 1957), short-stature children report feelings of belonging and acceptance within their families (Abbott et al., 1982; Rotnem, 1986). Sibling relationships have not been systematically studied, but jealousy of siblings over the special attention given to the short-stature child has been documented (Mitchell et al., 1986). Various authors have equated short stature with a chronic illness, and have cautioned against the disproportionate use of parental resources to attend to the short-stature child (Rotnem et al., 1979; Rotnem, 1986).

Summary and Synthesis of the Current Literature

Many of the original stereotypes concerning short-stature children have been reinforced. In particular, the description of nonaggressive, passive youngsters has been underscored, as has the notion of a vulnerable population at risk for various psychosocial difficulties. In particular, consistent dissatisfaction with body-image, physical adequacy, and popularity are reported by both parents and the children themselves. Parents tend to view their children as more psychologically and socially vulnerable than do the children themselves, and as having poorer social skills. It is unclear whether this concern is related to their child's short stature or "normal" parental sensitivity to their child's well-being. What has not been supported is the once generally accepted assumption that short-stature children have universally low self-esteem and are socially withdrawn. The children make distinctions among various aspects of their personalities and competencies, and thereby retain an adequate and sometimes overinflated sense of self-esteem and self-worth. They are rated by peers to be no more or less popular, to be no more or less attractive or likable, and to attend age-appropriate activities in equal numbers. Even though short-stature children wish for greater popularity, they do not fit the picture of social isolation which was presented in the early literature. Juvenilization and peer teasing continue to be areas of interpersonal difficulty, however long-term stable friendships also provide social stability for the short-stature child. Mastery of the physical environment is problematic, and forces dependence upon both peers and parents for the short-stature child.

Another misconception to arise out of early psychosocial literature is that short-stature children are ostracized and rejected by their parents. Parents seem to make distinctions about the various facets of their child's competencies as do the children. When parents and children are able to focus their attention on nonphysical areas of competence, such as school achievement and sedentary activities, and a psychologically supportive familial environment is provided, lower rates of maladjustment and incompetence are reported.

Directions for the Future

Despite clear trends in the psychosocial literature, small sample sizes as well as overlapping and collapsed diagnostic categories have prevented the emergence of clear descriptive personality styles and behavioral traits in the short-stature population. Lack of standardized instrumentation and protocols have also prevented true between-study comparisons. Differences in operational definitions, such as the distinction between observed behavior and self-reported competence, also need to be clarified in order to compare findings between studies.

Various areas of research in this population have been neglected. Among them are: the impact of having a short-stature child on family functioning and sibling adjustment; the impact of parental attitudes on psychosocial competence; critical ages for medical intervention which maximize psychosocial adjustment; the establishment of psychosocial criterion regarding the initiation of medical treatment; and the impact of the medical world (including treatment effects) upon the lives of families of short-stature children. Systematic evaluation of psychosocial interventions in this population have also not been attempted to date.

Though at first glance short stature is not life threatening or physically a prohibitive illness, by definition 3% of our pediatric population fall below the growth chart percentiles and are therefore considered to be short stature. A diagnostic entity of this magnitude deserves continued research and medical exploration to further our understanding of this condition, and help these youngsters achieve more adequate developmentally appropriate psychosocial adjustment and competence.

References

Abbott, D., Rotnem, D., Genel, M. & Cohen, D.J. (1982). Cognitive and emotional functioning in hypopituitary short statured children. *Schizophrenia Bulletin, 8*(2), 310–319.

Achenbach, T. (1979). The child behavior profile: An empirically based system for assessing children's behavioral problems cand competencies. *International Journal of Mental Health, 7*, 24–42.

Achenback, T.M., McConaughy, S.H., & Howell, C.T. (1987). Child/adolescent behavioral and emotional problems: Implications of cross-informant correlations for situational specificity. *Psychological Bulletin, 101*, 213–232.

Apter, A., Galatzer, A., Beth-Nalachmi, N. & Laron, Z. (1981). Self-image in adolescents with delayed puberty and growth retardation. *Journal of Youth and Adolescence, 10*(6), 501–505.

Clopper, R., Mazur, T., MacGillivray, M., Peterson, R., & Voorhess, M. (1983). Data on virilization and erotosexual behavior in male hypogonadotropic hypopituitarism during gonadotropin and androgen treatment. *Journal of Andrology, 4*, 303–311.

Clopper, R.R., MacGillivray, M.H., Mazur, T., Voorhess, M.L., & Mills, B.J. (1986). Post-treatment follow-up of growth hormone deficient patients: Psychosocial status. In B. Stabler & L. Underwood (Eds.), *Slow grows the child. Psychosocial aspects of growth delay* (pp. 83–96). Hillsdale, NJ: Lawrence Erlbaum.

Dean, H.J., McTaggart, T.L., & Freesen, H.G. (1986). Long-term social follow-up of growth hormone deficient adults treated with growth hormone during childhood. In B. Stabler & L. Underwood (Eds.), *Slow grows the child. Psychosocial aspects of growth delay* (pp. 73–82.). Hillsdale, NJ: Lawrence Erlbaum.

Deck, L.P. (1968). Buying brains by the inch. *Journal Collegiate University Personality Association, 19*, 33–37.

Drash, P. (1969). Psychologic counseling: Dwarfism. In L.I. Gardner (Ed.), *Endocrine and genetic diseases of childhood* (pp. 1014–1022). Philadelphia: W.B. Saunders Co.

Drotar, D., Owens, R., & Gotthold, J. (1980). Personality adjustment of children and adolescents with hypopituitarism. *Child Psychiatry and Human Development, 11*(1), 59–66.

Exner, J.E. (1974). *The Rorschach: A comprehensive system I.* New York: Wiley.

Feldman, S.D. (1975). The presentation of shortness in every day life-height and heightism in American sociology: Toward a sociology of stature. In S.D. Feldman (Ed.), *Life styles. Diversity in American society* (pp. 437–442). Boston: Little Brown.

Glasser, A.J. and Zimmerman, I.L. (1967). Clinical interpretations of the Wechsler Intelligence Scale for Children. New York: Grune and Stratton.

Gordon, M., Crouthamel, C., Post, E.M., & Richman, R.A. (1982). Psychosocial aspects of constitutional short stature: Social compentence, behavior problems, self-esteem and family functioning. *The Journal of Pediatrics, 101*(3), 477–480.

Grew, R.S, Stabler, B., Williams, R.W., & Underwood, L.E. (1983). Facilitating patient understanding in the treatment of growth delay. *Clinical Pediatrics, 22,* 685–690.

Holmes, C.S., Hayford, J.T., & Thompson, R.G. (1982a). Parents' and teachers' differing views of short children's behavior. *Child: Care, Health and Development, 8,* 327–336.

Holmes, C.S., Hayford, J.T., & Thompson, R.G. (1982b). Personality and behavior differences in groups of boys with short stature. *Children's Health Care, 11*(2), 61–64.

Holmes, C.S., Karlsson, J.A., & Thompson, R.B. (1985). Social and school competencies in children with short stature: Longitudinal patterns. *Journal of Developmental and Behavioral Pediatrics, 6,* 263–267.

Holmes, C.S., Karlson, J.A., & Thompson, R.G. (1986). Longitudinal evaluation of behavior patterns in children with short stature. In B. Stabler & L. Underwood (Eds.), *Slow grows the child. Psychosocial aspects of growth delay* (pp. 1–12). Hillsdale, NJ: Lawrence Erlbaum.

Holmes, C.S., Sidler, A.K., Tsalilcian, E., & Karlsson, J.A. (October 1988) *Six year longitudinal follow-up of short children: I. Social, School and Activity Competence.* Presentation at the Genentech National Cooperative Growth Study, Symposium III, Palm Desert, CA.

Kusalic M., & Fortin, C. (1975). Growth hormone treatment in hypopituitary dwarfs: Longitudinal psychological effects. *Canadian Psychiatric Association Journal, 20*(5), 325–331.

Lewis, C., Johnson, S.B., Knuth, C., Silverstein, J. *The psychosocial adjustment of growth hormone deficient children: Parent, teacher, peer and self perceptions.* University of Florida, unpublished manuscript.

Lewis, V.G., Money, J., & Bobrow, N.A. (1973). Psychologic study of boys with short stature, retarded osseous growth, and normal age of pubertal onset. *Adolescence, 8,* 445–454.

Lewis, V.G., Money, J., & Bobrow, N.A. (1977). Idiopathic pubertal delay beyond age fifteen: Psychologic study of twelve boys. *Adolescence, 12*(45), 1–11.

Meyer-Bahlburg, H.F.L. (1985). Psychosocial management of short stature. In D. Shaffer, A.A. Ehrhardt, & L.L. Greenhill (Eds.), *The clinical guide to child psychiatry* (pp. 110–114). New York: The Free Press.

Mitchell, C.M., Johanson, A.J., Joyce, S., Libber, S., Plotnick, L., Migeon, C.J., & Blizzard, R.M. (1986). Psychosocial impact of long-term growth hormone therapy. In B. Stabler & L. Underwood (Eds.), *Slow grows the child. Psychosocial aspects of growth delay* (pp. 97–109). Hillsdale, NJ: Lawrence Erlbaum.

Murray, H. (1943). *Thematic Apperception Test.* Cambridge, Harvard University Press.

Mussen, P.H. & Jones, M.C. (1957). Self-conceptions, motivations and interpersonal attitudes of late- and early-maturing boys. *Child Development, 28*(2), 243–256.

Nowicki, S., & Strickland, B.R. (1973). A locus of control scale for children. *Journal of Consulting and Clinical Psychology, 40,* 148–155.

Phares, V., Compas, B.E., & Howell, D.C. (1989). Perspectives on child behavior problems: Comparisons of children's self-reports with parent and teacher reports. *Journal of Consulting and Clinical Psychology, 1,* 68–71.

Piers, E.V. (1969). *Manual for the Piers-Harris Children's Self-Concept Scale.* Nashville: Counselor Recordings and Tests.

Pless, I.B., & Satterwhite, B. (1973). A measure of family functioning and its application. *Society for Scientific Medicine, 7,* 613–621.

Pumroy, D.K. (1966). Maryland parent attitude survey: A research instrument with social desirability controlled. *Journal of Psychology, 64,* 73–78.

Rapaport, D., Gill, M.M., & Schaffer, R. (1974). The Rorschach Test. In R.R. Holt (Ed.), *Diagnostic Psychological Testing* (pp. 268–463).

Richman, R.A., Gordon, M., Tegtmeyer, P., Crouthamel, C., & Post, E.M. (1986). Academic and emotional difficulties associated with constitutional short stature. In B. Stabler & L. Underwood (Eds.), *Slow grows the child. Psychosocial aspects of growth delay* (pp. 13–26). Hillsdale, NJ: Lawrence Erlbaum.

Rosenfeld, R.G., Northcraft, G.B., & Hintz, R.L. (1982). A prospective, randomized study of testosterone treatment of constitutional delay of growth and development in male adolescents. *Pediatrics, 69,*(6), 681–687.

Rotnem, D., Cohen, D., Hintz, R.L., & Genel, M. (1979). When treatment fails: Psychological sequelae of relative "treatment failure" with human growth hormone replacement. *Journal of the American Academy of Child Psychiatry, 19*(3), 505–520.

Rotnem, D., Genel, M., Hintz, R.L., & Cohen, D.J. (1977). Personality development in children with growth hormone deficiency. *Journal of the American Academy of Child Psychiatry, 16,* 412–426.

Rotnem, D.L. (1986). Size versus age: Ambiguities in parenting short-statured children. In B. Stabler & L. Underwood (Eds.), *Slow grows the child. Psychosocial aspects of growth delay* (pp. 178–190). Hillsdale, NJ: Lawrence Erlbaum.

Schumacher, A. (1982). On the significance of stature in human society. *Journal of Human Evolution, 11,* 697–701.

Sine, J.O., Pauker, J.D., & Sines, L.K. (1963). *The Missouri Children's Picture Series Manual.* Iowa City.

Stabler, B. (1979). *Research and the experience of being short.* Symposium on psychosocial aspects of growth delay. University of Texas Medical Branch of Galveston.

Stabler, B. (1988). Potential psychological risks in growth hormone therapy. In L.E. Underwood (Ed.), *Human GH: Progress and challenges* (pp. 172–178). New York: Marcell Dekker, Inc.

Stabler, B., Whitt, J.K., Moreault, D.M., D'Ercole, A.J. & Underwood, L.E. (1980). Social judgements by children of short stature. *Psychological Reports, 46,* 743–746.

Steinhausen, H.C., & Stahnke, N. (1977). Psycoendocrinological studies in dwarfed children and adolescents. *Archives of Diseases in Childhood, 51,* 778–783.

Wechsler, D. (1974). *Wechsler Intelligence Scale for Children—Revised.* New York: Psychological Corporation.

Wilson, P.R. (1968). The perceptual distortion of height as a function of ascribed academic status. *Journal of Social Psychology, 74,* 97–102.

Young-Hyman, D. (1986). Effects of short stature on social competence. In B. Stabler & L. Underwood (Eds.), *Slow grows the child. Psychosocial aspects of growth delay* (pp. 27–45). Hillsdale: Lawrence Erlbaum.

5
Assessing the Effects of Replacement Hormone Treatment on Psychosocial and Psychosexual Behavior in Growth Hormone Deficient Individuals

Richard R. Clopper

While the behavioral characteristics of very short children have received enough scientific attention that trends are beginning to emerge (Lee & Rosenfeld, 1987; Mazur & Clopper, 1987), the effects of endocrine treatments on psychosocial and psychosexual development remain difficult to characterize. The purpose of this chapter is to discuss the types of problems facing researchers in this area and to present a critical review of the few available reports that have bearing on the question of treatment effects in growth hormone deficient patients. This discussion focuses primarily on assessment of the impact of growth hormone and/or sex hormone treatment on psychosocial and psychosexual development and functioning in growth hormone deficient children and adolescents. The effects of treatment during the prenatal and adult periods will not be discussed.

The issues considered herein are not an exhaustive compendium of those faced by behavioral scientists as they attempt to assess the behavioral impact of endocrine treatments. Rather, these issues are some that are frequently encountered and/or often neglected which have important bearing on the interpretability of study results.

Defining the Task

On the face of it, assessing the behavioral effects of an endocrine therapy appears to be a straightforward empirical problem. Pick an appropriate dependent behavioral variable, measure it before and after the treatment, and compare the pretreatment and posttreatment data. Implementing such a plan in a way that yields clear results is considerably more complicated.

Supported in part by U.S. Public Health Service Grant # HD 19760 and funds of the Children's Growth Foundation of Western New York.

Description of Replacement Regimens

Typical Growth Hormone Regimen

By the time patients begin growth hormone (GH) replacement therapy, they have experienced several outpatient physical examinations including careful height measurements, a 1 to 3 day admission for provocative GH testing, and a variety of blood tests and x-rays. Once the test results are available and GH deficiency is diagnosed, each child returns to the hospital to learn how to administer the GH. In this training session, which is usually conducted by a nurse specialist, the child and at least one parent or guardian learns how to: correctly dilute the powdered hormone, prepare the syringe, draw-up the correct dose of GH, select and prepare the injection site, and give the injection using sterile procedures. The person giving the injections practices several times using sterile water and an orange or a practice injection pad until the basic technique is mastered. Then the parent or guardian gives the first GH injection to the child with the coaching of the nurse. The injection schedule is reviewed and questions answered before the family leaves with a several month supply of hormone to administer at home. The patient then begins a protracted period of three or more GH injections per week. While older patients are encouraged to give their own injections, a parent usually takes on this responsibility. The patient continues on this regimen with outpatient check-ups every few months to monitor growth. Since GH deficiency is not life threatening after the neonatal period, and since its growth-promoting effects substantially end with the end of the pubertal growth spurt, GH replacement usually ends during the late teenage years when the skeleton matures.

Two types of GH have been used for replacement therapy. Originally, treatment was with human GH extracted from donated pituitary glands (pGH). Since late 1985, production of a bioengineered form of human GH has been possible using recombinate DNA technology (rGH). Both forms have been effective in promoting linear growth.

Typical Sex Hormone Regimen

If pubertal onset does not occur, sex hormone replacement is begun. Typically boys are placed on long-acting intramuscular testosterone (T) preparations which require one injection every two to four weeks. Girls are usually placed on oral estrogen (E) and/or progesterone (P) preparations similar to birth control pills. These regimens induce somatic pubertal development but not necessarily fertility. Since sex hormones are not essential for life, replacement therapy continues as long as the patients wish to maintain their sexual development.

When puberty is delayed due to a deficiency of pituitary gondotropin (Gn), replacement therapy is still usually with gonadal sex steroid (T or

E/P). Replacing the pituitary gonadotropic hormones, follicle-stimulating hormone (FSH), and luteinizing hormone (LH) is considerably more expensive and requires frequent injections. The pituitary gonadotropins (LH and FSH) along with T, E, and P are required for fertility.

Review of Literature

Behavioral Correlates of Growth Hormone Treatment

Four published studies provide data on the behavioral correlates of GH treatment. All four reports examined groups of children and adolescents who received pGH for varying schedules and durations of treatment.

Money and Pollitt (1966) longitudinally studied 17 cases (11 boys/6 girls) of "dwarfism" of mixed etiology as they underwent treatment with pituitary-derived GH. Some parents were also studied. The patients were between the ages of 4 and 18 years at last follow-up and each experienced less than 4 years of intermittent pGH therapy. Using a structured interview, data were collected and tabulated regarding various attitudes, fantasies, social and romantic activities, and their experience with GH treatment. The authors reported that all of the patients evidenced a lag in their psychosocial and psychosexual development. As a group they tended to show inhibited and immature behavior. Furthermore, the authors suggested that GH treatment precipitated a "readjustment syndrome" in which patients had to reexamine their expectations for themselves and the changing behavioral expectations of others. This was not always a pleasant experience and could even lead to discontinuing treatment. These authors also suggested that there was an inverse relationship between the degree of a child's "psychomaturational lag" and the success the child's parents exhibit with treating the child in an age-appropriate manner.

Kusalic, Fortin, and Gauthier (1972) reported on a group of 11 GH recipients (9 boys/2 girls) who were evaluated before and after a 6-month period of GH therapy followed by 6 months of no treatment. The patients ranged in age from 5 to 21 years at the beginning of treatment and most had multiple pituitary deficits. Using a psychoanalytically oriented interview and projective tests they concluded that their patients exhibited immaturity, inhibited aggression, and low self-esteem prior to GH therapy. After treatment, Kusalic et al. (1972) reported a shifting to a "weak release of aggressive drive," emotional disequilibrium, and depressive symptomatology. Patients also harbored unrealistic expectations for treatment. Parents appeared unable to significantly modify their overprotective attitudes.

Rotnem, Cohen, Hintz, and Genel (1979) studied 11 children (9 boys/2 girls) using structured interviews and a projective test before and after 1 year of continuous GH treatment. The patients ranged in age from 7 to 22 years at follow-up. Some patients had isolated GH deficiency while others had multiple pituitary deficits. These authors reported that their sample exhibited "mild" features of a depressive nature which tended to increase over the year of treatment. Also, children in their sample tended to become angry as treatment progressed. Rotnem's group attributed this phenomenon to disappointment over slower growth than the patients expected to occur. Parent expectations for treatment and their preparation of the child for beginning treatment also seemed to be important for outcome. Behavioral inhibition, withdrawal, and low self-esteem were reported as characteristic of roughly half the sample (5 of 11) after treatment.

While assessing GH treatment was not the objective of their study, Holmes, Karlsson, and Thompson (1986) reported longitudinal and cross-sectional psychometric data on 47 patients with short stature, 17 of whom were GH deficient. The remaining subjects were diagnosed as having Turner's syndrome (TS) or constitutional delay (CD) of growth. Presumably the TS and CD patients were not on GH therapy. The follow-up period was 3 years. Using a standardized psychometric test and data from each child's growth record, these authors examined the relationships among age, sex, diagnosis, and the parent-reported scores on the behavioral competency and behavior problem scales of the Child Behavior Problem Checklist (Achenbach, 1979). With the exception of the school competency trends among GH-deficient children, mean scores on all scales remained within one standard deviation (SD) of the test means during the follow-up period. Their multivariate analysis suggested that all groups experienced a temporary decline in their school and social competency during the early teenage years. Cross-sectional analysis of age effects suggested improved competency in the middle to late teenage years. School competency scores were generally poorer (greater than −1 SD) among GH-deficient females than males, and for younger (mean age at follow-up = 12.8 yr) GH-deficient patients of both sexes relative to both the test mean and the older (mean age at follow-up = 17.5 yr) GH-deficient patients. Cross-sectional analysis by age suggested that behavior problem scores improved for all groups over time.

These published reports, therefore, indicate that pGH replacement in GH-deficient individuals is not always associated with improved behavioral functioning. Interview and projective test data suggest that GH-deficient patients may be prone to immature, inhibited, withdrawn, and/or depressed behavior prior to the beginning of GH therapy and that a large percentage of these patients continued to have similar problems following treatment. Furthermore, a few patients experience difficulty adjusting to the treatment and may even elect to discontinue it. Without the benefit of control or comparison groups, the earliest studies attributed patients' adjustment difficulties to factors such as parental and/or peer infantalization and disappointment with treatment effects (slower than expected growth). The one published report that used standardized psychometric measures suggests that GH treatment was associated with a relative decline (possibly temporary) in school performance during the early adolescent years. However, social competence and behavior problems were not significantly different from the test norms.

Ongoing Studies with Relevance for the Study of Treatment Effects

The following studies are of children and teenagers who received treatment with pGH and were then switched to rGH following a 10-month or longer hiatus from all GH treatment (see Variation in Treatment Regimens later in chapter). These reports are from ongoing studies that have been recently summarized at professional meetings rather than in print.

The Holmes group continued to follow their short-stature patients for an additional 3 years (Holmes et al., 1988). Their six-year follow-up data on 36 of their original study children (13 with GH deficiency) indicates that, though within the normal range, GH deficient patients were not as competent in school as the CD short-statured (medically healthy group) at follow-up. Social competency declined significantly over the 6-year study period for all short males (GH deficient and CD), while the social competency of females generally improved. Extracurricular activities also declined for GH and CD males but not for females over the study period. At the 6-year follow-up mark, all groups had mild elevations on the behavior problem scales. However, elevated scores were only correlated with height for the men (all diagnoses) in the sample ($r = -0.40$ to -0.65). These correlations suggest that short stature in older (mean = 18.1 years) teenage men is associated with poorer than average behavioral adjustment regardless of etiology.

Another study with bearing on the effects of GH treatment is that of Siegel, Koepke, Bedway, Postellon, Hopwood, and Clark (1988). As with Holmes et al. (1988), the intent of this study was not the assessment of treatment effects per se. Instead, Siegel's group was interested in clarifying the cognitive and academic achievement characteristics of GH-deficient children. Siegel and Hopwood (1986) previously reported cross-sectional standardized test data indicating that more than half of their 42 GH-deficient patients (30 boys/12 girls) were underachieving in math and/or reading relative to expectation based on IQ. By examining additional psychometric data on self-concept, visual-motor ability, and characteristics of the parent-child relationship, the authors concluded that the most likely explanation of underachievement was a cognitive deficit (82%) or poor performance related to poor self-concept (12%). Siegel et al. (1988) were able to retest 27 of their original patients seven years later while most continued on GH treatment. The average age at follow-up was 17 years. Univariate analysis of pretreatment and posttreatment achievement scores indicate that reading performance decreased while math performance increased over the follow-up period. Thirty-seven percent of patients were underachieving relative to IQ at follow-up, 74% of whom were also underachieving at pretest. Attentional factors appeared to be most predictive of patient underachievement.

Additional indirect evidence on the effects of GH treatment is provided by an extension of a study designed to assess the behavioral and attitudinal impact of the FDA ban on distribution of pGH and the announcement of risk of exposure to a degenerative neurologic disease (see Risk of Iatrogenic Neurologic Disorder later in chapter). Mazur et al. (1987) assessed the behavioral adjustment and attitudes toward GH treatment of 33 GH-deficient patients (29 boys/4 girls) and at least one parent as the patients resumed GH treatment with rGH. The majority of patients treated with pGH elected to resume treatment with rGH as soon as it became available. Twenty were retested with a standardized test of behavioral adjustment after the patients had received an average of 1.5 years of continuous treatment with rGH (Clopper et al., 1988). The mean age at follow-up was 15.7 years. Findings indicated that parental reports of behavioral problems decreased significantly over the course of treatment. However, parents also perceived their teenagers as being significantly less competent at follow-up, particularly in the areas of schoolwork and activities. Social competency scores tended to increase slightly. While individual patients varied in the severity of their parents' adjustment ratings, both the pretreatment and posttreatment mean scores were all within the normal range for the test.

To summarize, three continuing follow-up studies have looked at adolescent patients who received both pGH and rGH replacement. Though not originally designed to examine treatment effects, they provide additional insights by virtue of their longitudinal assessments during GH treatment. Using standardized psychometric tests, all three studies recently reported evidence of declining school competence during their 1.5- to 7-year follow-up periods. Furthermore, Siegel et al. (1988) found that reading performance declined more than math performance. Extracurricular activities also declined in the two studies assessing this variable while social competence improved slightly over a short follow-up period (1.5 years) but decreased over 7 years of follow-up.

Behavioral Correlates of Sex Hormone Replacement in Growth Hormone Deficiency

Martin and Wilkins (1958) reported interview data on 26 patients (20 male/6 female) with clinical evidence of GH deficiency and lack of sexual development "past the age of puberty." Pituitary deficits were idiopathic in 19 cases and tumor related in 7 cases. In this pretherapeutic GH era, none of the 26 patients received GH replacement. Sex steroid replacement was given to all patients beginning in the late teenage to young adult years and three patients also were tried on varying doses of chorionic gonadotropin (LH like) for up to 9 months. In the male patients, testosterone (T) replacement was followed by somatic maturation although testosterone-stimulated hair growth (virilization) was judged to be "considerably below normal." Testosterone therapy was judged beneficial by patients who cited more stamina, improved secondary sex characteristics and a growth spurt as specific benefits. Psychosexually, erections increased in frequency and nocturnal emissions began in some cases. The frequency of these phenomena appeared to fluctuate with the injection cycle. Frequency was highest just after an injection and appeared to wane just prior to the next injection. The patients were described as "avoiding" dating, petting, and intercourse. Women were begun on a combination of estrogen and low-dose androgen until somatic maturation was complete. They were then switched to "periodic therapy." The authors judged that sex hormone therapy was more effective both somatically and behaviorally in women than men since the women appeared to engage in more dating and, unlike the men, one woman married and reported satisfactory intercourse.

Meyer-Bahlburg and Aceto (1976) reported on the psychosexual behavior of 11 idiopathic hypopituitary patients (9 male/2 female) using a structured interview technique. Four patients had isolated GH deficiency while six had multiple pituitary deficits including Gn. The remaining case had normal Gn but primary testicular failure thought secondary to his mosaic genetic endowment (46, XY; 47, XXY; 48, XXXY). All were over 16 years of age at follow-up. While most patients received GH treatment, the duration was variable and sex steroid replacement (T or E) was given in only five cases at follow-up. One patient had discontinued T treatment because of the pain of injections. The results of this survey suggest that sexual behavior was late in onset and that frequencies of sexual behavior were low, particularly interpersonal eroticism. The authors felt that the two women in the sample were somewhat better adjusted behaviorally and experienced more complete pubertal development than did the treated males. Ten of the 11 patients were heterosexual while one male was effeminately homosexual/transexual in terms of gender identity.

Money and co-workers produced a series of clinical reports describing the psychosexual development of five men treated for panhypopituitarism secondary to a pituitary tumor (Money & Clopper, 1975), nine men with isolated GH deficiency (Clopper, et al., 1976), and six men with idiopathic multiple pituitary deficits including GH and gonadotropin (Gn) deficiencies (Money et al., 1980). Using transcripts of tape-recorded structured interviews with patients and parents collected over 3 to 25 years, Money's group reported that men deficient in both GH and Gn (Money et al., 1975; Money et al., 1980) reported delayed onset of ejaculation, masturbation, dating, necking, petting, intercourse, sexual fantasy, and marriage. Gender identity was clearly heterosexual. While T replacement (typically after age 15) was followed by increases in the reported frequency of solitary sexual behavior and to a lesser extent dating, the more intimate interpersonal forms of sexual behavior were infrequently reported well into the third decade of life. Furthermore, nonerotic social activities such as social group membership and frequency of activities with friends appeared infrequent as well. Similar reports were made by the isolated GH-deficient group even though they experience spontaneous, though slightly delayed, pubertal onset (Clopper et al, 1976). Only a few of the isolated GH group experienced simultaneous GH (pituitary-derived GH therapy) and sex steroids (endogenous T). A normal control group was not studied.

The initial reports by Money's group were extended by comparing four groups of men with: (1) isolated GH deficiency; (2) GH and Gn deficiency; (3) isolated Gn deficiency; and, (4) no pituitary deficit but primary testicular failure (anorchia) on the same variables and with comparable Kinsey norms (Gebhard & Johnson, 1979) when norms were available (Clopper, 1981). Gender identity was again heterosexual. Between-group trends were not statistically significant, aside from delays for all four groups in the onset of all behaviors except erection relative to available norms, and significantly lower frequencies of courtship and marriage among all of the pituitary-deficient groups relative to the anorchic group. Study group sizes were quite small (8–10 men per group).

Noting that male patients with GH deficiency and Gn deficiency frequently express dissatisfaction with the degree of their virilization after T replacement and that Gn replacement is a more completely physiologic therapy for these men, Clopper et al. (1983) reported retrospective interview data on four such men who switched from androgen to Gn therapy. In this pilot study, three men between the ages of 23 and 30 years were switched from T replacement to Gn (LH and FSH) replacement. A fourth patient, age 15, was switched from the weak androgen fluoxymesterone to Gn during his final year of GH replacement. The older three men did not receive concurrent GH replacement. After two years of continuous Gn replacement the current status of each man's psychosexual behavior was assessed using the interview methods of Money et al. (1975) and Clopper (1981). Psychosexual development and functioning on prior androgen therapy could only be assessed retrospectively. Detailed data on the degree of patient virilization and pubertal development as well as periodic measurement of sex steroid levels in the blood had been obtained prospectively over the course of Gn replacement. The results indicated uniform reports of the onset of new psychosexual behavior (solitary and/or social) and increases in the frequency of most previously established psychosexual behaviors after switching to Gn therapy. In addition, physician ratings of the degree of virilization each man experienced on Gn treatment dramatically increased over similarly obtained ratings prior to the change in therapy. All four men were heterosexual. A longitudinal, double blinded, placebo-controlled clinical trial of the relative behavioral, endocrinological, and somatic (virilization) effectiveness of T and Gn replacement in GH- and Gn-deficient men is currently in progress.

To date, published reports on the behavioral effects of sex hormone replacement in GH-deficient patients have relied exclusively on self-report interview data. Since treatment regimens (both GH and sex hormone) and patient characteristics varied markedly from study to study, results must be considered tentative. In men, the most consistent findings are that psychosexual behavior (with the exception of erection) may begin later than usual in GH-deficient individuals and that T replacement is followed by increases (or initial onset in some cases) in a variety of private or solitary psychosexual phenomena (e.g., erection, fantasy, and masturbation). Present evidence suggests that the more intimate interpersonal sexual behaviors (e.g., necking, petting, and coitus) remain relatively infrequent in GH-deficient men even after long periods (10 years) of T replacement. This later trend is consistent with reportedly low rates of marriage summarized elsewhere in this volume (see Dean chapter). The effects of Gn replacement are just now beginning to be explored. Data on women are so rare that interpretation is not warranted.

Pitfalls and Continuing Challenges

At least two sets of complicating issues must be addressed in future studies if our understanding of treatment effects is to advance. These same issues must also be considered while interpreting results from existing studies. One set of issues centers on endocrine variables while the second centers on the methodological and design issues inherent in the behavioral study of children.

Endocrine Treatment Variables

Rhythmic Secretion and Goodness-of-Fit

Endocrine treatments tend to be "replacement" regimens rather than "pharmacologic " regimens. That is, the administered hormone is given with a dose and frequency that attempts to *approximate* physiologic levels of the hormone in the healthy child. "Approximate" is a key word here. While clearly beneficial to GH-deficient individuals in terms of linear growth, the goodness-of-fit is poor between the pattern of serum concentrations obtained by the exogenously administered hormone and the dynamic pattern of serum concentrations seen in naturally occurring secretion. The maximum plasma GH level achieved by a replacement injection appears to be higher and of longer duration than the spontaneous cycle of peaks seen in normal subjects (Brook, 1989). Consequently, the net effect of a replacement injection of GH appears be a smoother, less variable serum concentration of GH immediately following an injection than that seen in normals. In the very deficient child, this peak is likely to be followed by a very low level of GH until the next injection 10 or more hours later depending upon the injection schedule used. It remains highly possible that this problem leads to chronic under replacement in terms of total exposure to GH over the course of a child's total growth period.

Similar to GH replacement, typical sex steroid (T or E) replacement regimens also produce patterns of serum concentrations that mute the dynamic fluctuations seen in normal secretion. Perhaps the most noticeable example is seen in the pattern of serum T induced by T replacement with one of the long-acting T esters (e.g., T enanthate) which are currently the treatment of choice in male hypogonadal states. These T preparations induce serum T concentrations in or above the normal range shortly after injection which then gradually decline over the next 1 to 4 weeks depending upon the dosage and the ester used. The spontaneous pattern of normal T secretion shows multiple fluctuations of up to threefold in magnitude over the course of one day (see Griffin & Wilson, 1985).

The goodness-of-fit problem is even more complicated when assessing the behavioral effects of sex hormone replacement therapy if the T or E deficit is the result of a deficit in the pituitary regulatory hormones that usually stimulate gonadal sex hormone production and fertility. Note that in the case of sexual infantilism due to hypogonadotropinism, which can occur along with GH deficiency, the typical treatment is replacement T or E in males or females, respectively. Such gonadal steroid replacement is *not* a completely physiologic endocrine replacement since the gonadotropins (FSH and LH) themselves are not replaced. FSH and LH have their own patterns of rhythmic secretion that are only partially subject to gonadal steroid control (Griffin & Wilson, 1985; Odell, 1989).

Another goodness-of-fit problem is the question of hormonal interaction as a contributor to behavioral as well as somatic effects. While one action of a hormone tends to be emphasized over others, hormones typically contribute to more than one physiologic process. For example, the role of GH in linear growth occurs simultaneously with its role in carbohydrate metabolism. It is, therefore, reasonable and probably prudent to consider the importance of the hormonal milieu to which a GH or sex hormone therapy is added. While still largely uncharacterized for behavioral phenomena, evidence already exists that multiple hormonal influences are important for normal somatic growth. For example, the normal growth spurt at puberty appears to be dependent upon not only gonadal steroids (T or E) but also GH, adrenal sex steroids, and insulin-like growth factor-I (Bourguignon, 1988). Similar interactions may also contribute to behavioral outcome. For example, the data of Clopper et al. (1983) and more recently Gooren (1988) on psychosexual functioning during GN versus androgen replacement in GH-deficient and GH-sufficient males, respectively, raises the possibility that both GN and T may be required for normal psychosexual functioning.

As can be seen, the goodness-of-fit problems inherent in studies of GH and sex hormone treatment must temper the interpretation of study findings until the behavioral importance of the cycles themselves are clear. These problems are not routinely addressed.

Compliance with Treatment

Assessing the effects of a hormone treatment requires knowing that the prescribed treatment actually produces plasma concentrations of the replaced hormone in the physiologic range. Since it is impractical to continuously monitor the replaced hormone level in a patient's plasma, investigators are forced to rely on patients' compliance to assure the success of adequate replacement. Growth hormone and sex hormone regimens are long-term chronic treatments for deficient individuals. Furthermore, both types of hormone replacement are less critical for good general health than are, for example, insulin replacement for juvenile onset diabetes mellitus or cortisol replacement in adrenal insufficiency. Consequently, compliance with GH or sex hormone therapy may appear to patients to be less necessary for maintaining their health. As a result, patients' motivation for good compliance may be less resolute. Since good compliance is central to assessing treatment effects, this source of error must be considered in the study of the effects of treatment. Adequate compliance has been assumed in the majority of studies reported earlier.

Variation in Treatment Regimens

There have been two sequential and distinct eras in the history of GH replacement therapy. The differences between the regimens of the two eras complicate the interpretation and comparison of behavioral data from the two time periods by adding additional sources of variation.

Prior to the widespread availability of bioengineered GH (rGH) in late 1985, GH therapy regimes were different from those available today because of a severe shortage of purified human pGH. Treatment was limited to only those children with severe short stature who tested deficient on two provocative tests of GH release. Also, treatment could only continue until a predetermined height was attained (often 5' 6" or less for males) even if puberty had not yet intervened to stop linear growth. In addition, many of the pituitary-derived GH recipients were periodically taken off GH therapy for several months each calendar year. When rGH solved the supply problem, it became possible for the first time to treat GH-deficient children continuously and indefinitely. Some children received only one of the regimens while other children began on pGH treatment and switched to rGH. Potential differences in the treatment expectations of children on the two regimens has not been assessed even though evidence from the earlier period of GH treatment suggests that treatment expectations may be important in assessing behavioral outcome (Kusilac et al., 1972; Rotnem et al., 1979; Grew et al., 1983).

A second difference between the pGH and rGH regimens is that the method of administration has changed and indeed may still be evolving. Most pGH was given by intramuscular injection three times per week. More recently different injection schedules (e.g., daily) and subcutaneous rather than intramuscular injection procedures have been used. The general effect of these differences on behavioral outcome and the specific effects on compliance and treatment expectation remain unexplored.

Risk of Iatrogenic Neurologic Disorder

Another difference is that some unknown number of patients receiving pGH may have been accidentally exposed to a lethal degenerative neurologic disease, Creutzfeldt-Jakob disease (CJD), which is believed to be due to a slow viral infection that may not manifest itself for decades after exposure (Gibbs et al., 1985; Brown et al., 1985). Treatment with rGH does not carry this risk. Some of the patients given pGH have also been treated with rGH following the 1985 FDA ban on the distribution of pituitary-derived GH. Many switched to rGH as soon as it became available (Mazur et al., 1987). The acute phase of CJD is marked by dementia, cerebellar and visual dysfunction, and movement disorders (Brown et al., 1979; Brown, 1980). The early neurologic effects of this virus are unknown and definitive diagnosis can be made only after death via brain histology (Gibbs et al., 1985). Consequently, this potential biasing effect must be considered in the selection of samples for future study and in the interpretation of data on the older cohort of pGH recipients.

Etiologic and Diagnostic Differences

One source of diagnostic variation in samples of GH-deficient patients is the severity and type of deficiencies present in each subject. In fact, the standard stipulative definition of growth hormone deficiency (e.g., provocative test values < 10 ng/ml) has recently been called into question by evidence of poor growth with normal (> 10 ng/ml) GH responses to provocation (Rose et al., 1988), normal GH response to stimulation but abnormal nocturnal GH secretory patterns (Bercu & Diamond, 1986), and a distribution of GH responses to stimulation among normally growing children that overlaps the classical deficient range (Rose et al., 1988). Since patients appear to be less homogeneous than originally thought with respect to the severity and type of GH deficiency based on current testing procedures and since the behavioral implications of these differences are currently undocumented, it becomes more important to document these differences in order to clearly define treatment effects.

A second source of diagnostic variation arises from the fact that some GH-deficient children have multiple hormone deficiencies. GH is one of five anterior pituitary peptide hormones. Adrenocorticotropic hormone (ACTH), thyrotropic (TSH), and the gonadotropins (FSH and LH) can also be deficient. ACTH and TSH, respectively, are treated with replacement oral cortisone and thyroid hormone (cf. Underwood & Van Wyk, 1985). Replacement of Gn is discussed previously in this chapter.

A third source of diagnostic variation is the correlated treatment variables found when GH or sex hormone deficiency is secondary to a CNS tumor, surgery, or radiation rather than occurring on an idiopathic basis. The treatment of pituitary or hypothalamic tumors typically involves brain surgery and/or radiation to prevent tumor regrowth (Preece, 1982). Posttreatment intellectual deficits have been reported in tumor-related GH deficiency (Clopper et al,, 1977; Galatzer et al., 1981) as well as in children who received cranial radiation for the treatment of acute lymphoblastic leukemia (Meadows et al., 1981; Krammer et al., 1988) some of whom also experience GH insufficiency after treatment (Blatt et al., 1984).

Perinatal trauma has also been implicated as a possible contributing factor to GH-deficient states (Rona & Tanner, 1977; Craft et al., 1980). Once again severity, duration, and type of perinatal insult probably influence the behavioral development of these children both through direct effects on the developing nervous system and indirectly through the familial environment (see Annecillo chapter for example of extreme environmental factors). However, perinatal trauma is seldom reported, much less systematically assessed in the existing behavioral literature.

Methodological and Design Issues

The assessment of the effects of endocrine treatments on behavioral variables presently occurs largely at the descriptive level of analysis. While one might consider the institution of GH or sex hormone replacement as an experimental manipulation, it is not the same as the classical rigidly controlled studies required to establish cause and effect relationships between variables. Nonetheless, the immediate task is one of establishing ways to study the effects of GH and sex hormone treatment on behavior, given the limitations of this type of human research.

Study Designs

The overall structure of a research project including the subject selection and data collection procedures determine the design of the project. While there are many different aspects to designing a study and good resources exist on this topic (e.g., Winer, 1971; Barlow & Herson, 1984; Stevens, 1986), the following considerations are particularly relevant to studies examining the behavioral concomitants of endocrine treatment regimens in children.

Since endocrine replacement therapies tend to be long-term chronic treatments and, since the physiologic activity of the hormones administered have specific time courses for their action, the way time is incorporated into a study design has significant bearing on the interpretability of the results. Prospective or forward-looking designs have distinct advantages over retrospective or backward-looking designs. Note that the improvement in psychosexual functioning reported by men after switching from androgen treatment to Gn replacement (Clopper et al., 1983) was based upon retrospective recall of each man's psychosexual functioning on androgen replacement two years prior to their assessment. The extent to which their recall was inaccurate or their experience with Gn replacement, per se, distorted their perception of their past behavior on androgen is unknown. Consequently, a conclusion of an enhancement of their psychosexual functioning on Gn relative to T is weakened by uncertainty introduced by the retrospective data. In the same study, prospectively collected endocrine data for the two treatments provided much clearer evidence that Gn replacement stimulated testicular function and improved virilization over the androgen-only state.

Another way the temporal aspects of treatment can enter a study design is through the choice of longitudinal or cross-sectional data collection strategies. This choice is also relevant to the accuracy of assesssing developmental phenomena where behavioral performance matures over time with the appearance of qualitatively different behaviors at different times in the life cycle. A cross-sectional design estimates treatment or developmental effects over time by comparing multiple samples of differing ages or durations of treatment on the dependent variables of interest. The longitudinal alternative follows one sample through time by repeatedly assessing the same subjects as they experience the treatment. While the cross-sectional approach provides economy in terms of data collection time and expense, it poorly assesses developmental phenomena that may interact with the treatment effects in significant ways. For example, Holmes et al. (1986) reported three-year longitudinal data on two groups of short children of mixed etiology, a younger group (mean initial age 9.7 years) and an older group (mean initial age 14.4 years). Examination of the three-year longitudinal trends for the two groups suggested that short children experienced a temporary decrease in their social and school competencies during early teenage years but rebounded during later teenage years. As Holmes et al. (1986, p. 9) correctly point out, this apparent trend needed to be interpreted cautiously since the age comparison was essentially cross-sectional since different samples of children were assessed at different ages. Their caution appears well founded, since their 6-year longitudinal data (Holmes et al., 1988) indicated an overall decline in social competence for males through the mid to late teenage years rather than the temporary decline suggested by the earlier cross-sectional analysis.

A subset of the cross-sectional versus longitudinal study design question is selection of within-subject, repeated measure designs versus randomized group comparison designs. In the context of studying treatment effects, the within-subject designs—where one set of subjects is followed over time (prospectively) and the same variables are assessed in each subject on multiple occasions—have several advantages over the group comparison designs where two different groups of subjects are assessed, one before and one after treatment. First, error in the estimation of treatment effects due to differences between individual subjects is minimized because each subject is compared with himself/herself over time. As a result, within-subject repeated measure designs tend to be statistically more powerful in terms of detecting differences and require fewer subjects to demonstrate effects (Stevens, 1986). However, they are also subject to confounding by order effects or carryover effects if more than one treatment is given. These problems are particularly troublesome in the area of sex hormone replacement and cognitive functioning since sex hormones are believed to affect cognitive performance (Rubin et al., 1981).

The problems with repeated measure designs are illustrated in the following example. GH patients with multiple pituitary deficits receive several hormone replacements simultaneously and possibly sequentially. Consider a male patient with GH and thyroid deficiencies that are recognized and treated simultaneously throughout childhood. In teenage, his Gn deficiency is diagnosed when puberty does not occur and T replacement is added to the overall regimen. In addition, suppose he was given a short course of a weakly androgenic steroid such as oxandrolone to boost his growth rate during a brief period of decelerating growth on GH and thyroid replacement alone. In this case, if a longitudinal prospective assessment of the effects of T replacement were begun at the time T was initiated, the carryover of an effect of the oxandrolone treatment could not be separated from the effect of T alone. Carryover effects can be minimized though not necessarily eliminated by estimating the dependent variables during each treatment and during a no-treatment period that falls between the two replacement treatments. There is also an order effect problem in the above example since it is possible that the effect of T might be different if it follows oxandrolone. Order of treatment problems sometimes can be eliminated by counterbalancing treatments such that half the subjects receive one treatment first while the other half receive the second treatment first. Again a no treatment or placebo period between treatments is needed to isolate the effects of the two therapies. These strategies are often not an option without significant preplanning and coordination by the treatment team.

Another design question is the issue of including and selecting a control or comparison group. In research with people on the effects of replacement treatment, true control groups are usually not possible. In the first place withholding treatment from a deficient person is unethical and inhumane. Secondly, randomly selected healthy control subjects are not perfect matches since they are not deficient. Furthermore, replacement therapies provide only an imperfect approximation of the healthy state. However, healthy control subjects, like other comparison groups, provide important reference data with which to assess the statistical and pragmatic significance of documented effects in the patient sample. Without such comparative standards treatment-associated changes in behavior can not be clearly interpreted. For example, Money and Pollitt's (1966) descriptive prospective study of GH-treated, GH-deficient patients suggested that their patients exhibited a "psychomaturational lag" and that the lag was related to the inabilty of the parents to treat the patients in accordance with their chronologic age. In the absence of a reference group it is not possible to decide how different the study group is from healthy normal-statured controls or if the noted association holds true for other healthy short children as well.

Several strategies for selecting comparison groups are worth keeping in mind. Matching comparison subjects with each patient adds strength to a study by minimizing differences between subjects based on factors such as sex, age, socioeconomic status (SES), and IQ. The closer the match between patients and the comparison subjects the more precisely the treatment effects can be estimated. None of the studies reviewed herein used a matched control group design. Currently, Siegel's group (personal communication) is using the sibling control variation in her continuing study of academic achievement in GH-deficient patients. In this approach, the closest sibling in age is also studied in order to control variation in outcome that is associated with familial factors (e.g., genetic endowment or SES). Even though perfect age and sex matches are often not possible within any given family, comparisons between the patient and sibling control groups should improve power to detect treatment-related effects over time.

The norms established by standardized tests provide a convenient comparison group. Assuming reasonably good validity and reliability, patients' standardized scores provide a means of representing each patient's relative position in the reference population. Age and sex norms are especially important in studies of children in order to allow for the effects of normal maturation. The more recent studies reviewed earlier demonstrate the utility of well-standardized psychometric tests (Holmes et al., 1986, 1988; Clopper et al., 1988; Siegel et al., 1988) for assessing the behavioral characteristics of GH-deficient children undergoing hormone replacement. These studies also appear to mark a shift in measurement strategies from interviews to standardized psychometric tests for assessing behavioral variables.

The assessment of treatment effects does not always require population norms, however. For example, treatment effects can be assessed within subjects using suitably precise and objective measures. Barlow and Hersen (1984) provide an excellent presentation of a variety of experimental designs that are appropriate for studying individual patients over time. The basic strategy is to assess each patient before a treatment is begun and reassess the patient during the treatment and after the treatment has ended. It is then possible to demonstrate the degree of association between the treatment and the measured behavior in each case. This type of single case design is particularly useful in clinical research.

Reducing Measurement Error

With the exception of Wechsler and Stanford-Binet IQ tests, early studies of the behavioral characteristics of GH-deficient patients and the behavioral correlates of their treatment relied almost exclusively on measuring behavioral variables through interview data (Mazur et al., 1987). While providing important and necessary exploratory data, interview data are highly subject to error due to factors such as intersubject variation in the comparability of patients' understanding of each question, rapport with the interviewer, patient perceptions of the social desirability of the accurate response, and the subjective quality of patients' reports about themselves. Interviews also do not routinely provide comparative data or norms. As a result, the early investigations gave important directions for further investigation but only broad characterizations of complex behavioral phenomena since the measures used provided poor resolution and specificity.

Kusalic et al. (1972) and Rotnem et al. (1979) added projective psychometric tests to their interview assessment. The projective tests only partially improved the precision of the behavioral data. They provided a clearer indication that poor self-esteem, anger/aggression, and depression might by important aspects of Money's psychomaturational lag hypothesis (Money et al., 1966). However, while improving the standardization of the stimulus (e.g., ink blots or photographs) and the scoring procedure, the projective test scores and their interpretation remain quite subjective and difficult to relate to other populations of children.

Population comparisons became much easier with the development of a well-constructed behavioral checklist with norms on clinical as well as nonclinical populations of children (Achenbach, 1979; Achenbach & Edelbrock, 1983). Use of these instruments allowed Holmes et al. (1988) to discern specific types of behavior problems and competencies that appear inversely correlated with the height of the short young men in her sample. Norms on psychosexual behavioral development remain especially difficult to obtain due in large part to the sensitivity of the subject and attendant taboos that hinder epidemiologic study of normal psychosexual behavior. For example, in an attempt to use normative data Clopper (1981) was only able to find norms for four of ten types of typical psychosexual behavior (Gebhard & Johnson, 1979).

The addition of physiological measures to a behavioral assessment battery provides several important strengths. To date, these types of measures have not been included in the study of the psychosocial and psychosexual correlates of replacement treatment. Physiologic measures tend to be more easily quantified and scaled than even some psychometric tests. They also have the advantage of relating complex behavioral phenomena to concurrent physiologic events and thereby establishing sets of convergent behavioral measures. Unfortunately, even the rather easily obtained, noninvasive measures such as changes in heart rate, respiration, blood pressure, or skin conductance remain largely unrelated to the types of psychosocial and psychosexual behavior that have been the focus in past studies of GH treatment. Also, measuring physiologic phenomena is often more expensive than interview of psychometric assessments due to a requirement of specialized biomedical instrumentation for physiologic recording.

The potential usefulness of including physiologic measures of treatment effects is currently under study by Clopper (unpublished data). In this study, GH- and Gn-deficient men are cycled through periods of treatment with T, Gn, and saline in order to assess the relative effectiveness of T and Gn for stimulating their pubertal psychosexual development. Each man is repeatedly assessed before, during, and after each treatment period. The behavioral outcome measures are self-reported frequencies of specific psychosexual behavior and physiologic measures of sleep erection phenomena recorded during each treatment period. The study is currently in progress so that only preliminary results are available. At this time it appears that the sleep erection cycle is differentially influenced by the treatments and that the physiologically assessed variables show more consistent differences per treatment than do the self-report measures.

Multivariate Assessment

The most straightforward assessment of a replacement treatment regimen is a multivariate one. Multiple background, diagnostic treatment, and behavioral outcome variables need to be assessed simultaneously in order to avoid oversimplification and distortion of the interrelationships of the variables. However, increasing the number of variables requires increasing the number of observations of each variable in order to accurately detect and assess treatment effects (see Stevens, 1986). This is usually accomplished by increasing the number of subjects in the study. However, GH-deficient patients are rare (see Dean chapter). How then does an investigator obtain a sufficient number of observations to carry out a multivariate study?

One way is to join with other researchers in a cooperative research study involving multiple treatment sites. By agreeing in advance on a common study protocol it is then possible to study much larger samples of these rare individuals than any one center can do alone. Even more than single center studies, multicenter studies usually require outside funding. Governmental and corporate support provide important funding opportunities for such projects.

A second way to solve the number of observations problem when the population under study is infrequent is to collect multiple observations on the available subjects. Though offering an alternative to increasing the sample size, this approach is technically less desirable because it lacks the strength of replicating treatment effects across a larger group of subjects. While it is technically possible to replicate the treatment effect within each subject by cycling between replacement treatment and placebo treatment periods, it is not always possible due to the ethics of not treating and/or the willingness of patients to knowingly alternate between periods on and off replacement.

Summary and Recommendations

The assessment of the behavioral effects of replacement hormone therapy has been hindered by many factors. Progress has necessarily awaited solutions for such problems as a limited supply of GH, a very limited number of patients to study, and the availability of psychometrically sound survey instruments. Improvements in psychometric test construction for children have partially alleviated the behavioral measurement problem. As survey instruments, however, they do not provide a detailed picture of the factors contributing to the patterns they identify. The approval of rGH for therapeutic use solved the problem of limited supply and improved the problem of scarce subjects by increasing the number of patients that can be treated. GH deficiency is still a rare condition, however. Bioengineered rGH also provided the first opportunity ever to study patients under continuous and potentially chronic replacement therapy.

Accounting for the potentially important confounding effects such as differing: (1) diagnostic characteristics (e.g., isolated or multiple deficiencies), (2) etiologies (e.g., idiopathic, perinatal insult, tumor, or sleep-related deficiencies), (3) modes of administering a regimen (e.g., frequency of injections and type of injection), (4) degrees of compliance with prescribed regimen, and (5) treatment expectations are difficult but not impossible tasks for new carefully designed studies. However, a significant hurdle remains. The potential importance of inherent differences in the replacement regimen versus the naturally occurring pattern of GH secretion can not be overlooked, particularly since the cyclical pattern of release of the pituitary-gonadal axis has recently been related to fluctuations in cognitive performance in women (Hampson, 1988; Kimura, 1988). A prospective comparison of patients beginning replacement and healthy control subjects would provide a good first step in evaluating this problem.

Assessing the behavioral effects of replacement hormone treatment in children and adolescents is clearly a multivariate problem that has received relatively little attention. As with any line of new investigation, the available reports have focused on descriptive or phenomenological analysis of the behavioral characteristics of patients receiving GH and/or sex steroid replacement. In general, behavioral assessments have progressed from relying on interview methods to the use of standardized psychometric questionnaires with well-established norms. Assessments of cognitive functioning are an exception to this trend since psychometrically sound IQ tests have been available for decades. Longitudinal study designs have been used from the beginning while the use of comparison groups and well-normed self-report instruments have appeared within the last 5 years. However, truly multivariate studies, in which a large set of relevant variables are simultaneously considered, are just beginning. This type of design and analysis requires either multiple observations on the same subjects during different treatments or large samples of subjects in order to obtain reliable results. Longitudinal multicenter studies offer a chance to systematically collect the necessary data sets to evaluate the significance of many factors that can not now be ruled out as trivial. Relating self-report measures of behavioral adjustment to less subjective physiologic measures such as heart rate, blood pressure, respiration, or EEG indices may provide a new generation of convergent behavioral measures for multivariate analyses.

The available literature on GH replacement does not provide data on the behavioral effects of GH per se. It does provide an increasingly detailed picture of behavioral development coincident with GH treatment. The original characterization of ''psychomaturational lag'' has been refined by subsequent work to suggest that, as a group, children undergoing GH treatment, while not grossly maladjusted, are prone to experience problems with academic achievement and social adjustment. There is conflicting evidence that GH replacement is associated with behavioral improvement. Puberty appears to be associated with more difficulty than the prepubertal years even for GH-deficient patients on treatment. However, it is unclear how much this trend is attributable to the typical problems of the pubertal period, to an effect of the GH deficit, or perhaps to the treatment itself. While two reports suggest that psychosexual adjustment is better in women than men, the number of women studied is so small that generalizations are unwarranted. The available evidence on men suggests delays in the onset of solitary sexual behavior (except erection) until sex hormone therapy is begun and continued delays in the onset of sociosexual behavior and marriage even after sex hormone replacement. Their degree of virilization on replacement T may also be deficient. However, interpretation of the psychosexual data is complicated by the lack of adequate normative information and the difficulties encountered with studying sexual development in human subjects.

As can be seen from this chapter, the time is ripe for new advances in our understanding of the behavioral correlates of GH and sex hormone replacement therapy. Careful multivariate prospective studies of large samples provide an exciting opportunity to clarify our understanding of the biobehavioral relationships associated with the treatment of GH and sex hormone deficiencies.

References

Achenbach, T.M. (1979). The child behavior profile: An empirically based system for assessing children's problems and competencies. *International Journal of Mental Health, 7*:24–42.

Achenbach, T.M., & Edelbrock, C. (1983). *Manual for the child behavior checklist and revised child behavior profile.* Burlington, VT: University of Vermont.

Barlow, D.H., & Hersen, M. (1984). *Single case experimental designs* (2nd ed.). New York: Pergamon Press.

Bercu, B.B., & Diamond, F.B. (1986). Growth hormone neurosecretory dysfunction. *Clinical Endocrinology and Metabolism, 15*:537–590.

Blatt, J., Bercu, B.B., Gillin, J.C., Mendelson, W.B., & Poplack, D. (1984). Reduced pulsatile growth hormone secretion in children after therapy for acute lymphoblastic leukemia. *Journal of Pediatrics, 104*:182–186.

Bourguignon, J. (1988). Linear growth as a function of age at onset of puberty and sex steroid dosage: Therapeutic implications. *Endocrine Reviews, 9*:467–488.

Brook, C.G.D. (1989). Growth hormone deficiency: Features, assessment, and management. In L.J. DeGroot (Ed.), *Endocrinology* (2nd ed.) (pp. 351–362). Philadelphia, PA: W.B. Saunders Co.

Brown, P. (1980). An epidemiologic critique of Creutzfeldt-Jakob disease. *Epidemiologic Reviews, 2*:113–135.

Brown, P., Cathala, F., Sadowsky, D., & Gajdusek, D.C. (1979). Creutzfeldt-Jakob disease in France: II. Clinical characteristics of 124 consecutive verified cases during the decade 1968–1977. *Annals of Neurology, 6*:430–437.

Brown, P., Gajdusek, D.C., Gibbs, C.J., & Asher, D.M. (1985). Potential epidemic of Creutzfeldt-Jakob disease from human growth hormone therapy. *New England Journal of Medicine, 313*:728–731.

Clopper, R.R. (1981). *Erotosexual behavior in men treated for growth hormone and/or gonadotropin deficiencies.* Unpublished doctoral dissertation, The Johns Hopkins University.

Clopper, R.R., Adelson, J.M., & Money, J. (1976). Postpubertal psychosexual function in male hypopituitarism without hypogonadotropinism after growth hormone therapy. *Journal of Sex Research, 12*:14–32.

Clopper, R.R., Mazur, T., MacGillivray, M.H., Peterson, R.E., & Voorhess, M.L. (1983). Data on virilization and erotosexual behavior in male hypogonadotropic hypopituitarism during gonadotropin and androgen treatment. *Journal of Andrology, 4*:303–311.

Clopper, R.R., Mazur, T., Mills, B., MacGillivray, M., & Voorhess, M. (October, 1988). *Longitudinal psychometric assessment of behavioral adjustment on growth hormone therapy.* Presentation at the Genentech National Cooperative Growth Study, Symposium III. Palm Desert, CA.

Clopper, R.R., Meyer, W.J., Udvarhelyi, G.B., Money, J., Aarabi, B., Mulvihill, J.J., & Piaso, M. (1977). Postsurgical IQ and behavioral data on 20 patients with a history of childhood craniopharyngioma. *Psychoneuroendocrinology, 2*:365–372.

Craft, W. H., Underwood, L.E., & Van Wyk, J.J. (1980). High incidence of perinatal insult in children with idiopathic hypopituitarism. *Journal of Pediatrics, 96*:397–402.

Galatzer, A., Nofar, E., Beit-Halachmi, N., Aran, O., Shalit, M., Roitman, A., & Laron, Z. (1981). Intellectual and psychosocial functions of children, adolescents and young adults before and after operation for craniopharyngioma. *Child Care, Health and Development, 7*:307–316.

Gebhard, P.H., & Johnson, A.B. (1979). *The Kinsey data: Marginal tabulations of the 1938–1963 interviews conducted by the Institute for Sex Research.* Philadelphia, PA: W.B. Saunders Co.

Gibbs, C.J., Joy, A., Heffner, R., Franko, M., Miyazaki, M., Asher, D.M., Parisi, J.E., Brown, P.W., & Gajdusek, D.C. (1985). Clinical and pathological features and laboratory confirmation of Creutzfeldt-Jakob disease in a recipient of pituitary-derived human growth hormone. *New England Journal of Medicine, 313*:734–738.

Gooren, L.J.G. (1988). Hypogonadotropic hypogonadal men respond less well to androgen substitution treatment than hypergonadotropic hypogonadal men. *Archives of Sexual Behavior, 17*:265–270.

Grew, R.S., Stabler, B., Williams, R.W., & Underwood, L.E. (1983). Facilitating patient understanding in the treatment of growth delay. *Clinical Pediatrics, 22*:685–690.

Griffin, J.E., & Wilson, J.D. (1985). Disorders of the testes and male reproductive tract. In J.D. Wilson & D.W. Foster (Eds.), *Williams textbook of endocrinology* (7th ed.) (pp. 259–311). Philadelphia: W.B. Saunders Co.

Hampson, E. (November, 1988). *Contributions of gonadal hormones to human cognitive and motor skills: Evidence from the menstrual cycle.* Paper presented at the Annual Meeting of the Society for Neuroscience, Toronto.

Holmes, C., Karlsson, J.A., & Thompson, R.G. (1986). Longitudinal evaluation of behavior patterns in children with short stature. In B. Stabler & L.E. Underwood (Eds.), *Slow grows the child: Psychosocial aspects of growth delay* (pp. 1–12). Hillsdale, NJ: Lawrence Erlbaum.

Holmes, C., Sidler, A.K., Tsalikian, E., & Karlsson, J.A. (October, 1988). *Psychological status of children with short stature: Six-year longitudinal follow-up.* Presentation at the Genentech National Cooperative Growth Study, Symposium III, Palm Desert, CA.

Kimura, D. (November, 1988). Hormonal influences on cognitive/motor function in post-menopausal women: The effect of hormone replacement. Paper presented at the Annual Meeting of the Society for Neuroscience, Toronto.

Kramer, J.H., Norman, D., Brant-Zawadzki, M., Ablin, A., & Moore, I.M. (1988). Absence of white matter changes on magnetic resonance imaging in children treated with CNS prophylaxis therapy for leukemia. *Cancer, 61*:928–930.

Kusalic, M., Fortin, C., & Gauthier, Y. (1972). Psychodynamic aspects of dwarfism: Response to growth hormone treatment. *Canadian Psychiatric Association Journal, 7*:29–34.

Lee, P.D.K., & Rosenfeld, R.G. (1987). Psychosocial correlates of short stature and delayed puberty. *Pediatric Clinics of North America, 34*:851–863.

Martin, M.M., & Wilkins, L. (1958). Pituitary dwarfism: Diagnosis and treatment. *The Journal of Clinical Endocrinology and Metabolism, 18*:679–693.

Mazur, T., & Clopper, R.R. (1987). Hypopituitarism: Review of behavioral data. In D.M. Styne & C.D.G. Brook (Eds.), *Current concepts in pediatric endocrinology* (pp. 184–205). New York: Elsevier.

Mazur, T., Clopper, R.R., MacGillivray, M.H., Voorhess, M.L., Tubb, L., Smith, A., & Mills, B.J. (1987). Human growth hormone and the FDA ban: Psychological impact. *Program and Abstracts of the Annual Meeting of the American Pediatric Society.* Anaheim, CA.

Meadows, A.T., Gordon, J., Massari, D.J., Littman, P., Fergusson, J., & Moss, K. (1981). Declines in IQ scores and cognitive dysfunction in children with acute lymphocytic leukaemia treated with cranial irradiation. *Lancet, 2*: 1015–1018.

Meyer-Bahlburg, H.F.L., & Aceto, T. (March, 1976). *Psychosexual status of adolescents and adults with idiopathic hypopituitarism.* Presented at the Annual Meeting of the American Psychosomatic Society, Pittsburgh, PA.

Money, J., & Clopper, R.R. (1975). Postpubertal psychosexual function in postsurgical male hypopituitarism. *Journal of Sex Research, 11*:25–38.

Money, J., Clopper, R.R., & Menefee, J. (1980). Psychosexual development in postpubertal males with idiopathic panhypopituitarism. *Journal of Sex Research, 16*:212–225.

Money, J., & Pollitt, E. (1966). Studies in the psychology of dwarfism: II. Personality maturation and response to growth hormone treatment in hypopituitary dwarfs. *The Journal of Pediatrics, 68*:381–390.

Odell, W.D. (1989). In L.J. DeGroot (Ed.), *Endocrinology,* Vol. 3 (2nd ed.) (pp. 1860–1872). Philadelphia, PA: W.B. Saunders Co.

Preece, M.A. (1982). Diagnosis and treatment of children with growth hormone deficiency. *Clinical Endocrinology and Metabolism, 11*:1–24.

Rona, R.J., & Tanner, J.M. (1977). Aeitology of idiopathic hypopituitary dwarfism: Hormone deficiency in England and Wales. *Archives of Diseases of Childhood, 52*:197–208.

Rose, S.R., Ross, J.L., Uriarte, M., Barnes, K.M., Cassorla, F.G., & Cutler, G.B. (1988). The advantage of measuring stimulated as compared with spontaneous growth hormone levels in the diagnosis of growth hormone deficiency. *New England Journal of Medicine, 319*: 201–207.

Rotnem, D., Cohen, D.J., Hintz, R.L., & Genel, M. (1979). Psychological sequelae of relative "treatment failure" for children receiving human growth hormone replacement. *Journal of Child Psychiatry, 18*:505–520.

Rubin, R.T., Reinisch, J.M., & Haskett, R.F. (1981). Postnatal gonadal steroid effects on human behavior. *Science, 211*:1318–1324.

Siegel, P.T., & Hopwood, N.J. (1986). The relationship of academic achievement and the intellectual functioning and affective conditions of hypopituitary children. In B. Stabler & L.E. Underwood (Eds.), *Slow grows the child: Psychosocial aspects of growth delay* (pp. 57–72). Hillsdale, NJ: Lawrence Erlbaum.

Siegel, P.T., Koepke, T., Bedway, M., Postellon, D., Hopwood, N., & Clark, K. (August, 1988). *Growth hormone deficient children: A seven year follow-up of intellectual, behavioral and academic functioning.* Presentation at the Annual Meeting of the American Psychological Association, Atlanta, GA.

Stevens, J. (1986). *Applied multivariate statistics for the social sciences.* Hillsdale, NJ: Lawrence Erlbaum.

Underwood, L.E., & Van Wyk, J.J. (1985). Hormones in normal and aberrant growth. In R.E. Williams (Ed.), *Textbook of endocrinology* (7th ed.). Philadelphia: W.B. Saunders Co.

Winer, B.J. (1971). *Statistical principles in experimental design (2nd ed.).* New York: McGraw-Hill.

6
Demographic Outcome of Growth Hormone Deficient Adults

HEATHER J. DEAN

Clinical and Anecdotal History of Untreated Growth Hormone Deficient Adults

General Tom Thumb was married to Lavinia Bump on February 10, 1963, in New York City. The groom was 95 cm tall, his wife was 80 cm tall or 12 to 13 standard deviations (SD) below the mean height for adult men and women. Their bodies were proportionate and they were sexually mature. Both were products of consanguinous marriages (McKusick & Rimoin, 1967). This inherited form of short stature was labeled ateleiotic dwarfism by an English physician, Hastings Gilford, after the Greek term "ateleios" meaning deficient. Several years later, the connection between a pituitary affection and this condition was made. Over the next 60 years, before the discovery and isolation of growth hormone in 1956, many case reports appeared in the literature describing adults in their sixth and seventh decades of life who fit the clinical description of ateleiotic or hypopituitary dwarfism (Faurbye, 1946; Hawley, 1963; Hewer, 1944; Goodman et al., 1968).

The affected individuals had immature doll-like facies with soft and wrinkled facial skin, Figure 6.1. The infantile features and midfacial bony hypoplasia made their facial features so characteristic that they tended to look alike. Their voices were high pitched. Infantile body proportions with truncal obesity was described in all of them. Micropenis and bilateral cryptorchidism were present in those males with associated gonadotropin deficiency. In those with normal gonadotropin secretion, puberty occurred spontaneously but was often delayed to as late as age 27 years. Sexual development, once it occurred, proceeded normally. The average age of menarche was 16 years and the average age of body hair growth in males was 19 years. The final adult height was 80 to 130 cm for the females and 95 to 150 cm for the males, approximately 6 to 13 SD below the present-day mean height for adult males and females. The females had normal fertility. Pregnancy was complicated by cephalopelvic disproportion since the off-spring had a normal birth weight and length regardless of the GH status of

FIGURE 6.1. Familial hypopituitarism in three of five siblings. (From McKusick & Rimoin (1967), "General Tom Thumb and Other Midgets," in *Scientific American*, 217. Copyright 1967 by Scientific American, Inc. Reprinted with permission.)

the fetus. Delivery by Caesarean section was the rule. Life span of these adults was apparently normal.

There is limited information regarding the psychosocial outcome of these individuals. P.T. Barnum may have exploited the stature of several of them in his circus, but it appears that they had satisfactory, gainful employment. "Tom Thumb is said to have delighted Queen Victoria and other European rulers with his easy and witty remarks . . . His wife taught school at the age of 13 before going into the circus" (McKusick & Rimoin, 1967). In contrast, the sexually immature individuals often lived at home with their parents throughout their adult lives (McKusick & Rimoin, 1967; Faurbye, 1946; Hawley, 1963; McArthur et al., 1985). Three affected siblings from an inbred Canadian Hutterite family, who refused growth hormone treatment when it became available in 1968, did not want to change their status within the community (McArthur et al., 1985). They were reported to be well accepted and free of significant psychological problems. Given their unique social environment, this experience, and that of General Tom Thumb, may not be representative of the level of psychosocial functioning of all hypopituitary adults. As described in previous chapters, modern authors have reported widespread problems in childhood including social isolation, younger peer group association, negative self-concept, and poor academic achievements despite normal intelligence (Mazur & Clopper, 1987). It is generally assumed that these problems antecede social difficulties in adult life.

Height Outcomes of Treated GH-Deficient Adults: Published Reports

With the advent of GH therapy in 1967, and proof of its efficacy in promoting linear growth (Guyda et al., 1975), national programs were established in Canada, Australia, Great Britain, Japan, Europe, and the United States. In all GH trials, it was assumed a priori that GH treatment would allow normal growth and increased adult height with amelioration of the psychosocial problems experienced by these patients. The remainder of this chapter will be a review of the published final adult height results and a brief review of the follow-up studies of psychosocial functioning, particularly as they relate to the demographic outcomes.

Reports of final adult height first appeared in 1981 (Burns et al., 1981; Bourguignon et al., 1986; Dean et al., 1985; Galatzer et al., 1987; Joss et al., 1983; Lenko et al., 1982; Van den Broeck et al., 1988). A summary of these published final adult height results is presented in Table 6.1. Although the growth hormone dosage schedules vary considerably, there is surprising agreement in the results. Most subjects, males and females, organic and idiopathic, achieved a disappointing overall mean height 2 to 3 SD below the normal population mean. The group had gained at least 2 SD over their predicted height without treatment but failed to achieve average adult stature. The subjects with gonadotropin deficiency achieved a final height approximately 1.0 SD greater than those with spontaneous

TABLE 6.1. Final heights of adults with hypopituitarism treated with growth hormone during childhood.

Reference	Number of subjects	hGH units/week	Standard deviation score for final height		
			All	Males	Females
Burns, 1981	42*	15–20	−2.3	−2.1	−3.3
Lenko, 1982	16	12	−2.3	−1.7	−3.3
Joss, 1983	13*	15–18	−2.4	NA	NA
Dean, 1985	116	6	−3.0	−3.1	−2.8
Bourguignon, 1986	22	18	NA	−2.5	NA
Clopper, 1986	39	NA	−2.2	NA	NA
Galatzer, 1987	42*	NA	NA	−2.5	−2.9
Van den Broeck, 1988	19*	8–24	NA	−1.7	NA
present study	40	6	−2.0	−2.1	−1.9

* idiopathic GHD only
NA = not available

82   H.J. Dean

onset of puberty due to their relative eunuchoid body proportions (Burnes et al., 1981; Bourgignon et al., 1986; Galatzer et al., 1987).

Height Outcomes of Treated Growth Hormone Deficient Adults: Prospective Studies

Since completion of our first follow-up study in 1983 (Dean et al., 1985), the Canadian collaborative Therapeutic Trial has included compulsory clinical follow-up and retesting of GH status 3 years after discontinuing GH therapy. This follow-up provides an opportunity not only to collect accurate demographic data and biochemical confirmation of GH deficiency by repeat stimulation tests, but also to inform subjects of the current status of Creutzfeld-Jakob disease and leukemia in association with GH therapy (Watanabe et al., 1988). The criteria for inclusion in this analysis were (1) follow-up during the years 1983 to 1987 inclusive (i.e., after the previous follow-up study); (2) treatment with growth hormone for at least one year; and (3) age at follow-up 18 years or greater. The demographic data on the 40 subjects seen in follow-up is shown in Figure 6.2 and Table 6.2. Final heights ranged from 139 to 170 cm. The mean final standard deviation score (SD score) for height of the 29 males was −2.1 and for the 11 females was −1.9. The subjects with GH deficiency secondary to removal of a craniopharyngioma were taller at diagnosis (data not shown) and maintained this advantage after GH treatment (p < 0.005). Since the

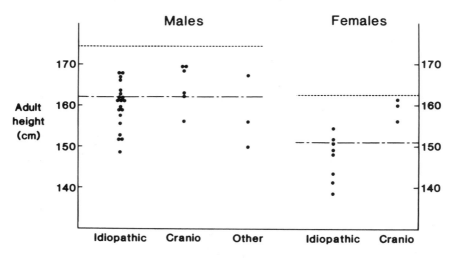

FIGURE 6.2. Final adult height of subjects at follow-up 1983–1987, by etiology and gender (abbreviations: cranio = craniopharyngioma; ----50 percentile for height for normal adult population; -.-.-. 3rd percentile for height for normal adult population).

TABLE 6.2. Demographic follow-up study during the years 1983 to 1987.

	Range (40)	X ± SD (40)	Sex		Etiology		
			M (29)	F (11)	Idio (28)	Cranio (8)	Other (4)
Age (yrs.) at:							
Start of treatment	5–21	12.6 ± 4.1	12.6 ± 4.4	12.4 ± 3.6	12.8 ± 4.3	12.1 ± 4.3	14.5
End of treatment	15–26	19.3 ± 3.0	19.9 ± 3.1*	17.8 ± 2.1*	19.1 ± 3.1	19.4 ± 2.9	19.6
Follow-up	18–33	23.2 ± 3.7	24.3 ± 3.5	22.7 ± 2.1	23.6 ± 3.5	24.8 ± 2.8	23.8
SD-score for height at follow-up	−0.1 to −4.1	−2.0 ± 1.0	−2.1 ± 1.0	−1.9 ± 1.2	−2.3 ± 0.9**	−1.3 ±0.8**	−2.0

( ) number of subjects
* $p < 0.05$
** $p < 0.005$

mean pretreatment SD score for height for all untreated subjects is approximately -4 to -6 (Dean et al., 1985), the growth hormone therapy has recovered only 2 to 4 SD of height. It is noteworthy that the mean age at start of treatment has decreased from 13.6 to 12.6 years and the mean final SDS height has increased from -3.0 to -2.0 in the current follow-up study compared to our earlier study (Dean et al., 1985). The possible reasons for permanent loss of at least 2 SD include the degree of growth failure at diagnosis, concomitant pituitary hormone deficiencies, frequency of GH administration and dose of GH. It is conceivable that the mean final height will increase further in the future with daily injections, doses of biosynthetic GH that are twofold to threefold larger than were previously used with pituitary GH, and earlier initiation of therapy. It will then be possible to reevaluate the psychosocial outcome without the confounding variable of persistent short stature.

In the evaluation of the effect of GH on final height, it is critical that the study group not be contaminated with children with constitutional delay of growth and development (CGD). This concern has been underscored by the recent controversy regarding the "best" test to determine GH status. There appears to be a physiological blunting of GH release in the immediate prepubertal years following provocative stimuli and under physiological conditions such as sleep. Therefore, it may be difficult to differentiate between CGD and GH deficiency using conventional clinical and biochemical criteria. For this reason, we have encouraged repeat GH testing 3 years after termination of GH therapy to reconfirm the diagnosis of GH deficiency. Seventeen (43%) of the 40 follow-up patients in 1983-1987 consented to retesting their GH status. The majority had an L-dopa propranolol provocative test. Sixteen of the seventeen subjects had a peak GH less than 5 ug/L, the defined limit for GH deficiency. One boy, with idiopathic GH deficiency, who, at diagnosis at age 12 years 8 months was 2.3 SD below the mean height for his chronological age, discontinued his GH after 3 1/2 years, having achieved a height of 155.7 cm. Three years later at follow-up at age 19 1/2 years, he was 161 cm (SD score-2.0) and had a normal peak GH response to L-dopa/propranolol of 10.3 μg/L and a normal peak sleep GH level of 16.6 μg/L. The frequency of this "transient" GH deficiency remains unknown. Burns et al. (1981) retested 25 subjects with total GH deficiency, using insulin-induced hypoglycemia and confirmed the diagnosis in all of them. Thus, despite the current climate of uncertainty regarding the definition of GH deficiency in childhood, it appears that over 95% have permanent GH deficiency using unconventional diagnostic critera. Follow-up retesting of GH-deficient subjects will be important in future outcome studies to ensure that increased height is due to GH therapy alone and not compounded by a misdiagnosis of GH deficiency in a child with constitutional delayed growth and development.

Psychosocial Outcome of Treated Growth Hormone Deficient Adults: Social Status

In 1985, the Canadian MRC Therapeutic Trial reported the first analysis of objective measures of psychosocial functioning in adult life, namely, employment and marital status (Dean, 1985). Ninety-six had completed their formal education and were considered to be in the work force. Forty-five percent of these subjects were 18 to 24 years of age and 23% of those 25 to 40 years of age were unemployed compared to national rates of 21.2% and 9.4% respectively (p < 0.001). The distribution of full-time (84%), part-time (10%), and seasonal (6%) employment was similar to the distribution in the general population. The distribution of subjects in various occupational categories was similar to the general population. Most subjects were employed in clerical, sales, and service-related jobs.

Only 15 of the 116 subjects were married. One 30-year-old woman was divorced. There were 6 biologic children in the 15 families and one child in the nonmarried group. Overall, the rate of unemployment was threefold higher and the rate of marriage fivefold lower than the general population (Figures 6.3 and 6.4). There was no relationship between employment or marital status and height, sex, education, place of residence, etiology (i.e., idiopathic versus organic) or other hormone deficiencies (i.e., isolated versus panhypopituitarism). There was a suggested negative relationship between marital status and gonadal status but this did not reach statistical significance.

Twenty-four (21%) had received formal psychosocial counseling. Seventy (73%) of the 96 nonstudents were living at home with their parents. Sixty-seven (58%) of the whole group had a driver's licence compared to 78% in the comparable age group across Canada. Approximately one third of the group complained of difficulty in purchasing age-appropriate clothing. Less than 20% of the group participated in extracurricular community group or activities.

Similar findings, using structured interviews regarding employment and marital status, have been reported by Clopper et al. (1986), and Mitchell et al. (1986) and are summarized in Table 6.3. The mean age, gender, percentage of isolated GH deficiency and mean duration of treatment were similar in the three studies. The final adult height (−3.0 SDS) of the Canadian group was similar to the group reported by Mitchell (157.5 cm). However, the mean adult height of the group reported by Clopper was −2.2 or 0.8 SDS greater than the other two studies. This increased mean height was not associated with any difference in percent employment or marriage. The rates of employment and marriage are consistently lower than expected from regional norms in these studies. In two of the studies (Dean et al., 1985; Clopper et al., 1986) there was a trend to lower marital rate for

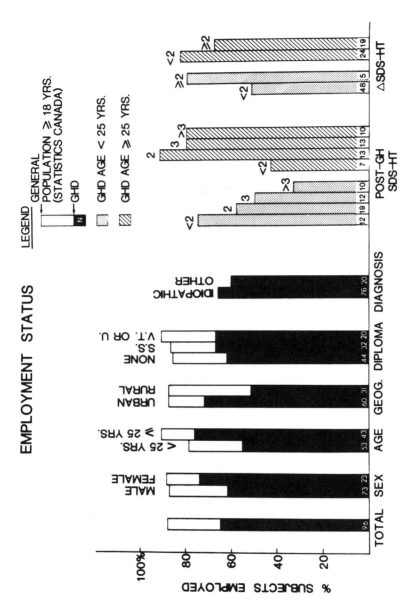

FIGURE 6.3. Employment status of adults with growth hormone deficiency (GHD) (*solid bars*) in work force compared with general population (open bars), where possible. Of 11 treatment variables studied only diagnosis, posttreatment standard deviation score (SDS) height, and increase in SDS height are illustrated. The only statistically significant relationship is with age ($P < 0.05$). Numbers at base of bars indicate number of subjects. Numbers above bars indicate the SDS-ht (standard deviation score for height). Subjects younger than 25 (*stippled bars*) and subjects at least 25 years old or greater (*hatched bars*). SS = secondary school; VT = vocational training; U = university. (Dean et al., 1986).

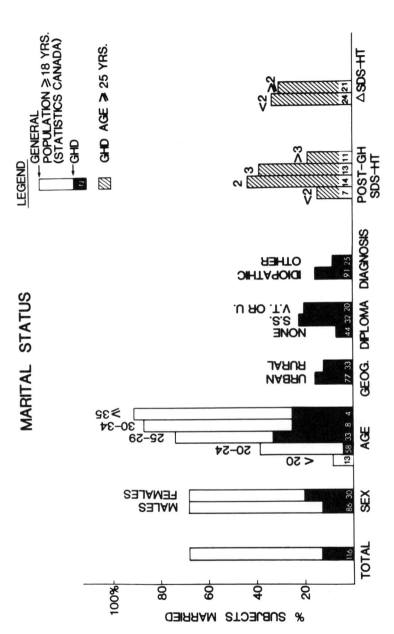

FIGURE 6.4. Marital status of adults with growth hormone deficiency (GHD) (*solid bars*) compared with marital status of general population (*open bars*), where possible. Only 96 patients who completed their education have a final diploma and only 45 patients older than 25 years have been included for posttreatment standard deviation score (SDS) height. The only statistically significant relationship is with age ($P < 0.01$). Numbers at base of bars indicate number of subjects. Numbers above bars indicate the number of SDS-ht as in Figure 6.2. The hatched bars indicate subjects at least 25 years old or greater (Dean et al., 1986).

TABLE 6.3. Psychosocial outcome variables of GH deficient adults

Reference	Country	Study population	Males %	Mean age (yrs.)	Isolated GHD %	Mean duration of treatment (yrs.)	% of subjects employed full time (*)	% of subjects married
Dean, 1985	Canada	116	74	24.5	63	6	62	13
Clopper, 1986	U.S.A.	39	82	23	46	7	54	10
Mitchell, 1986	U.S.A.	58	76	26	71	NA‡	57	30
Galatzer, 1987	Israel	42	55	NA	33	NA	100**	29†

* excluding students and housewives
** 22/30 employed subjects served in the army
† in age group 25–34 years
‡ NA = not available

the adults with multiple hormone deficiency compared to those with isolated GH deficiency. There is no obvious explanation for the higher marital rate in the Mitchell study.

One of the recurring themes in all of these studies was juvenilization. Over 50% of subjects complained that they looked and were treated by others as younger than their actual age. Mitchell et al. reported that the majority of subjects believed that they looked 5 to 9 years younger than their actual chronological age. The true prevalence of this problem is probably higher as there is a high level of denial (Mitchell et al., 1981).

The psychosocial functioning of GH-deficient adults in Israel differs from the reports from USA and Canada (Galatzer et al., 1987) (Table 6.3). Of the 30 Israeli subjects who had achieved their final adult height and completed their education, all were employed. Twenty-two (73%) of these subjects served in the Israeli army in noncombatant units. The vocational outcome may have been affected by compulsory military service. However, even after excluding the 20 students and soldiers, all of the other subjects were employed. The demographic outcome was similar to the worldwide experience since the final adult heights were 2.5 to 2.9 SD below the mean height for adults (Table 6.1). Laron and colleagues attribute their better vocational outcome to continuous multidisciplinary professional care provided by an integrated team of health professionals which includes specialized endocrine psychologists and social workers (Laron et al., 1986). One of the objectives of their team approach is the integration of the "patient within the family and community according to his needs and capabilities" (Laron et al., 1986). In this series, the rate of marriage in the subjects aged 25 to 30 years was only 33% of the rate of marriage in the general population. The rate of marriage of the younger subjects was 82% of the general population. The authors attribute this higher rate of marriage in the younger subjects to better psychological counseling in recent years.

The average adult height of the subjects in all of these studies was 2 to 3 SD below the average height for the general population. Thus, despite an improvement in relative height by approximately 2 to 4 SD with treatment, the subjects were significantly short in adult life. This persistent short stature despite GH therapy during childhood must not be overlooked in the interpretation of all of these studies. The psychosocial outcomes are compared to national normative data for a population with a normal distribution of height. The GH-deficient adult population is clearly different in height.

Summary

It is obvious that although GH therapy has improved the final height of GH-deficient individuals, their predicted ultimate stature is not within the normal range. Furthermore, in most series, serious psychosocial difficulties are implied by the low rates of employment and marriage. Whether there is a cause and effect relationship between the demographic outcome and psychosocial outcome remains speculative. There are now long-term prospective studies in progress in many countries to evaluate the effects of new GH treatment schedules and psychological counseling on final outcome—both physical and emotional.

Directions for Future Research

If we accept the premis that the ultimate goal of GH treatment is to enhance psychosocial well-being by increasing physical height, it is imperative that all therapeutic trials of GH monitor psychosocial parameters as well as height. If we cannot prove that GH therapy improves psychosocial outcome then we must question the rationale for GH therapy in children with short stature due to GH deficiency. Serious consideration must be given to future protocols that carefully assess psychosocial functioning in a longitudinal manner and that evaluate intervention strategies in children experiencing difficulty with peer relationships. All future GH therapeutic trials must be of sufficient length to determine final adult height and adult psychosocial status. The need for 8- to 10-year studies of GH treatment to the adult age group creates serious difficulty in securing funding and maintaining the interest of the investigators. However, this basic element of clinical GH research must be recognized in order to evaluate the true benefits of GH therapy.

The lack of a good control group remains a serious limitation of clinical GH research. The features of short stature, delayed puberty, daily injections, frequent medical intervention, and normal healthy longevity are not found together in any other chronic condition in childhood. Creative new approaches to overcome this limitation will be critical in the future.

References

Bourguignon, J.-P., Vandeweghe, M., Vanderschueren-Lodeweyckx, M., Malvaux, P., Wolter, R., Du Caju, M., & Ernould, C. (1986). Pubertal growth and final height in hypopituitary boys. A minor role of bone age at onset of puberty. *Journal of Clinical Endocrinology and Metabolism, 63,* 376–382.

Burns, E.C., Tanner, J.M., Preece, M.A., & Cameron N. (1981). Final height and pubertal development in 55 children with idiopathic growth hormone deficiency treated for between 2 and 15 years with human growth hormone. *European Journal of Pediatrics, 137,* 155–164.

Clopper, R.R., MacGillivray, M.H., Mazur, T., Voorhess, M.L., & Milb, B.J. (1986). Post-treatment follow-up of growth hormone deficient patients: Psychosocial status. In B. Stabler & L.E. Underwood (Ed.), *Slow grows the child.* Hillsdale, NJ: Erlbaum & Associates.

Dean, H.J., McTaggart, T.L., Fish, D.G., & Friesen, H.J. (1985). The educational, vocational, and marital status of growth hormone-deficient adults treated with growth hormone during childhood. *American Journal of Disease of Children, 139,* 1105–1110.

Dean, H.J., McTaggart, T.L., Fish, D.G., & Friesen, H.J. (1986). Long term social follow-up of growth hormone deficient adults treated with growth hormone during childhood. In B. Stabler & L.E. Underwood (Ed.), *Slow grows the Child,* Hillsdale, NJ: Erlbaum & Associates.

Faurbye, A. (1946). Pituitary dwarfism. *Acta Psychiatrica Scandinavica, 21,* 245–249.

Galatzer, A., Aran, O., Beit-Halachmi, N., Nofar, E., Rubitchek, J., Pertzelan, A., & Laron, Z. (1987). The impact of long-term therapy by a multidisciplinary team on the education, occupation and marital status of growth hormone deficient patients after termination of therapy. *Clinical Endocrinology, 27,* 191–196.

Goodman, G.H., Grumbach, M.M., & Kaplan, S.L. (1968). Growth and growth hormone. *New England Journal of Medicine, 278,* 59–68.

Guyda, H.J., Friesen, H.G., Bailey, J.D. et al. (1975). Medical research council of Canada therapeutic trial of human growth hormone. First five years of therapy. *Canadian Medical Association Journal, 112,* 1301–1309.

Hawley, R.R. (1963). A 67 year old pituitary dwarf. *Journal of Clinical Endocrinology, 23,* 1058.

Hewer, T.F. (1944). Ateleiotic dwarfism with normal sexual function. Results of hypopituitarism. *Journal of Endocrinology, 3,* 397–400.

Joss, E., Tuppinger, K., Schwartz, H.P. et al. (1983). Final height of patients with pituitary growth failure and changes in growth variables after long term hormonal therapy. *Pediatric Research, 17,* 676–679.

Laron, Z., Aran, O., Nofar, E., Beit-Halachmi, N., Pertzelan, A., & Galatzer, A. (1986). Psychosocial aspects of young adult growth hormone deficient patients previously treated with human growth hormone—a preliminary report. *Acta Pediatrica Scandinavica (Suppl), 325,* 80–82.

Lenko, H.L., Leisli, S., & Perkeentupa, J. (1982). The efficacy of growth hormone in different types of growth failure. Analysis of 101 cases. *European Journal of Pediatrics, 138,* 241–248.

Mazur, T., & Clopper, R.R. (1987). Hypopituitarism. Review of behavioral data. In D.M. Styne & C.G.D. Brook (Eds.), *Current concepts in pediatric endocrinology* (pp. 184–205). New York: Elsevier.

McArthur, R.G., Morgan, K., Phillips, J.A., Bala, M., & Klassen, J. (1985). The natural history of familial hypopituitarism. *American Journal of Medical Genetics, 22,* 553–566.

McKusick, V.A., & Rimoin, D.L. (1967). General Tom Thumb and other midgets. *Scientific American, 217,* 102–112.

Mitchell, C.M., Joyce, S., Johanson, A.J., Libber, S., Plotnick, L., Migeon, C.J., & Blizzard, R.M. (1986). A retrospective evaluation of psychosocial impact of long-term growth hormone therapy. *Clinical Pediatrics, 25,* 17–23.

Van den Broeck, J., Vandenschueren-Lodewegckx, M., & Eggermont, E. (1988). Prediction of final height in boys with non-tumorous hypopituitarism. *European Journal of Pediatrics, 147,* 245–247.

Watanabe, S., Tsunematsu, Y., Fujimoto, J., & Komiyama, A. (1988). Leukemia in patients treated with growth hormone. *Lancet i,* 1159.

7
A Theoretical Model for Classical Psychosocial Dwarfism (Psychosocially Determined Short Stature)

WAYNE H. GREEN

Classical psychosocial dwarfism (PSD) or psychosocially determined short stature offers an unusually clear example of the interaction between psyche and soma and their abilities to respond with exquisite sensitivity to the dynamic relationship between the child and his or her psychosocial environment. In this disorder, an inimical psychosocial environment, the most important feature of which is a severely disturbed primary caretaker/ child relationship, typically causes endocrine abnormalities, severe growth retardation, behavioral abnormalities, and developmental delays in the child. Abrupt improvement or worsening in the psychosocial environment of these children has dramatic and measurable consequences which often are reflected rather rapidly in a similar direction by changes in endocrine status, growth rate, and behavior. Consequently, PSD has provided us with an opportunity to make significant insights into some mechanisms through which the psychosocial environment may exert powerful influences on physical growth and psychological development through the mediation of the interposed central nervous system and its regulation of psychoneuroendocrine functioning.

This chapter summarizes the history of psychosocial dwarfism and then presents diagnostic criteria for classical psychosocial dwarfism. Following this, relevant clinical data and pertinent studies in animals and premature human infants will be reviewed and a theoretical model for psychosocial dwarfism will be presented. Other aspects of psychosocial dwarfism, particularly endocrinologic, behavioral, psychological, differential diagnostic, therapeutic, and formes frustes have been reviewed recently elsewhere (Campbell et al., 1982; Green et al., 1984; Green, 1986; Green et al., 1987; Green, 1989).

History

Although some similar children had been previously described, the first detailed reports of children exhibiting classical PSD were published in 1967 (Powell et al., 1967a; Powell et al., 1967b). A subgroup of children hospitalized for evaluation of severe growth failure and suspected idiopathic hypopituitarism showed significant and rapid improvement in their endocrine statuses and growth rates following hospitalization but with no other specific medical or hormonal treatment. This subgroup, now called psychosocial dwarfism, was further differentiated from the larger group by pathological psychosocial environments that were present prior to hospitalization and by unusual behavioral manifestations and developmental delays which improved following hospitalization.

These unexpected observations which led to its serendipitous discovery remain the sine qua non for confirmation of the diagnosis of classical PSD and necessitate three other assumptions. First, although children who develop PSD may have an underlying biological and/or psychological vulnerability and be a heightened risk for developing PSD, they are assumed to be inherently capable of physical growth and psychological development within the spectrum of normalcy. Second, it is an abnormal, inimical psychosocial environment that causes a child to develop PSD. Third, the syndrome of PSD is reversible, at least to a significant degree, upon the child's removal from the malignant environment and placement in a more favorable situation with no specific medical, hormonal, or psychiatric treatment, although removal of a child from the nuclear family to a completely new environment must be regarded as one of the most powerful psychiatric interventions.

Diagnostic Criteria for Classical Psychosocial Dwarfism (Psychosocially Determined Short Stature)

Diagnostic criteria are presented in Table 7.1.

Clinical Data of Etiological Relevance

Endocrinology of Psychosocial Dwarfism

The characteristic endocrinologic findings in PSD will be summarized only as they have been reviewed recently in considerable detail (Campbell et al., 1982; Green et al., 1987). Endocrine dysfunction, although typical of PSD, is heterogeneous. The only consistently abnormal findings are somatomedin levels in the range characteristically found in hypopituitary dys-

TABLE 7.1. Diagnostic criteria for classical psychosocial dwarfism (psychosocially determined short stature).

I. Onset usually between ages 2 and 4 years. The distribution is skewed toward the younger ages. There is some evidence that physical and behavioral abnormalities begin at about the same time.

II. Evidence of a severely disturbed relationship between primary caretaker and child which is of primary etiological importance. This is usually the mother-child dyad. Mothers exhibit heterogeneous psychopathologies. Fathers tend to be absent or have relatively little direct involvement with the PSD. Marital discord is very commonly reported. Child abuse may occur concomitantly.

III. Growth Disturbances
1. Birth weight is usually normal for gestational age.
2. Growth rate becomes markedly subnormal following an earlier period of relatively normal growth.
3. Psychosocial dwarfs are typically well below the third percentile in height.
4. Weight is usually within normal parameters for height. Malnutrition is *not* typically present in classical PSD. Malnutrition, however, may coexist and is not a disqualifier for the diagnosis. Some psychosocial dwarfs have a history of prior decreased weight for height (nonorganic failure to thrive).
5. Bone age is usually significantly retarded and approximates height age.
6. Growth lines (Harris, Park lines) are often evident on x-rays of the long bones.

IV. Endocrine Dysfunction
1. Fasting growth hormone is abnormally low in the large majority of psychosocial dwarfs in the growth inhibiting environment and immediately following removal from it, e.g., hospitalization.
2. Growth hormone levels following provocative stimulation soon after hospitalization are abnormally low in about 50% of the cases.
3. Somatomedin levels soon after admission to hospital are abnormally low in all of the relatively few cases reported.
4. Other endocrine abnormalities do not appear to be typical of PSD, although ACTH release or production may be mildly to moderately compromised in some cases and increased cortisol secretion rate has also been reported.
5. Delayed onset of puberty has been reported.

V. Behavioral and Developmental Disturbances
1. The most typical behavioral disturbances involve bizarre acquisition of food and water: polyphagia, gorging and vomiting, eating garbage or animal's food; and polydipsia, drinking from toilets, dishpans, or rain puddles.
2. Developmental or maturational lags: delayed gross motor milestones, delayed language development, immature articulation, IQ scores most frequently in the mildly retarded or borderline ranges.
3. Social immaturity, temper tantrums, poor peer relationships.
4. Social withdrawal, depression, apathy, elective mutism.
5. Sleep disorders: disturbances in sleep architecture on EEG tracings reported. Night roaming may also occur.
6. Pain agnosia and self-injury may occur.

VI. Disqualifiers
The symptoms are *not* caused primarily by other physical or mental disorders such as idiopathic hypopituitarism, organic failure to thrive, malnutrition, infantile autism, mental retardation, major depressive disorder, or another reactive attachment disorder of childhood.

TABLE 7.1. *Continued*

VII. Confirmation of Diagnosis

Following the psychosocial dwarf's removal from the inimical psychosocial environment (i.e., usually the primary caretaker) and placement in a more favorable setting, there must be a rapid and significant reversibility in symptoms without specific medical, hormonal, or psychiatric treatment. Growth hormone levels, if abnormal, may normalize within a few days, growth rate increases significantly and catch-up growth rates are frequent. Behavioral symptoms and developmental delays also improve although usually more slowly than physical symptoms. A temporary compensatory hyperkinesis consisting of increased motor activity, excessive exploratory curiosity, exhaustive verbal exchange, and overreacting to stimuli may occur and gradually return to normal levels. Serial IQ tests may show improvement.

Table modified from Green, W.H. (1986). Psychosocial dwarfism: Psychological and etiological considerations. In B.B. Lahey & A.E. Kazdin (Eds.), *Advances in Clinical Child Psychology,* Vol. 9 (pp. 245–278). New York: Plenum Press. With permission.

function but this has been reported in a small number of cases (a total of 17 with values given for only 9). No abnormal endocrine findings correlate invariably and directly with either growth failure or catch-up growth (Green et al., 1987).

The most typical and best studied endocrine abnormalities in PSD involve growth hormone. Fasting GH levels determined upon or soon after admission to the hospital were abnormally low in over 90% of the psychosocial dwarfs who were tested. Serum GH levels following provocative stimulation (e.g., insulin-induced hypoglycemia) soon after hospitalization and before increase in growth rates occurs are below normal in about 50% of cases (Green et al., 1987). GH levels rapidly reflect major positive or negative changes in the psychosocial environment and may normalize within a few days after placement in a new domicile. Therefore, GH tests must be done as soon as possible following intervention or the abnormalities may be missed. Once growth rate has increased, about 95% of psychosocial dwarfs tested have normal GH secretion following provocative stimulation (Green et al., 1987).

Four PSDs who were administered exogenous human GH while living at home showed minimal or no response (Tanner et al., 1971; Frasier & Rallison, 1972). Tanner (1973) suggested that this growth failure might be secondary to abnormal production and/or action of somatomedin (SM). Somatomedin may be considered the effector hormone of GH. Its production by the liver is stimulated by GH and its levels usually correlate better with growth rates than do GH levels (Laron, 1983).

Subnormal somatomedin levels were reported during the period of poor growth in 9 children with PSD. In 6 of these children, SM levels returned to normal coincident with increased growth rates. The other three children continued to have below normal SM activity even when they subsequently began growing at catch-up rates (Van Den Brande et al., 1975; D'Ercole et al., 1977). Van Den Brande et al. (1975) hypothesized that tissues had become hypersensitized to SM during the period when SM levels were in the hypopituitary range.

Although abnormalities in thyroid functioning have been reported, they are not typical of PSD. Thyroxine (T4) values are normal in the large majority of cases. In the few cases where abnormalities have been reported, they tend to be subnormal values occurring in significantly malnourished (not classical) psychosocial dwarfs. The low values were felt to reflect decreased thyroxine binding proteins secondary to malnutrition. They normalized when the malnutrition was eliminated (Krieger & Good, 1970).

The few studies of pituitary-adrenocortical axis function suggest that pituitary production or release of ACTH is mildly to moderately decreased in some cases of PSD. However, the studies considered different dosages adequate for provocative stimulation of ACTH secretion and measured different metabolites, making conclusions difficult. The administration of exogenous ACTH elicited normal adrenal cortex responses in the cases for which this was reported. Brasel (1973) reported that normalization of ACTH response to metyrapone provocation required up to two years in some cases; however, the lengthy time may be spurious because an insufficient amount of metyrapone may have been administered (Krieger & Mellinger, 1971).

Caloric Intake and Psychosocial Dwarfism

Malnutrition appears to play no part in the etiology of classical PSD. The confusion and controversy in the literature on this matter has occurred for several reasons. First, it is clear that malnutrition can cause endocrine dysfunction (GH, however, if abnormal, is usually elevated) and retarded growth. Second, infants with nonorganic failure to thrive or reactive attachment disorder of infancy secondary to pathological relationships with their primary caretakers usually do have a significant component of malnutrition present and some of them may fulfill diagnostic criteria for PSD at a later age. Third, some children with PSD are, in fact, additionally malnourished.

Powell and his co-workers noted that only one of the 13 children they reported appeared to be malnourished (Powell et al., 1967a). They further emphasized that the rapidity with which these children's growths accelerated or decelerated when the psychosocial environment was changed precluded malnutrition or malabsorption as a significant etiological factor.

A critical review of the literature, including a detailed analysis of longitudinal data of Saenger et al. (1977) correlating caloric intake, growth rate, hormonal levels, and psychosocial environment of a child with PSD, concluded that available evidence supported almost conclusively the existence of classical PSD with no concomitant malnutrition (Green et al., 1984).

Abnormal Radiographic Findings in Psychosocial Dwarfism

Two types of abnormal x-ray findings appear to be associated with psychosocial dwarfism. Widening of the cranial sutures has been reported during the period of characteristically rapid growth following placement in a favorable new environment (Capitanio & Kirkpatrick, 1969; Afshani et al., 1976; Gloebl et al., 1976). This widening is usually associated with a normal cerebrospinal fluid pressure but pseudotumor cerebri has been reported in four children, ages 20 to 61 months, with a clinical picture of deprivation dwarfism (Tibbles et al., 1972).

The other abnormality is more relevant to the etiology of PSD. A significantly increased incidence of growth lines (Harris, Park Lines) has been reported on x-rays of the long bones in psychosocial dwarfs (Patton & Gardner, 1963; Capitanio & Kirkpatrick, 1969; Gloebl et al., 1976; Hernandez et al., 1978). Growth lines are thought to be related to a temporary inhibition of growth followed by recovery with renewed or increased growth.

Hernandez and his co-workers (1978) reported on x-rays of the distal radii or knees of 23 children diagnosed PSD (age range 2 to 15 years; average 5.9 years) and 25 patients with idiopathic hypopituitarism (age range 1 month to 21 years; average 7.75 years). Twenty (87%) of the psychosocial dwarfs had growth lines compared to only 2 (8%) of the children with idiopathic hypopituitarism; this was a highly significant difference ($p < 0.0005$). The authors felt their data implied that the environment of the psychosocial dwarfs fluctuated, sometimes inhibiting and sometimes permitting growth.

Sleep Abnormalities, Nocturnal GH Secretion, and Psychosocial Dwarfism

Sleep studies are of particular interest in PSD as sleep is the most important physiological regulator of GH secretion. In children, GH is released primarily during slow wave sleep (SWS), especially Stage IV (Honda et al., 1969; Takahashi et al., 1968). Sleep disturbances which correlated with periods of poor growth have been reported in PSD (Powell et al., 1967a; Wolff & Money, 1973). Other studies have looked at the relationship between GH secretion and sleep stages.

Powell, Hopwood, and Barratt (1973) rehospitalized a 9.3-year-old female psychosocial dwarf about two months after her parents had separated and the mother and children had moved back to their former state. During this two-month period, the girl's growth rate had increased significantly as evidenced by weight gain and split cranial sutures on x-ray. Sleep EEGs were determined on hospital nights 5, 6, 26, 27, 28, and 40 and serum GH levels were determined through an indwelling catheter on nights 5, 6, and

40. Although sleep architecture was normal on all occasions, there was no significant GH secretion during slow wave sleep on nights 5 and 6. On night 40, however, there was a significant output (GH > 30 ng/ml) shortly after Stage IV sleep began. The authors noted that the decreased GH secretion on nights 5 and 6 was not related to any abnormality in sleep pattern; they suggested that normalization of GH secretion associated with SWS occurs more slowly than normalization of GH secretion in response to arginine and insulin provoked hypoglycemia. This is consonant with the caveat of Wise, Burnet, Geary, and Berriman (1975) that provocative tests of GH release may be normal although GH release under normal physiological conditions such as sleep may be inadequate. That different mechanisms control GH release is suggested by the fact that methysergide, a serotonin receptor blocker, diminishes GH response to insulin-induced hypoglycemia but enhances GH secretion during SWS.

Howse et al. (1977) determined GH secretion by indwelling catheter during EEG-monitored sleep in a 4-year-old male and two females, ages $4\frac{1}{2}$ years and 13 years, with PSD. The three children spent more time in SWS sleep (Stages III-IV) than did 14 normal, but short, children, although the difference was not significant and the children with PSD had briefer, diminished, and delayed GH secretion responses during Stage III-IV sleep. They also did not show the normal peak GH secretion during Stage IV sleep in the early sleep cycles; rather, their peak GH levels occurred during Stage III-IV of later sleep cycles and they were lower than peak GH levels of the short, normal children. The mean GH level of the psychosocial dwarfs during five hours of sleep was more than 2 standard deviations below the age-related mean established for normal short children. The authors postulated that "either a delay or reduction in sleep-mediated release of growth hormone releasing factor (GHRF), or alternatively, reduced pituitary responsiveness to GHRF, resulting from suboptimal GHRF priming action, in association with a low pituitary secretion potential" caused the delayed and abnormally low GH secretion in these children (Howse et al., 1977, p. 356).

Guilhaume, Benoit, Gourmelen, and Richardet (1982) studied the relationship between Stage IV sleep deficit and growth hormone secretion using polygraphic sleep recordings in four psychosocial dwarfs, a male age 12 months and 3 females ages 19, 28, and 36 months. Ten normal children of similar ages served as controls. Initial polygraphs were recorded within the first 3 nights for 3 cases and on the ninth night in the fourth case. The total sleep period (439.5 min) was significantly shorter ($p < 0.05$) in the 4 subjects than in controls (513.8 min). The psychosocial dwarfs had significantly more Stage I sleep than controls ($P < 0.01$) and tendencies for paradoxical (REM) sleep to be of shorter duration, and for increased intervening wakefulness, although only one subject had long periods of being awake. Stage III sleep was normal in the psychosocial dwarfs, however Stage IV sleep was totally absent in 3 of them and markedly less than the controls' Stage IV sleep in the fourth. As a consequence, SWS

(Stage III + Stage IV) made up only 11.9% of total sleep in the 4 subjects while it comprised 29% of the controls' total sleep. In addition, subjects had only 2.8 ± 0.7 SWS episodes per night compared to controls' 4.0 ± 0.4 SWS episodes. The psychosocial dwarfs also had shorter SWS episodes than controls (13.0 ± 4.8 min vs 31.1 ± 4.5 min) and only one subject had SWS episodes of 26 minutes or more while all controls did.

Follow-up sleep recordings after 21 to 105 days in the new environment and during catch-up growth revealed marked improvements in sleep. All 4 psychosocial dwarfs had Stage IV sleep and it was no longer significantly different from controls. While total sleep period increased and Stage I sleep decreased (i.e., normalized) in the subjects, they were still significantly different from controls ($p < 0.05$).

Guilhaume et al. (1982) also determined plasma GH levels following ornithine stimulation for 3 of the 4 subjects. They were subnormal in 2 subjects tested during the first 2 days of hospitalization and normal in the third subject on the ninth day of hospitalization. All 3 had normal GH levels when tested during the recovery period when they had normal sleep patterns and were exhibiting catch-up growth. While these findings are suggestive of a possible relationship between decreased duration of SWS and/or lack of Stage IV sleep and diminished GH secretion, unfortunately, GH levels were not determined during sleep.

Overall, the studies of nocturnal GH secretion in psychosocial dwarfs are discrepant. Most studies done very soon after hospitalization report abnormalities. While GH levels correlated with Stage IV sleep in one study, in another, subnormal GH output occurred with normal SWS. These studies may provide additional evidence that the GH secreting apparatus may be physiologically operational but that it can be overridden by CNS, particularly hypothalamic, mechanisms.

Depression, Endocrine Abnormalities, and Psychosocial Dwarfism

Because observers have repeatedly reported that many psychosocial dwarfs appear to be depressed or apathetic, there has been particular interest in the etiological relationship between depression and PSD. In 1975, Patton and Gardner suggested that states of depression might adversely affect GH secretion by acting through their effects on the cortex and anterior hypothalamus or median eminence and subsequent hypophyseal pathways. Endocrine changes similar to those seen in some patients with PSD (decreased GH secretion following insulin-induced hypoglycemia and increased cortisol secretion) have been reported in adult patients meeting unmodified research diagnostic criteria for major depressive disorder (MDD), endogenous subtype (for review see Stokes, 1988). Puig-Antich and his co-workers found similar abnormalities in prepubertal depressed children (Puig-Antich et al., 1979; Puig-Antich et al., 1981).

More recently, further study has shown the endocrine abnormalities in prepubertal children with MDD and PSD to be similar only superficially and specific differentiating features have been identified.

In 1985, Puig-Antich reported that cortisol hypersecretion occurred in prepubertal MDD children only occasionally and then when comparisons of circadian plasma cortisol patterns during the depressed and recovery periods were made. Overall, there were no significant differences in cortisol secretion among depressed subjects, nondepressed children with psychiatric problems, and normal children. As cortisol abnormalities do not appear characteristic of PSD, this clarification is of less significance than that concerning GH abnormalities.

Three major differences in GH secretion between children with endogenous MDD and children with PSD are apparent.

1. In PSD, GH secretion following stimulation by insulin-induced hypoglycemia rapidly normalizes upon removal from the inimical environment. On the other hand, children with endogenous MDD continue to mount inadequate GH responses to insulin-induced hypoglycemia for at least 4 months following recovery (Puig-Antich et al., 1984c).

2. Children with PSD who were monitored during sleep were reported to have decreased or normal nocturnal GH secretion whereas children with MDD secreted significantly more GH than either normal controls or nondepressed neurotic children (Puig-Antich et al., 1984a).

3. Following recovery, prepubertal children with MDD continue to secrete significantly greater amounts of GH during sleep than normal controls or nondepressed neurotic children for at least 3 months (Puig-Antich et al., 1984b). Although data are more limited for PSD, GH secretion during sleep tends to normalize with improvement and no evidence that abnormally high GH levels occur following recovery has been reported.

Thus, it appears that major depressive disorder, per se, is not a significant primary etiological cause of PSD. It remains to be clarified whether those children with PSD who appear to be depressed are, in fact, depressed with MDD or dysthymic personality disorder, or whether they exhibit similar behavioral symptoms that are reactive to the inimical environment and that usually ameliorate rapidly upon placement in an improved environment. Some children with PSD may be truly depressed; the endocrinologic complexities of those prepubescent children with a dual diagnosis, PSD and MDD, remain to be studied and clarified.

Interaction of Environment and Behavior in Psychosocial Dwarfism: Primary Caretaker/Child Relationship in Psychosocial Dwarfism

Drash and his co-workers (1968) noted that 9 of 14 children diagnosed as having PSD had suffered extreme emotional rejection by one or both of their parents and felt that this figured prominently in the etiology of their symptoms. Two children had been physically abused, another had been locked in a room for prolonged periods, and a fourth was isolated in a crib. The children's aggressivity and hostility were felt to reflect their parental models. Many of the children seemed to have never had a close emotional relationship with any adult. The parents often were completely unable to manage their children and other caretakers often found it very difficult to control these children while they were still living with their parents. Fortunately, the children's behavior usually improved following placement in a new domicile.

Ferholt and his co-workers (1985) published a detailed study of the personality and behavioral characteristics and psychodynamics of 6 males and 4 females (age range 3 years 2 months to 16 years 5 months; mean age $7\frac{1}{2}$ years) whom they diagnosed as having psychosomatic dwarfism. The authors reported that all 10 of their subjects had severely disturbed parent-child relationships. The parents held unremittingly negative feelings and disparaging thoughts about their children. Although the children exhibited an intense longing for their parents, and seemed positively attached to them, all had an underlying and unmitigated hatred for their parents, particularly their mothers, and experienced them as rejecting, unfair, and selfish. The children's behaviors appeared to be highly influenced by the presence or absence of their mothers in particular. Two children who were withdrawn and inactive while at home with their mothers became markedly more animated after their mothers departed. Also, the severe behavioral abnormalities occurred primarily at home with their parents and usually lessened or substantially normalized while at school or visiting neighbors and following hospitalization. These remarkably rapid behavioral improvements upon leaving the inimical home environment often obfuscated the extremely disturbed parent (mother)-child relationships and made correct diagnosis difficult. When improved or asymptomatic children were returned to an unimproved home situation, their symptoms promptly returned. This was even true for two children who had been asymptomatic for over a year in foster care (Ferhold et al., 1985). The authors did not uncover a history of repeated parent-child separations, sensory deprivation, severe neglect, starvation, or physical abuse, although most children had suffered an injury at the hands of their parents at least once.

Perhaps the most important single study in shedding light on the etiology of PSD is that published by Saenger and his colleagues in 1977. This is a meticulously detailed longitudinal study of a 7-year-old male lasting nearly one year and providing objective data about the interaction of psychosocial environment (hospital vs home), primary relationships (presence or absence of parents and a specific nurse to whom he became especially attached), actual caloric intake, growth hormone and somatomedin levels, and the child's growth rate and changes in growth rate (acceleration or deceleration). This study will be discussed further in the theoretical model section.

Studies in the Rat

A series of studies focusing on the effects of experimentally imposed disturbances upon rat mother-pup interactions, published over the past decade by Schanberg and his co-researchers at Duke University Medical Center, has revealed markedly similar neuroendocrine and tissue abnormalities to those reported in psychosocial dwarfism (Roger et al., 1974; Butler & Schanberg, 1977; Butler et al., 1978; Kuhn et al., 1978; Kuhn et al., 1979; Schanberg et al., 1984). These studies provide the first evidence in the realm of comparative psychoneuroendocrinology for endocrine abnormalities analogous to those of PSD in another species.

The normal relationship between the mother rat and her pups was interrupted by either removing the pups from the mother and placing them in an incubator or by anesthetizing the mother and continuing to allow the pups access to her. Following either of these procedures, three specific abnormalities occurred in biochemical processes reflecting growth:

1. There was an immediate, marked decrease in levels of tissue (brain, heart, liver, kidney, lung, and spleen) ornithine decarboxylase (ODC). ODC sensitively and accurately reflects cell growth and development and is a good index of the biological activity of growth hormone.

2. Serum growth hormone (GH) levels were selectively decreased. Serum GH levels of pups deprived for between 1 and 6 hours decreased between 40% and 47% and were significantly lower than GH levels of control pups. Within 15 minutes of their return to their mothers, the serum GH levels of the deprived pups returned to normal and overshot those of control pups by about 150%; these elevated levels declined to normal and stabilized in about an hour.

3. There was a relative loss of tissue responsivity to GH. The normal marked increase in ODC levels in liver and brain tissue following injections of ovine GH was eliminated after maternal deprivation for as little as two hours. This effect was rapidly reversed. In as few as two hours after the pup was restored to its natural mother or a foster mother, normal tissue responsivity to GH administration, as evidenced by marked increases in ODC levels, occurred.

Through a series of rather elegant experiments using various combinations of anesthetized or normal lactating mother and foster mother rats, with or without their nipples ligated, and maternally deprived or control rat pups, the authors demonstrated convincingly that these abnormalities were directly related to the imposed maternal deprivation, specifically the lack of the tactile stimulation by the mother rat's licking and grooming her pups, and not related to any coincident nutritional deprivation. This specific type of maternal deprivation caused suppression of ODC induction in the liver and all peripheral tissues that were studied. A significant decrease in serum GH levels appeared to precede and to cause the suppression of ODC induction.

Nutritional deprivation was also found to be an important regulator of ODC activity in the rat pup. However, it operated independently of and through different mechanisms than maternal deprivation (Schanberg et al., 1984). Nutritional deprivation suppressed ODC induction only in the liver; ODC levels in other peripheral tissues appeared unaffected. The specific suppression ODC induction in the liver appears to be caused by either decreased availability of the glucose molecule itself or a decreased physiologic response triggered by glucose which is necessary for the liver to produce ODC following certain stimuli (see Green, 1986, for a more detailed review).

Researchers were able to reverse the effect of maternal deprivation in the rat pups and normalize their GH levels by using wet brushes to mimic the mother rat's licking. The authors also noted that the decrease in GH levels in rat pups following the maternal tactile deprivation was a fairly discrete neuroendocrine response. Serum prolactin, TSH, and serum corticosterone levels, reflective of ACTH secretion, showed no significant changes indicating that this was not the more general suppression of anterior pituitary function that usually occurs in the rat in reaction to stress (Schanberg et al., 1984).

More recently, it was reported that intracisternal administration of β-endorphin to 6-day-old rat pups markedly decreased brain, liver, heart, and kidney ODC activity. This ODC response was in the opposite direction from that occurring with peripherally administered β-endorphin (Bartolome et al., 1986). Naloxone blocked the decrease in brain ODC indicating that CNS opioid receptors participated in that process. The ODC changes parallelled strikingly those found in the maternally deprived rat pups suggesting that central β-endorphin may play an important role in mediating the physiological expression of neonatal isolation-induced stress. The authors suggested that endogenous β-endorphin might act as a growth inhibitory factor during early perinatal development by influencing CNS release of other hormones or neuromodulators (Bartolome et al., 1986). This would suggest that the mother rat's licking her pup might inhibit the production of β-endorphin in her pup's central nervous system. This inhibition of β-endorphin production, in turn, would permit the continued synthesis and/or release of sufficient amounts of GH to maintain the GH levels required for normal growth to proceed.

It thus appears that, in rat pups, growth as reflected by ODC levels in various tissues is influenced independently and differentially by both maternal behavior (licking of the pups) and nutrition. Deprivation of either preserves "maintenance" cell metabolism at the expense of growth through specific, nonoverlapping actions (Schanberg et al., 1984). In any particular individual, maternal or nutritional deprivation could occur singly or simultaneously depending on the specific circumstances present at any given time.

Studies of Premature Human Infants

The effects of various types of supplemental stimulation on the development of premature infants have been studied for the past 25 years. A recent review reported that supplemental tactile and kinesthetic stimulation has generally enhanced growth and development of preterm infants but that findings pertaining to weight gain have been inconsistent (Scafidi et al., 1986).

One recent study focuses specifically on the effects of tactile/kinesthetic stimulation in premature infants which appear related to the findings in the rat pup reviewed earlier (Field et al., 1986; Scafidi et al., 1986; Schanberg & Field, 1987). Twenty preterm infants were stimulated tactilely by stroking various regions of their bodies with a warmed palm for two 5-minute periods and kinesthetically by passive flexion/extension movements of their extremities for one 5-minute period. Three of these 15-minute sessions were administered daily for 10 weekdays. Subjects were compared to 20 controls in the same nursery who had no significant differences on 13 measures of perinatal data including obstetric and postnatal complications. All subjects were medically stable, were no longer receiving oxygen or intravenous feeding, and had been admitted to a transitional care "grower" nursery. Experimental subjects and controls did not differ significantly in the number of feedings per day or the volume or calories of formula ingested per day.

After 12 days, the treatment group averaged 8 g (47%) more weight gain per day than the control group (25.0 ± 6.0 g/day vs 17.0 ± 6.7 g/day; $P = 0.0005$) and also gained more weight per calorie of intake per kilogram of body weight (0.21 ± 0.04 g/calorie/kg vs 0.15 ± 0.04 g/calorie/kg; $P = 0.0005$). The experimental group also exhibited significantly more motor activity ($P = 0.04$), more alertness ($P = 0.04$) and higher Brazelton Neonatal Behavior Assessment Scores on four of the Lester Cluster Scores indicating more mature habituation ($P = 0.02$), orientation ($P = 0.02$), motor ($P = 0.03$), and range-of-state ($P = 0.03$) behaviors. On average, the treatment group was discharged 6 days sooner than the control group (18.4 vs 24.7 days after treatment onset; $P < .05$). The authors suggested that the apparent increased efficiency in converting caloric intake to weight gain was either secondary to a change in basal metabolism per se or an increase in weight gain efficiency secondary to increased activity (Field et al., 1986).

Follow-up data on a subgroup of these infants at both 8 and 12 months continued to show significant group differences at $P < .05$; the experimental group had a higher weight percentile and had higher scores on both the Bayley Mental and Motor Scales (Schanberg & Field, 1987).

A Theoretical Model for Classical Psychosocial Dwarfism (Psychosocially Determined Short Stature)

Green et al. (1984) critically reviewed the literature on psychosocial dwarfism and concluded that classical PSD does occur independently of malnutrition and than an inimical psychosocial environment, specifically the primary caretaker-child dyad, was responsible for the behavioral manifestations, endocrine dysfunction, and severe growth retardation typically found in PSD. The authors speculated that the psychosocial stresses affected the relays of the cerebral cortex to the hypothalamus causing intermittent and reversible inhibition of hypothalamic releasing factor(s) especially growth hormone releasing factor. Although one cannot generalize from findings in the rat pup to humans, these data and the suggestion of a related phenomenon in premature human infants support this conclusion.

Thus, severely compromising the normal interactive tactile grooming behavior between the mother rat and her pup by removing the pup or anesthetizing the mother rat was found to cause rapidly endocrine changes strikingly like those seen in PSD: decreased serum GH levels and a relative peripheral insensitivity to exogenous GH. These changes were independent of any nutritional deprivation.

The data on premature infants just reviewed is particularly interesting as the experimental subjects and controls did not differ significantly in the number of feedings or the amount or calories of formula ingested per day. The authors suggested the increased efficiency in converting calories to weight gain was either secondary to a change in basal metabolism per se or an increase in weight gain efficiency secondary to increased activity. These data could possibly reflect an interesting link between the rat data and data available in PSD. Whether, in fact, the control premature infants were evidencing phylogenetic continuity with the maternally deprived rat pups and had decreased growth secondary to decreased tactile stimulation has yet to be determined.

The growth in these premature infants is also strikingly similar to the data of Saenger et al. (1977) in which the linear growth and growth rate in a psychosocial dwarf increased following his attachment to a special nurse although his actual caloric intake per kilogram decreased. Similarly, his growth and growth rate both decreased rather dramatically when the nurse went on vacation although his caloric intake did not change significantly. Finally upon the nurse's return, growth and growth rate dramatically accelerated despite no significant increase in total caloric intake and an actual decrease, although not a significant one, in caloric intake per kilogram. Peak GH secretion following insulin-induced hypoglycemia during the two periods of poor growth was less that 7 ng/ml while during the period of accelerated growth it was 13 ng/ml.

In all three cases (the rat pup, the premature infant, and the psychosocial dwarf), these physiological changes could be hypothesized to be of survival value to the individual and species. Thus, when a severe enough deviation or disruption in a crucial mother-offspring behavioral pattern occurs, placing the offspring in jeopardy, a temporary, CNS mediated, metabolic shift of caloric expenditure from active growth to maintenance of the status quo would follow rapidly. Normal metabolic functioning would resume once normal mothering behavior was restored. It is intriguing that in all three examples, the triggering event intimately reflects deficits in essential maternal care for the particular developmental stage of the rat, infant, or toddler—either intimate physical, tactile contact or a positive emotionally nurturing and secure relationship with the primary caretaker.

With speculative hindsight, one could wonder whether this normally protective mechanism contributed significantly to the failure to thrive, marasmus, and death of infants in nurseries such as those described by Spitz (1945, 1946a, 1946b) where infants suffered from maternal deprivation, prolonged isolation and multiple and severe stimulus deprivation.

Taking into account the considerable evidence that there is a fluctuating rather than continuously adverse psychosocial environment and the remarkable rapidity with which growth begins or ceases depending on the comings and goings of people with whom the PSD is very emotionally involved, either negatively or positively, Green et al. (1984) proposed a hypothetical model detailing the relationship among the psychosocial environment, endocrine function, and growth. This model has been modified and expanded to include behavioral manifestations and is presented in Table 7.2.

Particularly important in understanding this model is the emphasis on three distinct phases or possibilities in the child-primary caretaker relationship. Typically, the psychosocial dwarf experiences a hostile, highly stressful negative relationship with the primary caretaker and growth is severely inhibited. Following a satisfactory change in the primary caretaker there is an intermediate period during which growth resumes at normal rates but enough time has not elapsed for a special, positive, emotionally nurturing relationship to develop and for the psychosocial dwarf to become firmly emotionally attached to the new caretaker. If all goes well, in time, this new primary caretaker-child relationship provides the necessary reliable emotional and physical security for this attachment to occur. Once this occurs, catch-up growth frequently begins and the former psychosocial dwarf truly begins to thrive.

Too often, the distinction between simply removing the psychosocial dwarf from his family and providing a substitute caretaker is not differentiated from the more important goal of developing a new emotionally secure relationship with a predictably reliable caretaker. Children, including psychosocial dwarfs, are quite adaptable and most will attach to an inter-

TABLE 7.2. Relationships among psychosocial environments, hormonal levels, growth, and behavioral/developmental/psychological manifestations.

Psychosocial environments	Hormonal levels (GH and somatomedin)	Growth	Behavioral/ developmental/ psychological manifestations
Growth-inhibiting environment (inimical primary caretaker-child relationship)	95% low fasting GH; about 50% low GH after stimulation (usually insulin). Somatomedin usually low	Severe growth retardation (<3rd percentile) with minimal or no evidence of active growth	Typical bizarre behaviors: polyphagia and polydipsia. Delayed milestones. IQs borderline or mildly retarded
New domicile, e.g., hospital, foster home, or relatives' home (removal of negative stress, i.e., primary caretaker)	95% normal GH levels after stimulation with insulin. Somatomedin levels are variable	Growth rate increases. Growth proceeds at more normal rates (non-catch-up growth)	Improvement, sometimes rapid, in behavior. Improved social relatedness
Development of a strong, emotionally nurturing, reciprocal relationship with the new primary caretaker in new environment. (addition of positive emotional factor)	Growth hormone levels within normal limits. Somatomedin levels normal or normalizing	Acceleration of growth rates may occur with subsequent catch-up growth in many cases	Continued improvement. Compensatory hyperkinesis may also occur and gradually subside. Serial IQ test scores may improve

Table modified from Green, W.H., (1986). Psychosocial dwarfism: Psychological and etiological considerations. In B.B. Lahey & A.E. Kazdin (Eds.), *Advances in Clinical Child Psychology*, Vol 9 (pp. 245–278). New York: Plenum Press. With permission.

ested, concerned caretaker if he or she spends enough time with the child but this does not always occur.

Hopwood and Becker (1979) observed that, if a psychosocial dwarf was placed in a new unsatisfactory environment, physical growth sometimes occurred but that there was minimal or no behavioral and emotional improvement. It has also been reported that although a psychosocial dwarf began growing following intervention, catch-up growth did not occur until he was actually hospitalized. These data suggest that the physical manifestations of PSD are caused by more adverse conditions than are the emotional and behavioral symptoms.

While the proposed model seems consonant with most of the available data, it must be noted that, to date, the exact relationship between growth and hormonal levels has not been satisfactorily elucidated. Hopwood and Becker's (1979) observation that placing a younger sibling with PSD in the same foster home with his brother who also had PSD caused an abrupt halt in the physical growth and emotional maturation of both of them and an immediate reexacerbation of regressive and destructive behavior also adds complications to a simple model, although classical conditioning, with each sibling representing, psychologically, a return to the situation with the inimical caretaker, could be invoked to explain this. A similar phenomenon is known to exist among opiate addicts; some addicts, long after they are no longer addicted to or physically dependent upon heroin, develop craving and withdrawal symptoms upon physically returning to the area where they used to obtain and use narcotics (Wikler, 1973). It would be fascinating if classical conditioning were responsible for causing alterations in CNS endorphins in both the psychosocial dwarfs and the ex-heroin addicts.

Conclusion

Classical psychosocial dwarfism provides us with, perhaps, the most spectacular opportunity yet known to study and understand some mechanisms through which particular adverse external influences may, through CNS modulated events, influence both physical growth and behavior. Additional important insights into these mechanisms are being made through studies in the premature human infant and the comparative psychoneuroendocrinology of the rat pup. As our knowledge in this area increases, important therapeutic gains in treating PSD and its formes frustes, other related growth problems and attachment disorders, nonorganic failures to thrive, and some behavioral disorders will almost certainly follow.

References

Afshani, E., Osman, M., & Girdany, B.R. (1973). Widening of cranial sutures in children with deprivation dwarfism. *Radiology, 109,* 141–144.

Bartolome, J.V., Bartolome, M.B., Daltner, L.A., Evans, C.J., Barchas, J.D., Kuhn, C.M., & Schanberg, S.M. (1986). Effects of β-Endorphin on ornithine decarboxylase in tissues of developing rats: A potential role for this endogenous neuropeptide in the modulation of tissue growth. *Life Sciences, 38,* 2355–2362.

Brasel, J.A. (1973). Review of findings in patients with emotional deprivation. In L.I. Gardner & P. Amacher (Eds.), *Endocrine aspects of malnutrition: Marasmus, kwashiorkor, and psychosocial deprivation* (pp. 115–127). Santa Ynez, CA: Kroc Foundation.

Butler, S.R., & Schanberg, S.M. (1977). Effect of maternal deprivation on polyamine metabolism in preweanling rat brain and heart. *Life Sciences, 21,* 877–884.

Butler, S.R., Suskind, M.R., & Schanberg, S.M. (1978). Maternal behavior as a regulator of polyamine biosynthesis in brain and heart of the developing rat pup. *Science, 199,* 445–447.

Campbell, M., Green, W.H., Caplan, R., & David, R. (1982). Psychiatry and endocrinology in children: Early infantile autism and psychosocial dwarfism. In P.J.V. Beumont & G.D. Burrows (Eds.), *Handbook of psychiatry and endocrinology* (pp. 15–62). Amsterdam, the Netherlands: Elsevier Biomedical Press.

Capitanio, M.A., & Kirkpatrick, J.A. (1969). Widening of the cranial sutures. *Radiology, 92*, 53–59.

D'Ercole, A.J., Underwood, L.E., & Van Wyk, J.J. (1977). Serum somatomedin-C in hypopituitarism and in other disorders of growth. *Journal of Pediatrics, 90*, 375–381.

Drash, P.W., Greenberg, N.E., & Money, J. (1968). Intelligence and personality in four syndromes of dwarfism. In D.B. Cheek (Ed.), *Human growth: Body composition, cell growth, energy, and intelligence* (pp. 568–581). Philadelphia: Lea & Febiger.

Ferholt, J.B., Rotnem, D.L., Genel, M., Leonard, M., Carey, M., & Hunter, D.E.K. (1985). A psychodynamic study of psychosomatic dwarfism: A syndrome of depression, personality disorder, and impaired growth. *Journal of the American Academy of Child Psychiatry, 24*, 49–57.

Field, T.M., Schanberg, S.M., Scafidi, F., Bauer, C.R., Vega-Lahr, N., Garcia, R., Nystrom, J., & Kuhn, C.B. (1986). Tactile/kinesthetic stimulation effects on preterm neonates. *Pediatrics, 77*, 654–658.

Frasier, S.D., & Rallison, M.L. (1972). Growth retardation and emotional deprivation: Relative resistance to treatment with human growth hormone. *Journal of Pediatrics, 80*, 603–609.

Gloebl, H.J., Capitanio, M.A., & Kirkpatrick, J.A. (1976). Radiographic findings in children with psychosocial dwarfism. *Pediatric Radiology, 4*, 83–86.

Green, W.H. (1986). Psychosocial dwarfism: Psychological and etiological considerations. In B.J. Lahey & A.E. Kazdin (Eds.), *Advances in Clinical Child Psychology*, Vol 9 (pp. 245–278). New York: Plenum Publishing.

Green, W.H. (1989). Reactive attachment disorders of infancy or early childhood. In H.I. Kaplan & B.J. Sadock (Eds.), *Comprehensive textbook of psychiatry/V* (5th ed.) (pp. 1894–1903). Baltimore: Williams & Wilkins.

Green, W.H., Campbell, M., & David, R. (1984). Psychosocial dwarfism: A critical review of the evidence. *Journal of the American Academy of Child Psychiatry, 23*, 39–48.

Green, W.H., Deutsch, S.I., & Campbell, M. (1987). Psychosocial dwarfism, infantile autism, and attention deficit disorder. In C.B. Nemeroff & P.T. Loosen (Eds.), *Handbook of clinical psychoneuroendocrinology* (pp. 109–142). New York: Guilford Press.

Guilhaume, A., Benoit, O., Gourmelen, M., & Richardet, J.M. (1982). Relationship between sleep stage IV deficit and reversible HGH deficiency in psychosocial dwarfism. *Pediatric Research, 16*, 299–303.

Hernandez, R.J., Poznanski, A.K., Hopwood, N.J., & Kelch, R.P. (1978). Incidence of growth lines in psychosocial dwarfs and idiopathic hypopituitarism. *American Journal of Roentgenology, 131*, 477–479.

Honda, Y., Takahashi, K., Takahashi, S., Azumi, K., Irie, M., Sakuma, M., Tsushima, T., & Shizume, K. (1969). Growth hormone secretion during nocturnal sleep in normal subjects. *Journal of Clinical Endocrinology and Metabolism, 29*, 20–29.

Hopwood, N.J. & Becker D.J. (1979). Psychosocial dwarfism: Detection, evaluation and management. In A.W. Franklin (Ed.), *Child abuse and neglect, 3*, 439–447.

Howse, P.M., Rayner, P.H.W., Williams, J.W., Rudd, B.T., Bertrande, P.V., Thompson, C.R.S., & Jones, L.A. (1977). Nyctohemeral secretion of growth hormone in normal children of short stature and in children with hypopituitarism and intrauterine growth retardation. *Clinical Endocrinology, 6,* 347–359.

Krieger, I., & Good, M.H. (1970). Adrenocortical and thyroid function in the deprivation syndrome. *American Journal of Diseases of Children, 120,* 95–102.

Krieger, I., & Mellinger, R.C. (1971). Pituitary function in the deprivation syndrome. *Journal of Pediatrics, 79,* 216–225.

Kuhn, C.M., Butler, S.R., & Schanberg, S.M. (1978). Selective depression of serum growth hormone during maternal deprivation in rat pups. *Science, 201,* 1034–1036.

Kuhn, C.B., Evoniuk, G., & Schanberg, S.M. (1979). Loss of tissue sensitivity to growth hormone during maternal deprivation in rats. *Life Sciences, 25,* 2089–2097.

Laron, Z. (1983). Somatomedin, insulin, growth hormone, and growth. In G. Chiumello & M. Sperling (Eds.), *Recent progress in pediatric endocrinology* (pp. 67–80). New York: Raven Press.

Patton, R.G., & Gardner, L.I. (1963). *Growth failure in maternal deprivation.* Springfield, IL: Charles C. Thomas.

Patton, R.G., & Gardner, L.I. (1975). Deprivation dwarfism (psychosocial deprivation): Disordered family environment as a cause of so-called idiopathic hypopituitarism. In L.I. Gardner (Ed.), *Endocrine and genetic diseases of childhood and adolescence* (2nd ed.) (pp. 85–98). Philadelphia: W.B. Saunders Co.

Powell, G.F., Brasel, J.A., & Blizzard, R.M. (1967a). Emotional deprivation and growth retardation simulating idiopathic hypopituitarism. I. Clinical evaluation of the syndrome. *New England Journal of Medicine, 276,* 1271–1278.

Powell, G.F., Brasel, J.A., Raiti, S., & Blizzard, R.M. (1967b). Emotional deprivation and growth retardation simulating idiopathic hypopituitarism. II. Endocrinologic evaluation of the syndrome. *New England Journal of Medicine, 276,* 1279–1283.

Powell, G.F., Hopwood, N.J., & Barratt, E.S. (1973). Growth hormone studies before and during catch-up growth in a child with emotional deprivation and short stature. *Journal of Clinical Endocrinology and Metabolism, 37,* 674–679.

Puig-Antich, J. (1985). Biological factors in prepubertal major depression. *Psychiatry Annals, 15,* 390–393, 397.

Puig-Antich, J., Chambers, W., Halpern, F., Hanlon, C., & Sachar, E.J. (1979). Cortisol hypersecretion in prepubertal depressive illness: A preliminary report. *Psychoneuroendocrinology, 4,* 191–197.

Puig-Antich, J., Goetz, R., Davies, M., Fein, M., Hanlon, C., Chambers, W.J., Tabrizi, M.A., Sachar, E.J., & Weitzman, E.D. (1984a). Growth hormone secretion in prepubertal children with major depression. II. Sleep-related plasma concentrations during a depressive episode. *Archives of General Psychiatry, 41,* 463–466.

Puig-Antich, J., Goetz, R., Davies, M., Tabrizi, M.A., Novacenko, H., Hanlon, C. Sachar, E.J., & Weitzman, E.D. (1984b). Growth hormone secretion in prepubertal children with major depression. IV. Sleep-related plasma concentrations in a drug-free, fully recovered clinical state. *Archives of General Psychiatry, 41,* 479–483.

Puig-Antich, J., Novacenko, H., Davies, M., Tabrizi, M.A., Ambrosini, P., Goetz, R., Bianca, J., Goetz, D., & Sachar, E.J. (1984c). Growth hormone secretion in prepubertal children with major depression. III. Response to insulin-induced hypoglycemia after recovery from a depressive episode and in a drug-free state. *Archives of General Psychiatry, 41,* 471–475.

Puig-Antich, J., Tabrizi, M.A., Davies, M., Goetz, R., Chambers, W.J., Halpern, F., & Sachar, E.J. (1981). Prepubertal endogenous major depressives hyposecrete growth hormone in response to insulin-induced hypoglycemia. *Biological Psychiatry, 16,* 801–818.

Roger, L.J., Schanberg, S.M., & Fellows, R.E. (1974). Growth and lactogenic hormone stimulation of ornithine decarboxylase in neonatal rat brain. *Endocrinology, 95,* 904–911.

Saenger, P., Levine, L.S., Wiedemann, E., Schwartz, E., Korth-Schutz, S., Pariera, J., Heinig, B., & New, M.I. (1977). Somatomedin and growth hormone in psychosocial dwarfism. *Pädiatrie and Pädologie* (Suppl.), *5,* 1–12.

Scafidi, F.A., Field, T.M., Schanberg, S.M., Bauer, C.R., Vega-Lahr, N., Garcia, R., Poirier, J., Nystrom, G., & Kuhn, C.G. (1986). Effects of tactile/kinesthetic stimulation on the clinical course and sleep/wake behavior of preterm neonates. *Infant behavior and development, 9,* 91–105.

Schanberg, S.M., Evoniuk, G., & Kuhn, C.M. (1984). Tactile and nutritional aspects of maternal care: Specific regulators of neuroendocrine function and cellular development. *Proceedings of the Society for Experimental Biology and Medicine, 175,* 135–146.

Schanberg, S.M., & Field, T.M. (1987). Sensory deprivation stress and supplemental stimulation in the rat pup and preterm human neonate. *Child Development, 58,* 1431–1447.

Spitz, R.A. (1945). Hospitalism: An inquiry into the genesis of psychiatric conditions in early childhood. *The Psychoanalytic Study of the Child, 1,* 53–74.

Spitz, R.A. (1946a). Hospitalism: A follow-up report of investigation described in Volume I, 1945. *The Psychoanalytic Study of the Child, 2,* 113–117.

Spitz, R.A. (1946b). Anaclitic depression: An inquiry into the genesis of psychiatric conditions in early childhood, II. *The Psychoanalytic Study of the Child, 2,* 313–342.

Stokes, P.E. (1988). Psychoendocrinology of depression and mania. In: A. Georgotas & R. Cancro (Eds.), *Depression and mania* (pp. 290–304). New York: Elsevier.

Takahashi, Y., Kipnis, D.M., & Daughaday, W.H. (1968). Growth hormone secretion during sleep. *Journal of Clincial Investigation, 47,* 2079–2090.

Tanner, J.M. (1973). Letter to the editor. Resistance to exogenous human growth hormone in psychosocial short stature (emotional deprivation). *Journal of Pediatrics, 82,* 171–172.

Tanner, J.M., Whitehouse, R.H., Hughes, P.C.R., & Vince, F.P. (1971). Effect of human growth hormone treatment for 1 to 7 years on growth of 100 children, with growth hormone deficiency, low birthweight, inherited smallness, Turner's syndrome, and other complaints. *Archives of Disease in Childhood, 46,* 745–782.

Tibbles, J.A.R, Vallet, H.L, Brown, B.St.J. & Goldbloom, R.B. (1972). Pseudotumor cerebri and deprivation dwarfism. *Developmental Medicine and Child Neurology, 14,* 322–331.

Van Den Brande, J.L., Van Buul, S., Heinrich, U., Van Roon, F., Zurcher, T., & Van Steirtegem, A.C. (1975). Further observations on plasma somatomedin activity in children. In R. Luft & K. Hall (Eds.), *Advances in metabolic disorders. Vol 8. Somatomedins and some other growth factors* (pp. 171–181). New York: Academic Press.

Wikler, A. (1973). Dynamics of drug dependence. Implications of a conditioning theory for research and treatment. *Archives of General Psychiatry, 28,* 611–616.

Wise, P.H., Burnet, R.B., Geary, T.D., & Berriman, H. (1975). Selective impairment of growth hormone response to physiological stimuli. *Archives of Diseases in Childhood, 30,* 210–214.

Wolff, G., & Money, J. (1973). Relationship between sleep and growth in patients with reversible somatotropin deficiency (psychosocial dwarfism). *Psychological Medicine, 3,* 18–27.

8
Intelligence (IQ) Lost and Regained: The Psychoneuroendocrinology of Failure to Thrive, Catch-up Growth, the Syndrome of Abuse Dwarfism, and Munchausen's Syndrome by Proxy

CHARLES ANNECILLO, JOHN MONEY,
AND CECILIA LOBATO

Reversible failure of mental growth together with reversible failure of statural growth, both associated with child abuse, was first ascertained through the study of longitudinal follow-up case records in the Johns Hopkins Psychohormonal Research Unit (Money, 1977; see also Williams & Money, 1980). The purpose of this chapter is to review the history of research on the developmental constancy of IQ, preliminary to the presentation of data on the reversibility of IQ impairment in the failure-to-thrive syndrome known variously as reversible hyposomatotropinism, psychosocial dwarfism, and abuse dwarfism. This syndrome is characterized by impairment of statural, intellectual, and social growth and maturation before rescue from abuse, and catch-up growth following rescue. The parents or guardians of the child with abuse dwarfism are themselves nosologically classified under Munchausen's syndrome by proxy.

IQ Constancy

IQ Change and the Middle Class

The study of IQ improvement in the syndrome of abuse dwarfism (Money et al., 1983b) is one of a few systematic longitudinal investigations that provide support for the notion that IQ can change and that it changes in association with environmental circumstances.

Supported by USPHS Grant #HD00325-31

Constancy versus inconstancy of IQ can be demonstrated only in a longitudinal study by repeated follow-up testing. Major studies of middle-class children have revealed high-positive correlations between IQs tested at two different ages. In general, comparisons of IQs tested at age 18 years with IQs tested previously from age 6 years upward show high-positve correlations (Bayley, 1949). These data have been construed to perpetuate the commonly held belief that IQ remains constant. Although studies reporting IQ change in repeatedly tested normal middle-class children are rare, data from the Fels study (McCall et al., 1973) reveal that IQ does change with wide variation in this population.

McCall et al. (1973) report IQ data on 80 children (38 males and 42 females) who were tested a maximum of 17 times between 2 1/2 and 17 years of age. Despite high-positive correlational stability in IQ between age groups, individual IQ patterns displayed an average longitudinal change of 28.5 IQ points. Shifts of more than 40 points occurred in one in seven children.

Patterns of declining IQs in children were associated with high parental disciplinary penalties. The higher IQ patterns were associated with parental encouragement of intellectual behavior with moderate disciplinary penalties. These parental correlates with IQ change are in agreement with the findings on intellectual growth impairment and recovery in the syndrome of abuse dwarfism wherein abuse is associated with low and declining IQ, whereas rescue into a benign environment is associated with elevation of IQ. However, the typical finding of recovery from severe IQ impairment in the abuse dwarfism study is very different from lack of impairment in the normal middle-class children in the Fels study. The mean IQ for the children in the Fels study was relatively consistent at approximately 118. By contrast, the children rescued from abuse in the abuse dwarfism study had a mean IQ of 66 in abuse and a mean IQ of 90 after various periods of unequivocal rescue (range: abuse, 36–101; rescue: 48–133).

IQ Change, Social Isolation, and Low Socioeconomic Status

Low and presumably declining IQ performance has been reported in various cross-sectional studies of groups of children living in various poor quality environments. Gordon (1923) studied IQ in a group of canal boat children in England who were socially, culturally, and educationally isolated from the general population. He found that the older the child, the lower the IQ; at approximately age 12, the average IQ was 60. Sherman and Key (1932) obtained similar results in a study of isolated Virginia mountain children.

Heber (1976) reported findings on the beneficial effects of an intervention program in black children at high risk for mental retardation born into very low socioeconomic conditions. Focusing upon the birth parents, prevention of mental retardation was achieved through implementation of a vocational and educational intervention program. This form of mental retardation has been labeled cultural-familial (Heber & Garber, 1975) and sociocultural (Heber, 1976), suggesting environmental etiology. Sowell (1977) suggests that the key factor accounting for the lower mean IQ in the black population at large (approximately 15 IQ points below the general population mean of 100) is the disproportionately low socioeconomic status. He provides evidence for low IQ associated with low socioeconomic status among many immigrant groups when they entered the United States. The IQ level of these immigrant groups rose with improvement in their socioeconomic status. It is assumed that IQ scores in the black population would increase with socioeconomic improvement.

IQ Change, Child Abuse, and Neglect

Prior to the 1970s, research dealing directly with an association between child abuse and mental retardation was scarce (Brandwein, 1973). Various studies during the 1970s provide evidence for a deleterious effect of child battering and abuse on intellectual growth (Appelbaum, 1977; Brandwein, 1973; Buchanan & Oliver, 1977; Sandgrund et al., 1974). The Milwaukee Project (Heber & Garber, 1975) further showed the deleterious effect of the neglect of developmental stimulation and the beneficial effect of intervention in preventing developmental mental retardation.

Constancy versus inconstancy of IQ can be demonstrated only in a longitudinal study, that is, by repeated follow-up testing. Comprehensive longitudinal investigation of intellectual impairment in institutionalized infants and children can be found in studies conducted by Skeels (1966) and Dennis (1973). The environments of intellectual growth retardation within the institutions in both studies warranted the description of institutional abuse and neglect. In brief, they were overcrowded and inadequately staffed. Each author's descriptions of their respective institutional environments are very similar and reminiscent of the more well-known institutional conditions described by Spitz (1946).

In 1966, Skeels published his outcome study of mental retardation as a sequel to infantile institutional abuse and neglect (Skeels & Fillmore, 1937). A study group of children was transferred to an enriched institutional environment where they subsequently displayed dramatic intellectual growth improvement as compared to a group of comparable children who continued in their regular living arrangement. Adult follow-up revealed various qualitative differences between the groups. The study group was far superior in educational and vocational achievement and was found to be leading a more normal independent life style. Dennis (1973) further demonstrated intellectual impairment as a sequel to institutional abuse and neglect. He found that the younger the institutionalized foundlings were when adopted into normal family life, the earlier the resumption of normal intellectual growth and the higher the ultimate level of the adult IQ.

Nature of the Syndrome of Abuse Dwarfism

The syndrome of abuse dwarfism is characterized by a domicile-specific impairment of statural growth and growth hormone secretion (Patton & Gardner, 1975). Along with various other pathological features of the syndrome, both impairments occur as sequelae of child abuse (Money, 1977). All impairments are reversible upon change of domicile, away from abuse into an environment of rescue. Taxonomically, the syndrome today is usually known as abuse dwarfism (Money, 1977) or psychosocial dwarfism (Reinhart & Drash, 1969). However, the syndrome has had various diagnostic labels: environmental failure to thrive (Barbero & Shaheen, 1967); deprivation dwarfism (Silver & Finkelstein, 1967); maternal deprivation and emotional deprivation (Powell et al., 1967a; Powell et al., 1967b); and reversible hyposomatotropinism (Money & Wolff, 1974).

Three primary impairments in the syndrome are deficits in physique age, mental age, and social age relative to chronological age. Physique age includes height age, identified as deficient when the height is below the third percentile for chronological age. Physique age includes also the age of onset of puberty, which is delayed if abuse is sufficiently prolonged. Intellectual development is retarded so that the mental age, represented by IQ, is deficient (Money et al., 1983b). Social maturation also is retarded so that social age, including academic achievement age and psychosexual age, is deficient.

There are three known hormonal impairments in the syndrome. Growth hormone impairment is associated with statural growth impairment resulting in a physique age discrepant with chronological age. Impaired gonadotropin (LH and FSH) secretion is associated with delayed puberty. Impaired response to adrenocorticotropic hormone (ACTH) reserve as tested by metyrapone stimulation is partial—total impairment would be lethal.

The pathognomonic characteristic of the environment of growth failure is child abuse that usually occurs in the parental home where the child is the victim of multiple practices of abuse and neglect. Recovery takes place after rescue. The earlier the rescue, the greater the amount of physical, mental, and social catch-up growth that can be achieved. An initial hospitalization for a short period of time, approximately two weeks, allows for the resumption of growth hormone secretion, and leads to the onset of catch-up statural growth, both of which are essential to establish the diagnosis of abuse dwarfism.

Various forms of unusual social behavior represent a developmental deficit reversible upon rescue from abuse. Eating may be from a garbage can, for example, and drinking from a toilet bowl; or there may be binges of polydipsia and polyphagia, possibly followed by vomiting. There can also be a history of such reversible behavioral symptoms as enuresis, encopresis, social apathy or inertia, crying spells, insomnia, eccentric sleeping and waking schedules (Wolff & Money, 1973), abnormal EEG and sleeping growth hormone levels (Guilhaume et al., 1982; Taylor & Brook, 1986), pain agnosia and self-injury (Money et al., 1972), all occurring only in the growth-retarding environment of abuse.

After rescue, impairments such as delayed statural growth and delayed puberty can reverse rapidly whereas other impairments, such as intellectual growth, remain protracted. However, the rate of intellectual catch-up growth is positively correlated with the rate of catch-up statural growth (Money et al., 1983a). Intellectual catch-up growth has been measured by an increase of as much as 84 IQ points (Money et al., 1983b).

IQ Data in the Syndrome of Abuse Dwarfism

Sample and Procedures

The sample comprised a group of 34 patients, 15 females and 19 males (27 white and 7 black), with a diagnosis of abuse dwarfism. Except for 4 patients at around the fourth percentile, all were below the third percentile in stature. All of the patients had a history of abuse documented in their medical and social service records.

The patients were all seen and evaluated in the Psychohormonal Research Unit and the Pediatric Endocrine Clinic at The Johns Hopkins Hospital during the period 1958–1975. From a total of 50 patients, 34 met the criteria of having at least one follow-up IQ testing for before and after rescue comparison. Some patients had more than one follow-up IQ test and more than one change of environment.

Each patient has a consolidated case history on file in the Psychohormonal Research Unit. Most of the intelligence testing for this study was conducted by members of the Psychohormonal Research Unit. Relevant information was abstracted and reduced for each patient in chronological order of IQ testing and history of abuse and rescue. Ratings of the patient's domiciliary environmental conditions determined whether abuse was present, equivocal, or absent.

IQ Change in Consistent Rescue

Table 8.1 shows the results of IQ change in 23 patients who had a baseline abuse IQ and a final follow-up IQ after varying periods of consistent rescue. As a group, these patients displayed a significant IQ increase (t test for correlated samples, $z = 5.5$, $p < 0.001$).

When children with the syndrome of abuse dwarfism are rescued, their

TABLE 8.1. IQ elevation after rescue ($N = 23$).

IQ	Before rescue	After rescue	Increase in IQ	Age before rescue	Age after rescue	Increase in age
Mean	66	90	24	7-7	12-8	5-1
SD	16	21	21	4-4	5-11	3-1

Note: Age in years and months.
$R = .78; p < .005$

rate of statural growth accelerates, and they go through a period of catch-up growth. The change in rate of mental growth, if similar to that of statural growth, would be reflected in a progressive increase in IQ. A sudden increase in IQ would, by contrast, signify a response to a sudden lifting of some sort of inhibiting constraint on the manifestations of intelligence. If the hypothesis of progressive catch-up intellectual growth is correct, then there should be a correlation between the amount of IQ increase and the amount of time spent free of abuse after having been rescued.

Duration of rescue and two other variables relevant to resultant IQ, baseline IQ, and variability in age, were subjected to a multiple regression analysis. The analysis revealed that these differences together account for 60% of the variance in IQ elevation ($r = 0.779, p < 0.005$). Time in rescue accounts for most of the variance (z beta $= 4.47$) as compared with baseline IQ level (z beta $= 3.02$), while baseline age acted as a moderator variable with a negative beta weight (z beta $= -2.59$). The moderating role of age means that the younger the age at the time of rescue, the greater the amount of IQ elevation during a comparable period of time after rescue. Overall, the regression analysis supports the hypothesis that postrescue growth of intelligence is not sudden, but progressive over time.

IQ Change in Younger Versus Older Age Groups

In order to further test the age hypothesis, namely, that the younger the age at the time of rescue, the greater the amount of intellectual catch-up growth and IQ elevation, two subsets of patients were assembled (Table 8.2). The table shows a mean elevation of IQ in each subset of younger and older patients (33 and 16, respectively) which, when combined, attained a high degree of significance ($F = 27.18, p < 0.001$). Though this significance applies regardless of age of rescue, it is also evident that the younger the age at rescue, the greater the gain in IQ. Analysis of variance showed that the difference between the two subsets reached significance at the level of $p < 0.1$ ($F = 3.46$), which in view of the small size of the sample is quite substantial. In addition, the effect of age at the time of testing reached significance at the level of $p < 0.05$ ($F = 5.39$), reflecting rather convincingly the fact that the younger patients had, in general, higher IQs. Because they were younger, one may infer, they had suffered less IQ im-

TABLE 8.2. Postabuse elevation of IQ relative to age at time of rescue.

N	Age at rescue Range	Mean	Rescue mean	Baseline IQ $\bar{x}$ & SD	Follow-up IQ $\bar{x}$ & SD	IQ elevation $\bar{x}$ & SD
7	2-4 to 5-5	3-9	3-11	71 ± 21	104 ± 11	33 ± 24
7	5-8 to 15-7	10-3	4- 1	63 ± 15	78 ± 16	16 ± 7

Note: Age in years and months.

pairment prior to being rescued and therefore were able to benefit more from catch-up growth.

IQ Change in Equivocal Versus Abuse or Rescue Environments

Some patients spent at least one period of their lives in conditions that qualified as equivocal, that is, as definitely neither abusive nor nonabusive, or an alternation of both. If they were IQ tested at the onset and conclusion of this period, regardless of how many other tests they had, then they qualified for inclusion in Table 8.3.

Table 8.3 shows a consistent trend confirming IQ impairment in an environment of abuse and IQ elevation after change to an environment of rescue. The new finding of this table is that after removal from the environment of abuse, even when the change was rated as equivocal and not as a full rescue, IQ deterioration ceased or showed a minor degree of elevation.

Statistical evaluation of Table 8.3 required two analyses of variance—one comparing the three environmental sequences that began with abuse, and the other comparing the three sequences in which the second condition was unchanged from the first. In both analyses, the main effect for

TABLE 8.3. Change in IQ relative to abuse, equivocal and rescue environments.

Row	Environmental	N	$\bar{x}$ years of test-test interval	$\bar{x}$ IQ baseline	$\bar{x}$ IQ follow-up	$\bar{x}$ IQ diff.
1	Abuse/Abuse	5	8-0	83 ± 16	75 ± 4	−8 ± 15
2	Abuse/Equivocal	6	4-11	61 ± 9	78 ± 7	17 ± 8
3	Abuse/Rescue	16	4-11	67 ± 17	92 ± 16	24 ± 18
4	Equivocal/Equivocal	8	4-6	75 ± 12	82 ± 13	7 ± 10
5	Rescue/Rescue	12	4-2	77 ± 17	96 ± 19	19 ± 13

Note: $N = 34$, but 13 patients qualified for inclusion in two different sequences because of multiple changes of domicile over widely spaced periods of their lives.
Analysis of variance:
Rows 1,2,3: trials ($F = 12.52, p < .01$)
  interaction ($F = 3.43, p < .05$)
Rows 1,4,5: trials ($F = 6.12, p < .01$)
  interaction ($F = 3.55, p < .05$)

trials was significant at the level of $p < 0.01$ ($F = 12.52$ and 6.12, respectively), despite the deviant effect of the abuse/abuse group. Correspondingly, in both analyses the interaction effect between IQ change from one environment to the other was significant at the level of $p < 0.05$ ($F = 3.43$ and 3.55, respectively).

A post hoc test of simple main effects using a pooled error term showed that the interactions were largely the result of the dramatic improvements shown by the abuse/rescue and rescue/rescue groups. These data yet again confirm the impairment of deterioration of intellectual growth in association with abuse environments and improvement of intellectual growth in rescue, even when rescue is equivocal.

Intelligence and Statural Growth Data in the Syndrome of Abuse Dwarfism

Sample and Procedure

In 32 of the foregoing cases, there were matching data on both intellectual and statural catch-up growth, so that the rates of catch-up growth could be compared (Money et al., 1983a). Each patient had been tested for at least one follow-up IQ after having been rescued and measured for height at the same age.

In order to have comparable units of growth in both height and intelligence, each was reduced to a ratio, the denominator of which was chronological age. The familiar derivation of IQ by dividing mental age by chronological age ($IQ = MA/CA \times 100$) was parallelled by the derivation of HQ (height quotient) by dividing height age by chronological age ($HQ = HA/CA \times 100$). The norms used for height age determination were those in the growth tables of Data from the National Health Survey of the National Center for Health Statistics (Hamill et al., 1977). The number of before and after comparisons of IQs and HQs exceeded the number of patients since most had multiple assessments.

Intellectual and Statural Catch-Up Growth Correlate

Clinical observation revealed that statural growth tended to recover more rapidly than intellectual growth after rescue. However, a positive correlation between the extent to which the two quotients changed existed regardless of how fast or slow the change took place.

The primary statistical strategy is a before and after comparison of the difference between the pairs of IQs and the paris of HQs obtained before and after rescue (Table 8.4). The correlation coefficient is 0.42 and is significant at the $p < 0.01$ level. The duration of follow-up varied from around four months (2 cases) to thirteen years (1 case) with a median of five

TABLE 8.4. Means and correlation of IQ and
HQ increments accrued during follow-up
(*N*=32).

Follow-up status	IQ (M±SD)	HQ (M±SD)
Before rescue	69 ± 17	55 ± 17
After rescue	88 ± 18	82 ± 11
Difference	19 ± 22	27 ± 18
	r(Diff.) = 0.42; p < 0.01	

IQ: Intelligence quotient
HQ: Height quotient

years, three months. The longer the follow-up, the more likely a variation
in growth rate because of, for example, the intervention of the pubertal
growth spurt, or the effects of multiple changes in domicile. Because of
such sources of variation, the strength of the correlation between the rates
of intellectual and statural catch-up growth, as reflected in the quotients, is
all the more noteworthy.

A separate statistical strategy was performed in order to assess the
possible confounding effect of the pubertal growth spurt. The correlation
between IQ and HQ comparisons prior to puberty (between early and late
prepubertal measurements) was significant at the $p < 0.01$ level.

Munchausen's Syndrome by Proxy

Impostoring, Sacrifice, and Atonement

The nosological term, Munchausen's syndrome by proxy, was coined in
1975 to apply to parents of children with abuse dwarfism (Money & Werlwas, 1976; Money, 1986). Munchausen's syndrome itself is a psycho-
pathological form of medical impostering in which more is known about
the history and origin of the presenting symptom than is disclosed in the
confabulatory and erroneous version that is given to the physician or clinic
staff. In Munchausen's syndrome, the symptoms are self-induced. In
Munchausen's syndrome by proxy, they are induced in the child, by the
parents or guardians (Meadow, 1977; Money, 1989).

The primary instigator of abuse in the syndrome of abuse dwarfism
would appear to be, according to present evidence, the mother, with the
father being acquiescent in what is, in effect, a folie à deux—a conspiracy
or collusion to abuse (Money & Werlwas, 1976). When parents who have
produced the symptoms of abuse seek medical attention for their child, the
apparent paradox is resolved by the proposition that treatment will insure
that the child will not die. Thus the program of abuse can be continued.
The dynamics of the paradox are characterized by the theme of sacrifice

and atonement. A sacrificial victim is an offering to atone for the transgressions of the penitent (Money, 1989).

Parents who abuse their child are, in effect, sacrificing their child. The transgression that is being atoned for in the sacrifice of the child is either unspeakable, unspoken of, or logically disconnected from the child's suffering. In an intensive investigation of two cases (Money et al., 1985), it was the mother who was heir to a transgression so great that it could not be atoned for, except by sacrifice. The transgression was the mother's own birth out of wedlock, which was, in one of the two cases, as a sequel to incest.

Addiction to Abuse

One of the self-justifications used by abusing parents is that their child instigates abuse. More fully explained, this claim may signify a failure of parent-child bonding, even from birth onward. Subsequently, it may signify also that the abused has become addicted to abuse. The response to receiving abuse is to stimulate more of it. The neurochemistry of this addiction may well be linked to the neurosecretion of brain endorphin which has a morphinelike effect. There is as yet no direct neurochemical evidence in support of this proposition.

The indirect evidence is from the clinic, namely, that children with the syndrome of abuse dwarfism exhibit signs of pain agnosia while they live under conditions of abuse (Money et al., 1972). Rescued and in the hospital, they regain an aversive response to pain in about two weeks. This phase is accompanied by a period of hyperactivity which can last for several months, and persist for as long as two years. It is marked by disputatiousness and noncompliance which exasperates other people and provokes their retaliation.

Discussion

The data on catch-up intellectual growth and IQ elevation in the syndrome of abuse dwarfism contradict the orthodox doctrine of IQ constancy. They also contradict the doctrine that IQ is determined either primarily or exclusively by heredity (see Jensen, 1969, 1973; Eysenck, 1974; see also rebuttal by Hirsch, 1975, 1981). The present data do not deny that heredity plays a role as a determinant of IQ, for heredity, expressed as DNA sequences in the genetic code, clearly is a factor in the neurobiology of intelligence. The neurobiology of intelligence encompasses far more than heredity, however. The syndrome of abuse dwarfism demonstrates unequivocally that the neurobiology of intelligence can be influenced by the same factors in the extrinsic, social environment that change the growth-regulating neurobiology of GHRH secretion from the hypothalamus which, in turn, regulates the secretion of growth hormone from the adjacent pituitary gland.

Thus, the syndrome of abuse dwarfism is of exceptional significance in providing a clinical model of an intrinsic neurobiological receptor mechanism inactivated by input from an extrinsic stimulus originating in the sociobehavioral environment, and having an effect on growth such as a genetic stimulus might be expected to have. Moreover, the effect is not restricted to somatic growth in stature, so-called organic growth, but applies also to growth of intellect and sociobehavioral maturity, both of which have traditionally been classified as psychological and nonorganic. The syndrome of abuse dwarfism closes the gap that, in conventional wisdom, separates organic from psychogenic, and shows that both are subject to the same principle of growth, growth arrest, and catch-up growth.

This is the context in which the syndrome of abuse dwarfism opens the gate of psychosomatic determinism (Green, 1986) to a new era of empirical research. There is a rapidly expanding body of knowledge on the influence of brain hormones, or derivatives of them, on learning and retention in animals (Bohas & de Weid, 1981). The two chief hormones are ACTH (adrenocorticotropic hormone) and MSH (melanin stimulating hormone). Biochemically, both are peptides composed of long sequences of amino acids, only short segments of which are necessary to affect learning and retention. It remains for future research on the syndrome of abuse dwarfism to ascertain the extent to which failure of ACTH, MSH, and growth hormone may participate in intellectual as well as somatic growth arrest. The neurobiology of tactile stimulation has been found, in experiments with newborn rats, to be linked to growth hormone secretion and somatic growth (Schanberg & Field, 1987).

In the future, other neurohormones and neurotransmitters will undoubtedly be implicated in the syndrome of abuse dwarfism. One of these others may prove to be β-endorphin or a related brain opiate. The basis for this hypothesis, already mentioned, is that in abuse dwarfism, during the period of abuse, a child becomes pain agnostic, with return of pain sensitivity soon after rescue.

The syndrome of abuse dwarfism confirms the error of dichotomizing genetic versus environmental determinants. Conversely, it confirms the wisdom of modern genetics in postulating a genetic norm of reaction which, for its proper expression, requires phyletically prescribed environmental boundaries (Money & Ehrhardt, 1972). The polarization of nature and nurture can be depolarized by adding a third term into the equation: nature + crucial period + nurture produces the phenotype. The phenotype may be behavioral, not only morphological (Money & Annecillo, 1987).

The pathognomonic characteristic of the environment of growth failure, in the syndrome of abuse dwarfism, is secret child abuse. It is without question that parental child abuse is as much a social evil as is cholera, or radiation from nuclear waste. The issue is whether child abuse should be classified as a crime and the parents prosecuted or whether it should be reclassified as a disease.

It is a recent phenomenon of the last two centuries that some manifestations of socially condemned behavior have been transferred from criminology to medicine and reclassified as syndromes of illness. Society's money and energy have then been directed toward diagnosis, prognosis, and treatment. The doctrine of prevention is correlated with the doctrine of etiology or cause. Indeed, the reclassification of criminality as pathology may be contingent on the discovery of its etiology and of the possibility of its cure or prevention.

In the abusing parents, Munchausen's syndrome by proxy is manifested in a theme of atonement by sacrifice. Etiologically the significance of this theme is that it provides only a necessary, but not sufficient, cause for abuse. What still is needed is an explanation of how forbidden and repugnant behavior becomes endorsed and practiced. For so complete a reversal of negative into positive, Solomon (1980) formulated and tested the theory of opponent-process learning. Opponent-process is seen in action when, for example, the initial panic and terror of a novice practicing a daredevil sport gives way to the exhilaration and ecstasy of the aficionado. This conversion is undoubtedly accomplished by a corresponding change in the neurochemistries of the learning brain. Deciphering them in their application to Munchausen's syndrome by proxy will require further advances in neuroscience.

References

Appelbaum, A.S. (1977). Developmental retardation in infants as a concomitant of physical child abuse. *Journal of Abnormal Child Psychology, 5,* 417–423.

Barbero, G.J., & Shaheen, E. (1967). Environmental failure to thrive: A clinical view. *Journal of Pediatrics, 71,* 639–644.

Bayley, N. (1949). Consistency and variability in the growth of intelligence from birth to eighteen years. *Journal of Genetic Psychology, 75,* 165–196.

Bohas, B., & de Wied, D. (1981). Actions of ACTH- and MSH-like peptides on learning, performance, and retention. In J.L. Martinez, Jr., R.A. Jensen, R.B. Messing, H. Rigter, and J.L. McGaugh (Eds.), *Endogenous peptides and learning and memory processes* (pp. 59–77). New York: Academic Press.

Brandwein, H. (1973). The battered child: A definite and significant factor in mental retardation. *Mental Retardation, 11,* 50–51.

Buchanan, A., & Oliver, J.E. (1977). Abuse and neglect as a cause of mental retardation: A study of 140 children admitted to subnormality hospitals in Wiltshire. *British Journal of Psychiatry, 131,* 458–467.

Dennis, W. *Children of the crèche.* New York: Appleton-Century-Crofts (1973).

Eysenck, H.J. (1974). *The inequality of man.* London: Temple Smith.

Gordon, H. (1923). Mental and scholastic tests among retarded children. *Educaton Pamphlet 44.* London: Board of Education.

Green, W.H. (1986). Psychosocial dwarfism: Psychological and etiological considerations. *Advances in Clinical Child Psychology, 9,* 245–278.

Guilhaume, A., Benoit, O., Gourmelen, M., & Richardet, J.M. (1982). Relationship between sleep stage IV deficit and reversible HGH deficiency in psychosocial dwarfism. *Pediatric Research, 16,* 299–303.

Hamill, P.V.V., Drizd, T.A., Johnson, C.L., Reed, R.B., & Roche, A.F. (1977). NCHS growth curves for children birth-18 years, United States. Vital and health statistics data from the National Health Survey, Series 11, Number 165, USPHS National Center for Health Statistics. Washington: U.S. Government Printing Office.

Heber, R.F. (1976). *Sociocultural mental retardation—A longitudinal study.* Paper presented at the Vermont Conference on the Primary Prevention of Psychopathology.

Heber, R.F., & Garber, H. (1975). The Milwaukee Project: A study of the use of family intervention to prevent cultural-familial mental retardation. In B.Z. Friedlander, G.M. Sterrit, & E.K. Girvin (Eds.), *Exceptional infant: Vol. 3. Assessment and intervention* (pp. 399–433). New York: Brunner/Mazel.

Hirsch, J. (1975). Jensenism: The bankruptcy of "science" without scholarship. United States Congressional Record, 122: No. 73, E2671–2672; No. 74, E2693–2695; No. 75, E2703–2705; E2716–2718, E2721–2722 (originally published in Educational Theory, 25, 3–27, 102).

Hirsch, J. (1981). To "Unfrock the charlatans." *Sage Race Relations Abstracts, 6,* 1–65.

Jensen, A.R. (1969). How much can we boost I.Q. and scholastic achievement? *Harvard Educational Review, 39,* 1–123.

Jensen, A.R. (1973). *Educability and group differences.* New York: Harper and Row.

McCall, R.B., Appelbaum, M.I., & Hogarty, P.S. (1973). Developmental changes in mental performance. *Monographs of the Society for Research in Child Development, 38* (3, Serial No. 150).

Meadow, R. (1977). Munchausen syndrome by proxy: The hinterland of child abuse. *Lancet, 2,* 343–345.

Money, J. (1977). The syndrome of abuse dwarfism (psychosocial dwarfism or reversible hyposomatotropinism): Behavioral data and case report. *American Journal of Diseases of Children, 131,* 508–513.

Money, J. (1986). Munchausen's syndrome by proxy: Update. *Journal of Pediatric Psychology, 11,* 583–584.

Money, J. (1989). Paleodigms and paleodigmatics: A new theoretical construct applicable to Munchausen's syndrome by proxy, child abuse dwarfism, paraphilias, anorexia nervosa, and other syndromes. *American Journal of Psychotherapy, 43,* 15–24.

Money, J., & Annecillo, C. (1987). Crucial period effect in psychoendocrinology: Two syndromes, abuse dwarfism and female (CVAH) hermaphroditism. In M.H. Bornstein (Ed.), *Sensitive periods in development: Interdisciplinary perspectives* (pp. 145–158). Hillsdale, NJ: Erlbaum.

Money, J., Annecillo, C., & Hutchison, J.W. (1985). Forensic and family psychiatry in abuse dwarfism: Munchausen's syndrome by proxy, atonement, and addiction to abuse. *Journal of Sex and Marital Therapy, 11,* 30–40.

Money, J., Annecillo, C., & Kelley, J.F. (1983a). Abuse-dwarfism syndrome: After rescue, statural and intellectual catchup growth correlate. *Journal of Clinical Child Psychology, 12,* 279–283.

Money, J., Annecillo, C., & Kelley, J.F. (1983b). Growth of intelligence: Failure and catchup associated respectively with abuse and rescue in the syndrome of abuse dwarfism. *Psychoneuroendocrinology, 8,* 309–319.

Money, J., & Ehrhardt, A.A. (1972). *Man and woman, boy and girl: The differenti-*

ation and dimorphism of gender identity from conception to maturity. Baltimore: Johns Hopkins Press.

Money, J., & Werlwas, J. (1976). Folie à deux in the parents of psychosocial dwarfs: Two cases. *Bulletin of the American Academy of Psychiatry and the Law, 4,* 351–362.

Money, J., & Wolff, G. (1974). Late puberty, retarded growth and reversible hyposomatotropinism (psychosocial dwarfism). *Adolescence, 9,* 121–134.

Money, J., Wolff, G., & Annecillo, C. (1972). Pain agnosia and self-injury in the syndrome of reversible somatotropin deficiency (psychosocial dwarfism). *Journal of Autism and Childhood Schizophrenia, 2,* 127–139.

Patton, R.G., & Gardner, L.I. (1975). Deprivation dwarfism (psychosocial deprivation): Disordered family environment as cause of so-called idiopathic hypopituitarism. In L.I. Gardner (Ed.), *Endocrine and genetic diseases of childhood and adolescence* (2nd ed.) (pp. 85–98). Philadelphia: W.B. Saunders Co.

Powell, G.F., Brasel, J.A., & Blizzard, R.M. (1967a). Emotional deprivation and growth retardation simulating idiopathic hypopituitarism. I. Clinical evaluation of the syndrome. *New England Journal of Medicine, 276,* 1271–1278.

Powell, G.F., Brasel, J.A., Raiti, S., & Blizzard, R.M. (1967b). Emotional deprivation and growth retardation simulating idiopathic hypopituitarism. II. Endocrinologic evaluation of the syndrome. *New England Journal of Medicine, 276,* 1279–1283.

Reinhart, J.B., & Drash, A.L. (1969). Psychosocial dwarfism: Enviromentally induced recovery. *Psychosomatic Medicine, 31,* 165–171.

Sandgrund, A., Gaines, R.W., & Green, A.H. (1974). Child abuse and mental retardation: A problem of cause and effect. *American Journal of Mental Deficiency, 79,* 327–330.

Schanberg, S.M., & Field, T.M. (1987). Sensory deprivation stress and supplemental stimulation in the rat pup and preterm human neonate. *Child Development, 58,* 1431–1447.

Sherman, M., & Key, C.B. (1932). The intelligence scores of isolated mountain children. *Child Development, 3,* 279–290.

Silver, H.K., & Finkelstein, M. (1967). Deprivation dwarfism. *Journal of Pediatrics, 70,* 317–324.

Skeels, H.M. (1966). Adult status of children with contrasting early life experiences: A followup study. *Monographs of the Society for Research in Child Development, 31* (3, Serial No. 105).

Skeels, H.M., & Fillmore, E.A. (1937). The mental development of children from underprivileged homes. *Journal of Genetic Psychology, 50,* 427–439.

Solomon, R.L. (1980). The opponent-process theory of acquired motivation. *American Psychologist, 35,* 691–712.

Sowell, T. New light on black IQ. *New York Times Magazine,* March 27, 1977, pp. 56–62.

Spitz, R.A. (1946). Hospitalism. *Psychoanalytic Study of the Child, 2,* 113–117.

Taylor, B.J., & Brook, C.G.D. (1986). Sleep EEG in growth disorders. *Archives of Disease in Childhood, 61,* 754–760.

Williams, G.J., & Money, J. (Eds.) (1980). *Traumatic abuse and neglect of children at home.* Baltimore: Johns Hopkins University Press.

Wolff, G., & Money, J. (1973). Relationship between sleep and growth in patients with reversible somatotropin deficiency (psychosocial dwarfism). *Psychological Medicine, 3,* 18–27.

9
Disorders of the Sex Chromosomes: Medical Overview

JUDITH LEVINE ROSS, M.D.

Turner Syndrome (45 XO, Gonadal Dysgenesis)

In 1938 Doctor Henry Turner described 7 patients with short stature, sexual infantilism, and a wide-carrying angle at the elbow (cubitus valgus) (Turner, 1938). The original pathologic diagnosis in Turner syndrome was noted to be gonadal dysgenesis, with fibrous streaks replacing normal ovarian tissue (Wilkens & Fleishmann, 1954). The missing X chromosome was discovered by Barr in 1959 (Ford et al., 1959). The most common features constituting the phenotype of females with Turner syndrome are short stature, sexual infantilism, and gonadal dysgenesis. A group of associated characteristic features have also been described and are detailed below.

Incidence

Turner syndrome (gonadal dysgenesis) occurs in approximately 1 out of 5,000 births (Evans, 1977). Most patients (60%) have a single X chromosome but mosaic karyotypes are also common where only some cells are missing the X chromosome or part of the X chromosome (Palmer & Reichmann, 1976).

The 45 XO karyotype is one of the most common chromosomal abnormalities. There is a considerable in utero loss of 45 XO embryos and fetuses. About 10% of all aborted fetuses have a 45 XO karyotype. Less than 3% of 45 XO fetuses survive to term (Hall, 1987).

Pathophysiology

The X chromosome abnormality can arise through several different mechanisms. It may be the consequence of nondysjunction or chromosome loss during oogenesis or spermatogenesis. Alternatively, this chromosomal anomaly can develop early in conception as a result of an error in mitosis such as anaphase lag or mitotic nondysjunction. The mitotic error is not associated with advanced maternal age. Studies of sex-linked traits such as the Xg blood groups demonstrate that the paternally derived X chromsome is lost more commonly than the maternally derived X chromosome (Soltan, 1968). The chromosome loss in gonadal dysgenesis can include all cells missing an X chromsome, deletions of the long or short arm of the X chromsome (46 XXq−, 46 XXp−), or a duplication of the long arm of the second X chromosome (46 XX isochromosome X). The genotype may be mosaic, where several populations of cells exist in a single individual. The most common mosaic pattern in patients with Turner's syndrome is 45X/46XX (Palmer & Reichmann, 1976).

In patients with Turner syndrome, the ovaries apparently form normally but become progressively atretic after the fourth month of gestation (Singh & Carr, 1966). By the ninth month of gestation, the ovaries are replaced by fibrous tissues without follicles. The oocyte complement apparently undergoes accelerated atresia so that few if any oocytes remain by birth (Singh & Carr, 1966). A complete second X chromosome seems to be a requirement for normal ovarian development in utero.

Clinical Aspects

The typical patient with 45 XO gonadal dysgenesis has the phenotypic abnormalities described in the following. Individuals with the same 45 XO karyotype, however, can have few or many of these traits. Patients who are mosaics (only part of cells are missing the X chromsome) or patients in whom a part of one X is missing may have fewer of the phenotypic characteristics.

The neonatal presentation for Turner syndrome is termed Bonnevie-Ullrich syndrome (Ullrich, 1949). These infants are born with lymphedema of the hands and feet and webbed necks as well as the other features of children with Turner syndrome.

Short Stature

Short stature is a very common feature in these patients (98%). The mean final height of adult women with Turner syndrome ranges from 142.6 to 147.3 cm (Palmer & Reichmann, 1976; Brook et al., 1974; Lenko et al., 1979; Lyon et al., 1980; Sybert, 1984; Demetriou et al., 1984; Park et al., 1983). In general, the children are born small for gestational age, have decreased growth velocity during childhood, and lack a pubertal growth spurt. The final height is significantly correlated with midparental height (Lenko et al., 1979). The etiology of the short stature does not seem to be related to growth hormone deficiency or abnormalities in adrenal or gonadal hormones (Ross et al., 1985; Lucky et al., 1979). Somatomedin C levels are decreased in these patients (Ross et al., 1985; Cuttler et al., 1980).

Sexual Infantilism Secondary to Gonadal Dysgenesis

These patients have immature-appearing, normal female external genitalia. The cervix and uterus are infantile. The ovaries are fibrous streaks of connective tissue that are parallel to the fallopian tubes and lack follicles. Although most women with Turner syndromes have streak gonads, a small minority do have primary follicles in the ovary. Rarely, women with Turner syndrome have occurrence of menarche and an attenuated period of menses. Conceptions have occurred in 11 known women with the 45 XO karyotype (Baudier, 1985). Patients with Turner's syndrome develop pubic and axillary hair spontaneously and have normal adrenarchal development (Lucky et al., 1979).

Somatic Abnormalities

The congenital lymphedema is thought to give rise to many of the phenotypic abnormalities listed herein. Patients with Turner syndrome have distinctive facies including micrognathia, rotated ears, a fishlike mouth, and an increased incidence of ptosis and strabismus. A high arched palate is commonly present. The neck is broad with a low posterior hairline. Webbing of the neck is present in 25% to 40% of the patients. The chest is broad, often with hypoplastic nipple development. Coarctation of the aorta, which is the most common cardiac defect, occurs in 10% to 20% of patients. Other cardiac anomalies include aortic stenosis and bicuspid aortic valve (Nora et al., 1970). Patients with Turner syndrome are at an increased risk for dissecting aortic aneurysms. Antibiotic prophylaxis is recommended for the individuals with cardiac abnormalities.

These females have an increased incidence of renal abnormalities (approximately 60% of patients) including rotated kidney, horseshoe kidneys, and double collecting systems (Litvak, 1978). The renal abnormalities put patients at an increased risk for ureteropelvic obstruction and urinary tract infections.

Other features of Turner syndrome may include cubitus valgus of the elbow, congenital edema of the hands and feet, pigmented nevi, hyperconvex nails, and scoliosis. Females with Turner syndrome have an increased incidence of autoimmune thyroid disease, idiopathic hypertension, diabetes mellitus, osteoporosis, and inflammatory bowel disease. The frequency of otitis media is increased and is related to the abnormal angle of Eustacian tube insertion in the palate. Hearing should be tested in children who have had frequent ear infections.

These patients, on average, have normal intelligence. The incidence of mental retardation is not increased. In general, the performance IQ is lower then the general population whereas the verbal IQ is slightly higher than the general population. Specific deficits in spatial, motor, and perceptual abilities have been described (McCauley et al., 1987).

Diagnosis

The diagnosis of Turner syndrome should be considered in a newborn with lymphedema and webbed neck, or in a female child presenting with short stature and/or delayed sexual maturation. A karyotype rather than a buccal smear should be sent because mosaic karyotypes are not accurately detected with the buccal smear. Children less than 9 or 10 years may have normal levels of luteinizing hormone (LH) and follicle stimulating hormone (FSH) (<10 mIU/ml). Only after age 10 do levels of LH and FSH rise to the castrate range (LH >25 mIU/ml, FSH >40 mIU/ml). Children diagnosed with Turner syndrome should be screened with a renal ultrasound, cardiac echocardiogram, and thyroid hormone levels because of the increased incidence of renal, cardiac, and thyroid abnormalities.

Treatment

Growth

Treatments used to stimulate growth have included oxandrolone (Urban et al., 1979; Joss et al., 1984), growth hormone (Rosenfeld et al., 1986; Raiti et al., 1986), and low-dose estrogen (100 ng/kg/day) (Ross et al., 1986; Martinez et al., 1987). These hormones have increased growth rates for 1 to 2 years in girls with Turner syndrome. There has been no growth treatment to date that is associated with a significant increase in adult height of women with Turner syndrome.

Estrogen Replacement

Estrogen is indicated as replacement treatment for the purpose of secondary sex characteristic development. The recommended age for commencing estrogen replacement therapy is between 13 and 15 years. A suggested treatment regimen involves starting the patient on 10 μg per day of ethinyl estradiol until breakthrough vaginal bleeding occurs. Patients then start a cycling regimen of ethinyl estradiol (20-35 μg/day) for the first 25 days of each month and progesterone (Provera 5-10 mg) from day 16 to 25 of the cycle. Alternatively, these women can be cycled monthly with estrogen and progesterone as is contained in oral contraceptives.

Klinefelter Syndrome (47 XXY, Gonadal Dysgenesis)

Dr. Harry Klinefelter first described 9 men with gynecomastia and small testes in 1942 (Klinefelter, 1986). The presence of an extra X chromosome in males with Klinefelter syndrome was reported in 1959 (Jacobs & Strong, 1959). The most common phenotypic features of males with Klinefelter syndrome include gynecomastia, small testes, azoospermia, and tall stature. Other associated characteristics have also been described and are detailed following.

Incidence

The 47 XXY karyotype is one of the most common chromosomal disorders, second in frequency only to Down's syndrome. Klinefelter syndrome occurs in approximately 1 in 600 live births (Gerald, 1976). About 10% of patients with Klinefelter syndrome have a mosaic (46 XY/47 XXY) karyotype (Steward et al., 1982). Males with a mosaic karyotype generally are less severely affected (Sarkar & Marimutho, 1983).

Pathophysiology

The additional X chromosome in Klinefelter syndrome results from either meiotic nondysjunction in the egg or sperm, or from mitotic nondysjunction in the xygote. The extra X chromsome is more frequently derived from the mother (Soltan, 1968) and is generally associated with advanced maternal age.

Klinefelter syndrome is the most common cause of testicular failure in men and is characterized by impaired spermatogenesis and testosterone synthesis. It is not known how the presence of the extra X chromosome damages the testes.

Clinical Aspects

The most typical adult features of Klinefelter syndrome are gynecomastia, small testes, azoospermia, and tall stature. Clinical recognition of Klinefelter syndrome at birth is very unusual. In general, boys with the XXY karyotype have decreased birth weight, birth length, and head circumference (Ratcliffe, 1982). In newborn males with Klinefelter's, the testes are usually normal in size but the incidence of undescended testes in one or both testes is increased (Ratcliffe, 1982). Only a minority of the newborns have decreased penile length (Ratcliffe et al., 1986).

Testicular Failure

The testes of prepubertal boys with Klinefelter syndrome have a reduced number of spermatogonia (Ferguson-Smith, 1959) and are frequently decreased in size in comparison to normal boys (Ratcliffe, 1982; Ratcliffe et al., 1986; Laron & Hochman, 1971). Penile growth during childhood is decreased (Ratcliffe, 1982; Ratcliffe et al., 1986). The mean age of pubertal onset in males with Klinefelter syndrome is generally normal (Ratcliffe, 1982; Topper et al., 1982). These boys develop pubic hair and some testicular and penile enlargement before age 14 years. The maximum testicular volume (3-5 ml) is reached at approximately 13 years of age (Ratcliffe et al., 1986; Topper et al., 1982). In midpuberty, there is evidence of testicular growth arrest, or actual decrease in testicular size due to the fibrosis and hyalinization of the seminiferous tubules. The testes of adult XXY males are characterized by the absence of germinal cells and the presence of clumped, hyperplastic Leydig cells. Rarely, 47 XXY adult males produce sperm and have been able to father children (Laron et al., 1982).

During childhood and even in early puberty, pituitary-gonadal function is relatively normal in 47 XXY individuals. Gonadotropin (LH, FSH) levels are within the normal range until age 12 but become elevated by ages 14 to 15 (Topper et al., 1982; Salbenblatt, 1985). Testosterone levels are generally normal until age 16 but are frequently decreased compared to normal adult males after age 16 (Topper et al., 1982; Salbenblatt, 1985). Increased estradiol levels are found throughout puberty. The ratio of estradiol to testosterone is frequently elevated, which may account for the frequent occurrence of gynecomastia (Salbenblatt, 1985).

Tall Stature

Boys with Klinefelter syndrome have increased height velocities (Steward et al., 1982; Ratcliffe, 1982; Ratcliffe et al., 1986; Schibler et al., 1974) and tend to be over the fiftieth percentile for height very early in childhood (Steward et al., 1982; Ratcliffe, 1982; Ratcliffe et al., 1986; Ferguson-Smith, 1959) despite having decreased weight and length at birth. The mean adult height of Klinefelter males is approximately 179 to 184 cm which is significantly increased in comparison to the normal male population (Schibler et al., 1974; Becker, 1972).

Leg length is disproportionately increased during early childhood (Ratcliffe, 1982; Ratcliffe et al., 1986) and increased fat deposition in the hips can be detected by 10 years of age (Ferguson-Smith, 1959). The eunuchoidal body habitus may therefore be apparent even before the initiation of puberty. Eunuchoidal body proportions are frequently observed in adult 47 XXY males (Schibler et al., 1974; Becker, 1972).

Gynecomastia

Gynecomastia is noted in about 30% to 40% of men with Klinefelter syndrome (Robinson et al., 1986) and the incidence of breast cancer is increased in this population (Price et al., 1980).

Associated Somatic Abnormalities

Males with Klinefelter syndrome have an increased incidence of kyphosis and scoliosis, varicose veins, obesity, germ cell tumors, diabetes mellitus, autoimmune thyroid disease, and cerebrovascular disease (Price et al.,

1980; Worstman et al., 1986; Nielsen, 1972). Neurologic abnormalities observed with greater frequency include ataxia and essential tremor (Boltshauser & Deonna, 1978).

Psychological Aspects

Boys with the XXY karyotype may have neuromuscular deficits in childhood (Robinson et al., 1986; Salbenblatt et al., 1987). They have a selective reduction in verbal IQ scores in comparison to performance scores, but mean intelligence quotients are in the average range (Robinson et al., 1986; Salbenblatt et al., 1987). These boys are more likely to have speech and language delays as well as school difficulties (Stewart et al., 1982; Robinson et al., 1986; Salbenblatt et al., 1987; Evans et al., 1986). Certain personality characteristics have been noted in some 47 XXY boys including poor self image, shyness, immaturity, and antisocial behavior (Steward et al., 1982; Caldwell & Smith, 1972).

Diagnosis

The diagnosis should be pursued in the prepubertal male with small testes, tall stature, or disproportionately long legs. The diagnosis should also be considered in the adolescent presenting with significant gynecomastia, small testes, or eunuchoidal body proportions. A karyotype rather than buccal smear should be sent to detect mosaicism.

Luteinizing hormone and follicle stimulating hormone levels are usually elevated after age 12 to 14 years but tend to be in the normal range in prepubertal boys. Testosterone levels will frequently be normal in prepubertal and midpubertal boys and may be decreased in young adult and adult XXY males. Patients diagnosed with Klinefelter syndrome should be screened with thyroid hormone levels because of the increased incidence of thyroid disease.

Treatment

Gynecomastia

The increased breast tissue can be surgically removed because of the psychological problems that can ensue and the increased risk of breast carcinoma.

Testicular Failure

Approximately two thirds of males with Klinefelter's syndrome have inadequate levels of androgen and require replacement therapy with testosterone. Adolescent males are generally started on 25 to 50 mg of testosterone enanthate intramuscularly (IM) every 4 weeks and the dosage is gradually increased. A recommended adult testosterone replacement dose is testosterone enanthate, 200 mg IM every 2 weeks.

XYY Males

The incidence of the 47 XYY karyotype is approximately 1 in 1,000 births (Welch, 1985). Birth weight and length are generally normal in these males. Growth velocity is increased from early childhood, so XYY males tend to be taller than normal males (Stewart et al., 1982; Evans et al., 1986; Welch, 1985). Puberty tends to occur normally (Robinson et al., 1986). Adolescent and adult XYY males have an increased incidence of both gonadal failure and elevated testosterone levels (Welch, 1985). The incidence of school problems and behavioral difficulties in these patients is increased (Stewart et al., 1982; Sarkar & Marimutho, 1983; Ratcliffe, 1982; Ratcliffe et al., 1986; Ferguson-Smith, 1959; Laron & Hochman, 1971; Topper et al., 1982; Laron et al., 1982; Salbenblatt et al., 1985; Schibler et al., 1974; Becker, 1972; Robinson et al., 1986; Welch, 1985; Netley, 1986) and some XYY males have decreased verbal IQ performance (Ratcliffe et al., 1986).

XXX Females

The incidence of the 47 XXX karyotype is between 1 in 400 and 1 in 600 births (Hamerton et al., 1975). The birth weight and length of these infants is normal. These patients have increased height and decreased weight in comparison to normal women (Stewart et al., 1982; Ratcliffe et al., 1986; Robinson et al., 1986). Sexual maturation may be slightly delayed but is generally normal (Ratcliffe et al., 1986; Robinson et al., 1986). Girls with the XXX karyotype have a generalized reduction in intellectual abilities with delayed language processing and poor language skills (Stewart et al., 1982; Ratcliffe et al., 1986; Evans et al., 1986).

References

Baudier, M.M., Chilhal, H.J., & Dickey, R.P. (1985). Pregnancy and reproductive function in a patient with nonmosaic Turner syndrome. *Obstetrics and Gynecology (Suppl), 65,* 605.

Becker, K.L. (1972). Clinical and therapeutic experiences with Klinefelter's syndrome. *Fertility Sterility, 23,* 568.

Boltshauser, E.M., & Deonna, T. (1978). Klinefelter syndrome and neurologic disease. *Journal of Neurology, 219,* 253.

Brook, C.G.D., Murset, G., Zachmann, M., & Prader, A. (1974). Growth in children with 45, XO Turner's syndrome. *Archives of Diseases of Childhood, 49,* 789.

Caldwell, P.D., & Smith, D.W. (1972). The XXY (Klinefelter's) syndrome in childhood: Detection and treatment. *Journal of Pediatrics, 80,* 250.

Cuttler, L., Van Vliet, G., Conte, F.A., Kaplan, S.L., & Grumbach, M.M. (1980). Somatomedin C levels in children and adolescents with gonadal dysgenesis: Differences from age matched normal females and effect of chronic estrogen replacement therapy. *Journal of Clinical Endocrinology and Metabolism, 60,* 1087.

Demetriou, E., Emano, S.J., & Crigler, J., Jr. (1984). Final height in estrogen-treated patients with Turner syndrome. *Obstetrics and Gynecology, 64,* 459.

Evans, H.J. (1977). Chromosome anomalies among live births. *J Med Genet, 14,* 309.

Evans, J.A., de von Flindt, R., Greenberg, C., Ramsey, S., & Hamerton, J.L. (1986). Physical and psychological parameters in children with sex chromsome anomalies: Further follow up from the Winnipeg cytogenetic study of 14,069 newborn infants. *Birth Defects, 22,* 183.

Ferguson-Smith, M.A. (1959). The prepubertal testicular lesion in chromatin-positive Klinefelter's syndrome. *Lancet I,* 219.

Ford, C.E., Jones, K.W., Polani, P.E., De Almeida, J.C., & Briggs, J.H. (1959). A sex-chromosome anomaly in a case of gonadal dysgenesis (Turner syndrome). *Lancet 1,* 711.

Gerald, P.S. (1976). Sex chromosome disorders. *New England Journal of Medicine, 294,* 707.

Hall, J. (1987). Turner syndrome: an update. *Growth, genetics, and hormones* (p. 4). Biomedical Information Corporation.

Hamerton, J.L., Canning N., Ray, M., et al., (1975). A cytogentic survey of 14,069 newborn infants. I. Incidence of chromosomes abnormalities. *Clinical Genetics, 8,* 223.

Jacobs, P.A., & Strong, J.A. (1959). A case of human intersexuality having a possible XXY sex-determining mechanism. *Nature, 183,* 302.

Joss, E., & Zuppinger, K. (1984). Oxandrolone in girls with Turner's syndrome. *Acta Paediatrica Scandinavica, 73,* 674.

Klinefelter, H. (1986). Klinefelter's syndrome: Historical background and development. *Southern Medical Journal, 79,* 1089.

Laron, Z., Dickerman, Z., Zamir, R., & Galatzer, A. (1982). Paternity in Klinefelter's syndrome—a case report. *Archives of Andrology, 8,* 149.

Laron, Z., & Hochman, H. (1971). Small testes in prepubertal boys with Klinefelter's syndrome. *Journal of Clinical Endocrinology and Metabolism, 32,* 671.

Lenko, H.L., Perheentupa, J., & Soderholm, A. (1979). Growth in Turner's syndrome: spontaneous and fluoxymesterone stimulated. *Acta Paediatrica Scandinavica (Suppl) 277,* 57.

Litvak, A.S., Rousseau, T.G, Wrede, L.D., et al. (1978). The association of significant renal anomalies with Turner's syndrome. *Journal of Urology, 120,* 671.

Lucky, A.W., Marynick, S.P., Rebar, R.W., Cutler, G.B., Jr., Glenn, M., Johnsonbaugh, R.E., & Loriaux, D.L. (1979). Replacements oral ethinyloestradiol therapy for gonadal dysgenesis: Growth and adrenal androgen studies. *Acta Endocrinol Logica, 91,* 519.

Lyon, A.J., Preece, M.A., & Grant, D.B. (1980). Growth curve for girls with Turner syndrome. *Archives of Diseases of Childhood, 60,* 932.

Martinez, A., Heinrich, J.J., Domene, H., Escobar, M.E., Jasper, H., Montuori, E., & Bergade, C. (1987). Growth in Turner's syndrome: Long-term treatment with low dose ethinyl estradiol. *Journal of Clinical Endocrinology and Metabolism, 65,* 253.

McCauley, E., Kay, T., Ito, J., & Treder, R. (1987). The Turner syndrome: cognitive deficits, affective discrimination, and behavior problems. *Child Development, 58,* 464.

Netley, C.T. (1986). Summary overview of behavioral development in individuals with neonatally identified X and Y aneuploidy. *Birth Defects, 22,* 293.

Nielsen, J. (1972). Diabetes mellitus in patients with aneuploid chromosome aberrations and in their parents. *Human Genetics, 16,* 165.

Nora, J.J., Torres, F.G., Sinha, A.K., et al. (1970). Characteristic cardiovascular anomalies of XO Turner syndrome, XX and XY phenotype and XO/XX Turner mosaic. *American Journal of Cardiology, 25,* 639.

Palmer, C.G., & Reichmann, A. (1976). Chromosomal and clinical findings in 110 females with Turner syndrome. *Human Genetics, 35,* 35.

Park, E., Bailey, J.D., & Cowell, C.A. (1983). Growth and maturation of patients with Turner's syndrome. *Pediatric Research, 17,* 1.

Price, W.H., Clayton, J.F., Wilson, J., Collyer, S., & DeMey, R. (1980). Causes of death in X chromatin positive males (Klinefelter's syndrome). *Journal of Epidemiology and Community Health, 39,* 330.

Raiti, S., Moore, W.V., Van Vliet, G., & Kaplan, S.L. (1986). Growth-stimulating effects of human growth hormone therapy in patients with Turner syndrome. *Journal of Pediatrics, 109,* 944.

Ratcliffe, S.G. (1982). The sexual development of boys with the chromosome constitution 47, XXY (Klinefelter's) syndrome. *Clinics in Endocrinology and Metabolism, 11,* 703.

Ratcliffe, S.G., Murray, L., & Teague, P. (1986). Edinburgh study of growth and development of children with sex chromosome abnormalities III. *Birth Defects, 22,* 73.

Robinson, A., Bender, B.G., Borelli, J.B., Puck, M.H., Salbenblatt, S.A., & Winter, J.S.D. (1986). Sex chromosomal aneuploidy: Prospective and longitudinal studies. *Birth Defects, 22,* 23.

Rosenfeld, R.G., Hintz, R.L., Johanson, A.J., et al. (1986). Methionyl human growth hormone and oxandrolone in Turner syndrome: preliminary results of a prospective randomized trial. *Journal of Pediatrics, 109,* 936.

Ross, J.L., Long, L.M., Cassorla, F.G., Loriaux, D.L., & Cutler, G.B., Jr. (1986). The effect of low dose estradiol on 6-month growth rates in patients with Turner syndrome. *Journal of Pediatrics, 109,* 950.

Ross, J.L., Long, L.M., & Cutler, G.B., Jr. (1985). Growth hormone secretory dynamics in Turner syndrome. *Journal of Pediatrics, 106,* 202.

Salbenblatt, J.A., Bender, B.G., Puck, M.H., Robinson, A., Faiman, C., & Winter, J.S.D. (1985). Pituitary-gonadal function in Klinefelter syndrome before and during puberty. *Pediatric Research, 19,* 82.

Salbenblatt, J.A., Meyers, D.C., Bender, B.G., Linden, M.G., & Robinson, A. (1987). Gross and fine motor development in 47,XXY and 47,XYY males. *Pediatrics, 80,* 240.

Sarkar, R., Marimutho, K.M. (1983). Association between the degree of mosaicism and the severity of syndrome in Turner mosaics and Klinefelter mosaics. *Clinical Genetics, 24,* 420.

Schibler, D., Brook, C.G.D., Kind, H.P., Zachman, M., & Prader, A. (1974).

Growth and body proportions in 54 boys and men with Klinefelter's syndrome. *Helvetica Paediatrica Acta, 29,* 325.

Singh, R.F., & Carr, D.H. (1966). The anatomy and histology of XO human embryos and fetuses. *Anat Rec, 155,* 369.

Soltan, H.C. (1968). Genetic characteristics of families of XO and XXY patients, including evidence of source of X chromosome in 7 aneuploid patients. *J Med Genet, 5,* 173.

Stewart, D.A., Netley, C.T., & Park, E. (1982). Summary of clinical findings of children with 47,XXY, 47,XYY, and 47,XXX karyotypes. *Birth Defects, 18,* 1.

Sybert, V.P. (1984). Adult height in Turner syndrome with and without androgen therapy. *Journal of Pediatrics, 104,* 365.

Topper, E., Dickerman, Z., Prager-Lewin, R., Kaufman, H., Maimon, Z., & Laron, Z. (1982). Puberty in 24 patients with Klinefelter syndrome. *European Journal of Pediatrics, 139,* 8.

Turner, H.H. (1938). A syndrome of infantilism, congenital webbed neck, and cubitus valgus. *Endocrinology, 23,* 566.

Ullrich, O. (1949). Turner syndrome and status Bonnevie Ullrich. *American Journal of Human Genetics, 1,* 179.

Urban, M.D., Lee, P.A., Dorst, J.T., et al. (1979). Oxandrolone therapy in patients with Turner's syndrome. *Journal of Pediatrics, 94,* 829.

Welch, J.P. (1985). *The Y chromosome, part B, clinical aspects of Y chromosome abnormalities* (pp. 323–343). New York: Alan R. Liss, Inc.

Wilkins, L., & Fleishmann, W. (1954). Ovarian agenesis: Pathology, associated clinical symptoms and the bearing on the theories of sex differentiation. *Journal of Clinical Endocrinology, 4,* 1270.

Worstman, J., Mosco, H., & Dufau, M.L. (1986). Increased incidence of thyroid disease among men with hypergonadotropic hypogonadism. *American Journal of Medicine, 80,* 1055.

10
Cognitive Development of Children with Sex Chromosome Abnormalities

BRUCE G. BENDER, MARY H. PUCK,
JAMES A. SALBENBLATT, AND ARTHUR ROBINSON

Sex chromosome abnormalities (SCA) are genetic disorders associated with a variety of developmental problems, including learning disorders (LD). The study of SCA provides a unique opportunity to increase understanding of genetic and developmental features of LD. Most research on the genetics of LD involves cross-sectional or retrospective study of children in order to trace familial patterns of transmission of problems of unknown cause. In contrast to such heterogeneous samples, children with LD and the same SCA share a single genetic factor. The possibility of identifying an LD subtype of specific genetic causation is thus greatly increased, especially when the study is prospective (Pennington et al., 1982). Longitudinal evaluations of children with SCA may elucidate the relationship between physical and cognitive development and help to establish the processes by which SCA is associated with LD. This knowledge in turn may increase understanding of the role of the sex chromosomes in normal cognitive development.

Chromosome Abnormalities

Before 1956, the precise number of chromosomes present in each cell had not been established. We now know that humans normally have 46 chromosomes divided into 23 pairs or homologs, two of which are sex chromosomes (Tjio & Levan, 1956). The chromosomal constitution of normal males and females is 46, XY and 46, XX, respectively (Figures 10.1 and 10.2).

Abnormalities of chromosome number (aneuploidy) result from an ab-

This study was supported in part by grant 5R01-HD10032 from the USPHS; grant RR-69 from the General Clinical Research Centers Program of the Division of Research Resources, NIH; and The Genetic Foundation.

normal segregation of the chromosomes (nondisjunction) occurring either in gametogenesis (meiotic nondisjunction) or as a postzygotic phenomenon (mitotic nondisjunction). In the former case, all cells of the conceptus will be aneuploid, whereas in the latter event there may be two or more populations of cells that will vary in their chromosomal consititution, a situation called mosaicism.

Sex Chromosome Disorders

Chromosome abnormalities occur in one of 200 newborns. Approximately half of these are SCA. The most common SCAs occur when a male has an extra X or Y chromosome or when a female has an extra X or, in the case of X monosomy, is missing all or part of an X: 47,XXY, 47,XYY, 47,XXX, 45,X, 46XXp−, 46,XXq−, respectively. Two or more extra sex chromosomes are rare. SCA mosaicism can include a variety of cell line combinations such as 45,X/46,XX; 45,X/46,XX/47,XXX, and so on.

With the exception of the 45,X karyotype (Turner syndrome), X and Y aneuploidy is rarely identified at birth because affected individuals appear normal. Aneuplody of the autosomes, on the other hand, is usually accompanied by readily identifiable physical and developmental problems, incuding mental retardation. Early studies of individuals with SCA have been necessarily limited to retrospective analyses of adult patients, a biased form of ascertainment. There developed, however, sufficient data to establish that newborns with SCA have an increased risk of impaired cognitive development. In the last two decades, newborn screening has

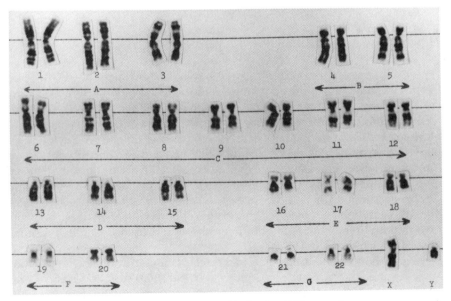

FIGURE 10.1. Normal male karyotype displaying XY sex chromosome constitution.

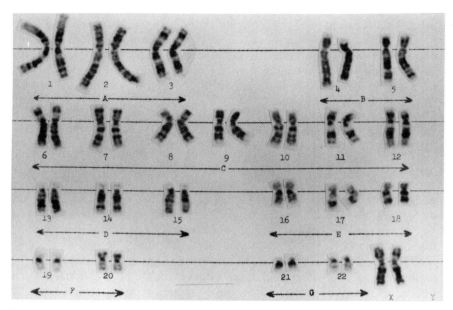

FIGURE 10.2. Normal female karyotype displaying XX sex chromosome constitution.

established incidence rates of SCA and has also provided unbiased samples for careful comparative study of physical, intellectual, and emotional development.

The Denver Study of Children with SCA

The Denver study (Robinson et al., 1979) is one of nine groups from five countries following children with SCA identified at birth (Robinson et al., 1979; Stewart, 1982). In Denver, amniotic membranes obtained from placentas of 40,000 consecutive newborns at two hospitals were examined during the 10-year period between 1964 and 1974 (Robinson & Puck, 1967). Any discrepancy between laboratory findings and the phenotypic sex of the babies was confirmed by chromosome analysis or karyotyping of peripheral blood cells. Of 68 infants with SCA thus identified, 46 became the basic sample for data presented in this chapter (Table 10.1). (Eight infants died in the neonatal period, seven children declined to participate, two initially joined the study but later chose to drop out, and five discontinued after moving out of state).

Regularly scheduled physical and developmental evaluations, psychologic testing, and extensive interviews with parents and children constitute the research protocol. School reports and academic evaluations are integrated into the clinical records. Siblings who share similar genetic and environmental background serve as controls. All family members partici-

TABLE 10.1. Denver study subjects.

Karyotype	Number of subjects	Mean age	Age range
47,XXY	14	12.0	10–19
47,XYY	4	10.5	10–13
47,XXX	11	15.5	12–19
45,X and Partial X Monosomy			
45,X	6	12.9	10–18
46,XXq-	2	15.0	12–17
45,X/46,X,r(X)	1	11.3	—
Female Mosaics			
45,X/46,XX/47,XXX	1	17.0	—
45,X/46,XX	4	17.9	16–22
46,XX/47,XXX	1	11.0	—
45,X/47,XXX	2	15.6	12–19
46,XY Controls	16	12.3	8–19
46,XX Controls	16	12.7	6–18

pated in the research program, thus reducing the special attention that could otherwise be experienced as discomfort by the identified child. In order to prevent overrepresentation of single characteristics shared by large families, only one male and one female sibling were included as controls from any one family.

Early development of these children has been reported elsewhere (Bender et al., 1983a; Bender et al., 1984a, 1984b; Eller et al., 1971; Pennington et al., 1980; Puck et al., 1975; Robinson et al., 1979c; Tennes et al., 1975; Tennes et al., 1977). The results reported here will summarize accumulated data pertaining to cognition and school experiences. The Denver sample of 48 children with SCA is one of the largest in existence. However, the number of subjects in each karyotype group remains small, and generalizations are made cautiously. Discussion includes results from other longitudinal studies of SCA.

Intelligence

Although only two children with SCA are clearly mentally retarded, the Full Scale intelligence score of the group of 38 nonmosaic propositi is significantly lower than that of the controls by an average of 14 points on the Wechsler Intelligence Scale for Children. Variability is considerable. Twenty-two scores are in the statistically defined average range of $\pm$ 1 standard deviation (SD) (86 to 115), 14 are below average, and 2 are above average. Among controls, 25 are average, 1 below, and 4 above.

Intellectual functioning varies among the groups as well as among the individuals. The Full Scale IQ distrubitions seen in Figure 10.3 suggest a normal distribution, with almost equal numbers of scores falling above and

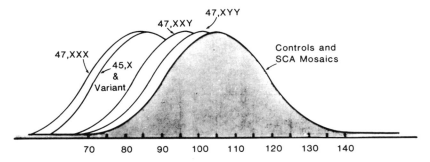

FIGURE 10.3. Estimated full-scale IQ distributions for SCA and control children.

below each mean, although the variability in the SCA groups is greater than that of the control groups. The 47,XXX and X monosomy groups each have mean IQs significantly lower than those of female controls, as determined by the Mann-Whitney U Test. Cognitive development of the 47,XXX girls has been of greatest concern, since six have IQs in the borderline retarded range. The 47,XXY and 47,XYY boys' scores are not significantly lower than those of male controls although they suggest a trend in that direction. The mosaic group and the controls have similar scores, a finding consistent with other developmental parameters (Bender et al., 1983a; Bender et al., 1984a; Robinson et al., 1982; Robinson et al., 1979).

Mean Verbal and Performance IQ scores (Figure 10.4) reflect more specifically the areas of strength and weakness discussed in the next section. Male propositi, both 47,XXY and 47,XYY, have relatively lower

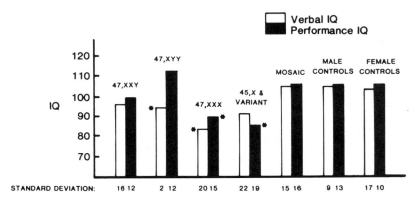

*INDICATES SIGNIFICANT DIFFERENCE FROM CONTROL GROUP OF SAME SEX (P<.05)

FIGURE 10.4. Mean verbal and performance IQs.

Verbal IQ scores, but only for the XYY group are they significantly lower than for controls and therefore indicative of weaker language skills. The 45,X and partial X monosomy group shows just the opposite pattern: a significantly decreased Performance IQ suggesting visual-perceptual problems. For the most generally impaired 47,XXX group, both Verbal and Performance IQs are significantly lower than those of controls.

Neuropsychological Studies

Cognitive and neuromotor measures were selected to evaluate specific functional areas of development not described by IQ tests. The resulting deficit patterns help differentiate the groups of children and clarify the developmental implications of each SCA. Language, auditory memory, spatial ability, and neuromotor skills are each described separately.

Language

Language function was evaluated by a speech-language pathologist blind to the identity of each child. Following the administration of a test battery for measurement of auditory perception, speech production, receptive language, and expressive language, the clinician rated each child's overall language skills on a continuum from unimpaired to severely impaired. Impaired children are those whose moderate to severe language deficits are believed to interfere with daily social and academic success (Figure 10.5).

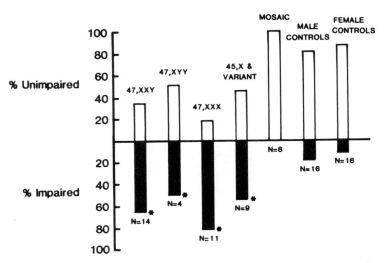

*INDICATES SIGNIFICANT DIFFERENCE FROM CONTROL GROUP OF SAME SEX (P<.05)

FIGURE 10.5. Language.

All SCA groups except the mosaics demonstrated significant language impairment relative to same-sex controls. The mosaic group did not differ from the controls. As reported earlier (Bender et al., 1983a), the 47,XXX group most frequently and most severely demonstrated incapacity to understand and to use language. The 47,XXY group, with better comprehension ability, nonetheless used language inefficiently, often requiring more time to process information even when the correct answer could be produced.

The 45,X and partial X monosomy group showed a marked increase in problems of speech production. Moderate to severe language dysfunction was seen in the three girls with IQ scores in the borderline and moderately retarded range. With one exception, the X monosomy girls have histories of chronic otitis media, a condition suspected of impeding language development (Paradise, 1981; Reichman & Healey, 1983).

Auditory Memory

The Auditory Sequential Memory subtest of the Illinois Test of Psycholinguistic Ability (Kirk et al., 1968), requiring the child to accurately repeat lists of digits of increasing length, was used to measure auditory short-term memory. "Impaired" children were those whose performance was below the tenth percentile for age (Figure 10.6). The boys with SCA were the only group with significantly impaired memory compared to controls. Fifteen of these 19 boys also have poor language and reading skills. The relationship of auditory short-term memory to academic progress is discussed later.

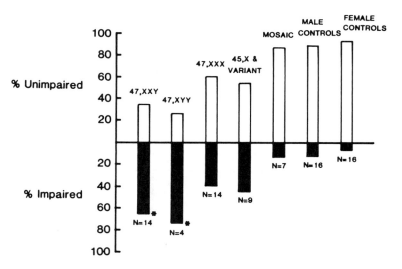

*INDICATES SIGNIFICANT DIFFERENCE FROM CONTROL GROUP OF SAME SEX (P>.05)

FIGURE 10.6. Auditory memory.

Studies of other dimensions of memory, such as short-term and long-term memory for verbal and figural information, are in progress.

Spatial Ability

With the exception of mosaics, girls with SCA scored below the tenth percentile more frequently than controls in this area. Diminished performance on the Spatial Relations Test from the Primary Mental Abilities Battery, identifying difficulty in completing a geometric figure, was consistently present in the X monosomy group. Five 47,XXX girls also had scores in the impaired range. This deficit was relatively rare among the other groups, occurring only in 3 of 18 boys, one mosaic girl, and one female control (Figure 10.7).

Neuromotor Skills

With the exception of mosaics, all groups with SCA had significant neuromotor dysfunction, as shown by scores from the Bruininks-Oseretsky Test of Motor Proficiency (Figure 10.8). Only one of 32 siblings was affected.

An apparent relationship exists between language and motor function and later learning skills (Pennington et al., 1980). Impaired language or motor skills occurred in 30 children with SCA. Half of the 38 nonmosaic subjects experienced both; all 19 have required special education. Six control siblings had delayed language or motor skills, but none had both.

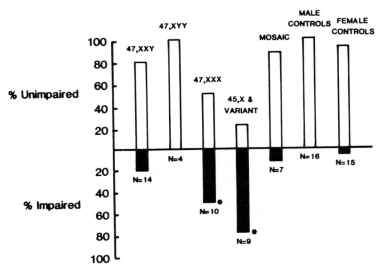

*INDICATES SIGNIFICANT DIFFERENCE FROM CONTROL GROUP OF SAME SEX (P<.05)

FIGURE 10.7. Spatial ability.

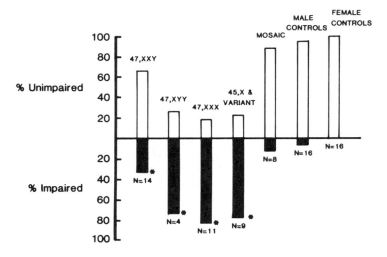

*INDICATES SIGNIFICANT DIFFERENCE FROM CONTROL GROUP OF SAME SEX (P<.05)

FIGURE 10.8. Neuromotor skills.

Special Education

In light of the frequency of language, memory, spatial, and neuromotor difficulties described in the previous section, it is not surprising that 82 percent of nonmosaic children with SCA are identified by school personnel as LD and receive partial or full-time special education, in contrast to 13% of control and mosaic children (Table 10.2). As reported elsewhere (Patten, 1983), diminished self-esteem is characteristic of LD children, anxiety and depression are common, and adaptive skills are minimal. Many are characterized by their awkwardness, labored self-expression, and limited capacity to communicate or compete with peers. The teachers who are responsible for individual academic evaluations and LD classroom placement are usually not informed of the SCA diagnosis. This serves as an unbiased verification of our clinic evaluations.

Twenty-three of the female LD propositi received part-time special education and spent the remaining time in regular classes, whereas eight were in full-time remedial classes (Table 10.2). Two girls, one 45,X and one 47,XXX, each with an IQ of 50, are in classes for mentally retarded children. LD male subjects (13%) required less full-time remediation than LD girls (40%), reflecting the girls' lower IQ scores and generally greater impairment. One mosaic girl required part-time special education in multiple subjects.

Trends of karyotype-specific learning problems for which special education was provided are listed in Table 10.2. The 16 LD boys required

TABLE 10.2. Current incidence of specific learning disorders requiring educational intervention in propositi and controls as reported by schools.

	No problem	Reading	Reading and language	Math	Math and language	All school subjects	Mental retardation	Number in part-time special education classes	Number in full-time special education classes
47,XXY (N = 14)	2	6	3	0	0	3	0	10	2
47,XYY (N = 4)	0	2	2	0	0	0	0	4	0
Male controls (N = 16)	14	1	0	0	0	1	0	2	0
47,XXX (N = 11)	2	1	3	0	3	1	1	6	3
45,X and variants (N = 9)	3	1	0	0	0	4	1	3	3
Female mosiacs (N = 8)	7	0	0	0	0	1	0	1	0
Female controls (N = 16)	14	0	0	0	1	1	0	2	0

assistance in reading and language skills, and three in the 47,XXY group were deficient in all school subjects. The girls' learning problems were more diverse. Four of nine LD girls in the 47,XXX group received help in reading and language only, whereas five were assisted in math, language, and other subjects. Six of the nine 45,X and partial X monosomy girls were weak in all subjects.

Without LD controls whose need for intervention has been neglected, the effectiveness of special education programs for these subjects is inconclusive. However, it is our clinical observation that most intervention has been beneficial to the LD children in the Denver study. At the same time, elimination of the LD has not been possible in most cases. The increased rate of LD placement for children with SCA relative to controls likely reflects the cumulative effects of longstanding cognitive impairment.

Summary and Discussion of Specific Karyotypes

Thirty-one of the 38 nonmosaic children with SCA have LD necessitating intervention. Other studies also report increased frequency of learning problems in children with SCA (Hier et al., 1980; Ratcliffe et al., 1982b; Stewart, 1982). Girls (47,XXX and 45,X and partial X monosomy) appear to be more severely affected than boys, with a mean IQ score more than one standard deviation below population norms and 30% in full-time special education. Eleven percent of the boys and none of the controls require full-time special education. Boys with 47,XXY and 47,XYY karyotypes have similar language and reading difficulties.

The mosaic children, who were followed in the same manner as nonmosaic propositi, are relatively unaffected and cannot be distinguished by this study from controls. This result refutes the agrument that a deleterious self-fulfulling prophecy creates problems for all identified children studied.

Although more than half of the nonmosaic children have LD, their cognitive characteristics are not identical. Karyotype-specific patterns are summarized and discussed in the following sections.

47,XXY

Nine of the 14 boys have impaired language and auditory memory skills, although only three have inadequate spatial ability. These are weak foundation skills that interfere with academic mastery. The great majority—12 of 14—have impeded ability to read. Seven—one half of our sample—shared the characteristics of a specific language-based dyslexia described by Vellutino (1979). Despite Full Scale IQs above 90, they have impaired language and verbal memory, reduced speed of cognitive processing, and a school history of reading failure. Dyslexia in 47,XXY boys is the most frequent LD of any propositi group and comes closest to a "pure" LD subtype associated with a specific SCA.

Impaired language (Bender et al., 1983a; Funderburk & Ferjo, 1978; Graham et al., 1981) and impaired reading skills (Annell et al., 1970; Pennington et al., 1982; Stewart et al., 1979; Theilgaard et al., 1971; Walzer et al., 1978) have been reported frequently among 47,XXY boys, although data regarding the associated skills are incomplete. Verbal IQ was found by some investigators to be lower than Performance IQ (Annell et al., 1970; Pasqualini et al., 1957; Ratcliffe et al., 1982; Theilgaard et al., 1971; Walzer et al., 1978), whereas comparable Verbal and Performance IQs were found by others (Barker & Black, 1976; Funderburk & Ferjo, 1978; Money, 1964; Robinson et al., 1982; Stewart et al., 1979). Significant Performance-Verbal IQ discrepancy was found to be indistinguishable from the tendency for many school-age boys to have slightly higher Performance IQ (Bender et al., 1983b). Problems limited to expressive language have also been reported (Robinson et al., 1979b). In the only blind study of language in SCA and control children, Bender, Fry, and co-workers (1983) observed reduced receptive and expressive language and slow language processing in 13 of 14 boys. Netley and Rovet (1982) also reported disorders of receptive language. The relationship of reduced auditory short-term memory (Graham et al., 1981; Stewart et al., 1979) to other deficits is not yet clearly defined. Future studies will examine speed of processing and other facets of memory such as immediate versus delayed and verbal versus visual.

Five of 14 47,XXY boys have neuromotor impairment including hypotonia, primitive reflexes, and decreased sensory-motor integration, coordination, speed, and strength at the gross and fine motor levels. The effect of these is compromised performance in writing skills, speed on timed activities, and ability to compete with peers in athletic games. The relationship between delayed motor development and learning difficulties is not totally clear, yet an association seems to exist; four of the five motor-impaired propositi have a language disability and are receiving special education. Decreased motor skills and signs of neurologic dysfunction occur with increased frequency in the general LD population (Pyfer & Carlson, 1972) and suggest variation in the development of the brain and central nervous system.

Inferences have been made about anomalous brain development in 47,XXY boys. Specifically, poor language skills may indicate relative impairment of left hemisphere functions (Graham et al., 1981; Neilsen, 1969; Neilsen et al., 1979). Utilizing a large sample of 33 unselected 47,XXY boys, Netley and Rovet (1982) reported an increase in nonright-handedness (i.e., left or mixed handedness) in propositi. In combination with previously published but sketchily described data showing visual field asymmetries (Stewart et al., 1982), the authors suggested that males with a supernumerary X may fail to establish left hemisphere dominance for language. Ratcliffe and Tierney (1982), in response, reported no increased nonright-handedness in a group of 32 47,XXY boys also identified through neonatal screening and noted that familial left-handedness may have contributed to the disparity in results. Bender, Puck, and co-workers (1983) similarly found predominant right-handedness in their 47,XXY sample and presented results of dichotic listening studies showing a right ear advantage in propsiti similar to that of controls and presumably indicating left hemisphere language dominance in both groups. Although studies of handedness and visual field or dichotic listening asymmetry are inconclusive, the hypothesis of atypical organication of hemispheric function in 47,XXY subjects remains an intriguing one awaiting further investigation using neuropsychologic and electrophysiologic techniques.

47,XYY

In this small sample of four boys, some interesting findings emerge. All four 47,XYY boys have received educational remediation in reading. Intelligence is uniformly within the average range as seen in their WISC-R IQs, although two had low average IQs on the WPPSI (Robinson et al., 1979c). Three of four have at least mildly decreased language and auditory memory skills. Neuromotor skills are impaired in three of four, as seen in standardized testing and confirmed on physical examination with reports of hypotonia and sensory motor dysfunction. Spatial skills are unimpaired and in two cases are superior.

The combined reading, language, memory, and neuromotor dysfunction in this group appears to be quite similar to that of the 47,XXY group and may indicate that the two groups have the same type of dyslexia. However, some differences between the groups seem to exist. The 47,XYY boys have a slightly higher Full Scale IQ and stronger spatial-perceptual skills. Only two have reduced speed of linguistic processing, in contrast to the almost unanimous finding of this deficit in the 47,XXY group.

Published results of cognitive studies of 47,XYY subjects have appeared less frequently than for 47,XXY subjects, and, consequently, less is understood about impairment in selected cognitive systems. The small sample size of this study precludes adequate comparison of the 47,XYY to other groups. However, results from this study, when combined with those of seven other studies of unselected 47,XYY boys worldwide (a total of 42 propositi), yielded some common findings (Bender et al., 1984b). Approximately half of the total sample had language, motor, and reading impairment and received special education. If ability patterns in 47,XYY boys prove to match those of 47,XXY boys, it may also follow that they have similar anomalies in the organization of cerebral hemispheres.

47,XXX

Mean ability levels of the 47,XXX girls are more impaired than those of any other group of propositi, with Full Scale IQs more than one standard deviation below test norms. One girl with a recent IQ score of 50 has demonstrated variable test performance over time, including a previous IQ score of 75, a reflection of her serious emotional dysfunction. While this group demonstrated increased frequency of deficits in all areas studied—language, memory, spatial, and neuromotor—memory and spatial skills appear to be the least affected. In contrast, 45,X and partial X monosomy girls show marked increase in spatial deficits, and 47,XXY boys frequently have low scores on auditory memory.

Eight of 11 girls are deficient in language and neuromotor skills and were placed in classes for multiple educational problems. Language disability is well recognized in 47,XXX girls. However, results in various studies show some inconsistencies. Significantly lower Verbal than Performance IQs were reported in other studies (Neilsen et al., 1982, Ratcliffe et al., 1982b; Stewart et al., 1982) but not found here. Rovet and Netley (1983) presented evidence that 47,XXX girls are more deficient in verbal than spatial skills. The authors also found evidence of deficient short-term memory for verbal and spatial information. Although this is consistent with our finding of increased memory deficits, it places primary emphasis upon memory impairment in the cognition of 47,XXX girls and indicates that verbal skills alone are affected. Data from Stewart and co-workers (1982) and Ratcliffe, Tierney, and associates (1982b) that the majority of their combined 21 47,XXX subjects received educational remediation in a variety of subjects agree with our findings but disagree with the conclusion of Nielsen and co-workers (1982) that compared with other propositi "they did relatively well at school" (p. 74). When viewed together, these studies leave unclear the extent to which language impairment is an isolated deficit or a specific deficit in addition to generally reduced intellectual ability and whether affected school subjects reflect primarily language problems (as in speech or reading) or below-average progress across all subjects.

Neuromotor deficits, although not described by other investigators, are frequent in our 47,XXX group. Delays in the early motor development of 47,XXX girls have been documented (Pennington et al., 1980; Tennes et al., 1977). Stanine scores from the gross and fine motor sections of the Bruininks-Oseretsky Test of Motor Proficiency indicate that eight are functioning in the lower 4% of children in their age groups. Dysfunction involving balance, equilibrium, visual-motor skills, and sensory-perceptual integration have all been observed, although the hypotonia seen in male subjects was not observed in 47,XXX girls. These deficits have been relatively ignored, perhaps because competitive athletic skills in girls are not emphasized by clinicians and parents or because other behavioral and educational problems demanded more immediate attention. The severity of their impairment, however, has had a strong developmental impact. As noted by Ayres (1982), the child with poor sensory integration may feel awkward and clumsy, play less skillfully than other children, and integrate information from eyes and ears poorly. This often results in inadequate academic skills, social isolation, and poor self-esteem. Clearly, the dysfunction must be traced to central nervous system development. Available information is inadequate to localize neuropsychologic dysfunction in 47,XXX girls. Generalized impairment and absence of asymmetric perceptual, motor, or neurologic findings suggest bilateral cerebral involvement. In contrast, Rovet and Netley (1982) speculate that impaired language suggests minimal right hemisphere involvement in language processing and immature sites in the left hemisphere.

45,X and Partial X Monosomy

While this group includes girls with more than one karyotype, each is missing part or all of an X chromosome in every cell and has physical stigmata and cognitive impairment consistent with Turner syndrome (Bender et al., 1984a). A specific deficit in spatial functioning has been identified in patients with Turner syndrome (Money & Alexander, 1966) and associated with reduced performance in orienting to left-right directions (Alexander et al., 1964), copying shapes (Silbert et al., 1977), handwriting (Pennington et al., 1982), and solving math problems (Garron, 1977). The present study has found impaired spatial skills to occur more frequently in this group than any other skill deficiency. Rovet and Netley (1980; 1982) found decreased success on a test of spatial rotation of three-dimensional figures in a group of 31 patients with Turner syndrome (11 to 18 years old) from an endocrine-gynecology clinic. Detailed analysis of test performance led the authors to conclude that the subjects, while using a rotation strategy similar to that of the chromosomally normal female control groups, did so more slowly, and that response speed may be the primary deficit underlying their reduced spatial skill. Response speed was not measured in the Denver study. However, the poor skill level measured on the untimed Spatial Relations Test indicates that, even with ample processing time, the 45,X and partial X monosomy girls experienced great difficulty on mental rotation tasks.

Unresolved is the question of whether spatial skills are impeded in X monosomy subjects or whether this specific and striking disorder is imposed upon a picture of generally reduced intelligence. Evidence of an association between mental retardation and Turner syndrome has been cited (Polani, 1960; Shaffer, 1962). Others argued that a specific reduction in Performance IQ on the Wechsler scales in the presence of normal Verbal IQ accounts for the misunderstanding that women with Turner syndrome have low general intelligence (Buckley, 1971; Garron, 1977), and at least two investigators have reported above-average language ability (Alexander & Money, 1965; Nielsen et al., 1977). We found a significantly reduced Performance IQ on the Wechsler Preschool and Primary Scale of Intelligence at 4 to 5 years of age but noted a trend toward a reduced Verbal IQ (Pennington et al., 1982). Subsequent administration of the WISC-R found both Verbal and Performance IQ to be lowered in the X monosomy group (Bender et al., 1984a). Others have confirmed the finding of impaired skills which lie outside the classification of "spatial" (Pennington et al., 1984). Rovet and Netley (1983) reported average scores but slow responses on a test of language comprehension. Waber (1979), in her neuropsychologic study of 11 subjects with Turner syndrome (ages 13 to 21 years), found lower Full Scale and Verbal but *not* Performance IQs. She also presented evidence of deficient visual memory, verbal fluency, and motor skills in these subjects. The results of her study are not definitive in light of the fact that (a) more than half of the subjects had 45,X mosaicism; (b) the karyotypes of the subjects were not determined but were implied by Barr body analysis; and (c) controls were matched with propositae for age and scores on four subtests of the Wechsler scales, the rationale for the latter being questionably based on previous finding of adequate verbal skills in Turner syndrome subjects (Garron, 1978).

Two factors may mediate the relationship between karyotype and cognitive patterns. The reduced speech and language ability seen in the dichotomous ratings reported here and previously (Bender et al., 1983a) is associated with increased chronic ear infection in the X monosomy group. Attention disorders and hyperactivity have also been observed with increased frequency (Bender et al., 1984a; Hier et al., 1980). Hyperactivity and distractibility may generally decrease test performance or exacerbate existing LD and have been associated with increased clumsiness (Taylor, 1980) and neurodevelopmental abnormality (Sandberg et al., 1978). All five attention-disordered propositae have neuromotor deficits characterized by poor sensory-motor function and decreased perceptual awareness of their bodies in space as well as language impairment. These combined results point toward neurologic dysfunction. However, definitive evidence of brain damage or disease has not been found. Silbert and co-workers (1977) suggested that spatially impaired patients with Turner syndrome "have a selective deficit in cortical functions that are lateralized to the right cerebral hemisphere." Money (1973), also drawing from the neuropsychologic literature on patients with brain injuries, localized the deficit in Turner syndrome to the right parietal lobe and indicated some similarity to patients with Gerstmann syndrome. Pennington, Heaton, Karzmark, and Pendleton (1984) conducted a neuropsychologic study of women with Turner syndrome that employed an algorithmic system to match deficits with four control groups, including normal women and women with left hemisphere, right hemisphere, or diffuse brain damage. They found that the overall performance ratings of the women with Turner syndrome were depressed to the same level as that of patients with known brain damage that could not be localized. Reduced scores on tests of spatial ability cannot singularly prove right hemisphere impairment because impaired spatial ability occurred in all of the control groups with brain damage in the Pennington study. Waber (1979), in like fashion, concluded that her test results reflected involvement of both cerebral hemispheres and suggested that a more productive approach of cognitive deficits in patients with Turner syndrome would attempt to understand potential alternations in brain development rather than to localize specific areas of damage.

Rovet and Netley (1982) also prefer a neurodevelopmental explanation rather than relating the cognitive limitations of individuals with SCA to those of patients with disease or injury to the brain. They argue that verbal functioning, primarily located in the left hemisphere of normal women, is more diffusely distributed between the right and left hemispheres of women with Turner syndrome. As a result, the right hemispheres do not develop a specialized capacity to process nonverbal information.

Phenotypic Variability

Variability characterizes children with SCA as it does children with normal chromosome constitution and children with other genetic disorders. The IQ range in the nonmosaic group is wide (50 to 122). Each karyotype group includes at least one person 18 years of age or older who is making plans, realistically, to enter college. While mean test scores are frequently discussed in studies of children with SCA, there is no single cognitive or behavioral profile that describes all children with the same SCA. Recognition of this variability and of the concept that "genetic" abnormality does not produce unalterable developmental disabilities is important for two reasons. First, school personnel who occasionally learn of a child's SCA and who read accessible literature on SCA may conclude that the child will have predictable and specific disabilities. Although the diagnosis of SCA indicates a potential limitation on the child's development, it is important to examine each child's abilities, disabilities, and specific life circumstances. Second, this variability makes more difficult the prediction of the future development of fetuses with SCA identified by intrauterine diagnosis. The informed genetic counselor will present to such expectant parents an appreciation for the many genetic and enviromental factors that interact with the SCA and does not offer a simple picture of learning disorders and related developmental problems.

Learning Disorders and Environment

The chromosomal abnormality is but one of many factors, both genetic and environmental, that together contribute to the development of children with SCA. The SCA increases risk of but does not invariably result in developmental disability. It is generally agreed that the environment and genotype interact to determine development of ability and behavior (Vale, 1980).

A "supportive" environment contributes to the amelioration of developmental problems associated with SCA (Robinson et al., 1982). Although environmental effects are difficult to control and measure in a developmental study involving a small number of children with SCA, we have attempted to do this by placing families of SCA and control subjects into a dichotomous category reflecting stability and instability and observing the relationship between these categorizations and developmental impairment. This is done on the basis of a composite assessment of socioeconomic status, parenting skills, and adverse stress conditions. "Dysfunctional" families, for example, might include those with punitive and erratic parents, those in which stressful events such as death of a family member have occurred, and those in which poverty has further impeded the success of its members. (The presence of any one of these factors does not necessarily result in a "dysfunctional" categorization.) Results indicate that children with SCA are significantly more likely to require full-time LD placement if they come from dysfunctional families, whereas control children are not equally affected by family dysfunction (Figure 10.9). Robinson, Bender, Puck, and Salbenblatt (1983) reported similar findings when examining emotional development. While the emotional development of all children worsened, that of children with SCA showed a greater decline than that of controls as family dysfunction increased. In short, emotional development and learning capability in children with SCA are closely associated with family stability, and these deteriorate more rapidly when exposed to an unsupportive environment than occurs with euploid children.

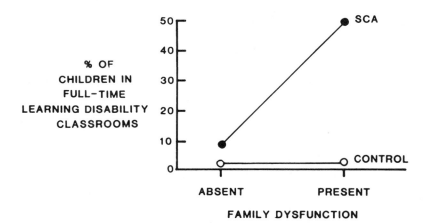

FIGURE 10.9. Learning disabilities.

Mechanisms for the Cognitive Expression of SCA

The question remains about what biologic mechanism in individuals with SCA affects the central nervous system and the development of higher cortical functions. Little evidence is available, and theories about the cause or causes remain speculative. Nonetheless, several possible explanations are available and can be examined.

An abnormality in the number of chromosomes constitutes a different problem from an abnormal gene or genes. A gene may have a very specific effect upon particular biochemical or neurodevelopmental system and may result in a specific disease such as phenylketonuria or, as more recently recognized, a learning disorder subtype such as dyslexia (Smith et al., 1983). A chromosome abnormality, on the other hand, is likely to have a more global and disruptive developmental effect. The fact that SCA results in less severe disability than autosomal abnormalities may be a result of the genetic inactivation of all but one X chromosome early in development (Lyon, 1962). However, the possibility remains that the early-developing nervous system may be affected before inactivation occurs or that part of the genetic material may be inactivated before other parts. Attempts have been made to describe spatial ability as a form of X-linked inheritance because it is favored in boys and men. This model cannot account for the spatial impairment in X monosomy girls, who like boys have only one X chromosome and might be expected to share their advantage on spatial tasks. It has also been discredited on other methodologic and theoretical grounds (Boles, 1980).

The quantitative effect of SCA may be understood at the cellular level. Polani (1977) cited evidence that additional chromosomes increase the length of the mitotic cycle, whereas the absence of one X chromosome shortens the cycle. He asserts that these changes in the cell cycle cause increases and decreases in the number of cells at the time of differentiation, resulting in a biochemical effect that may interfere with various aspects of physiologic growth. Barlow (1973) hypothesized that changes in the rate of cell division that occur in the presence of an extra sex chromosome may have a specific effect on the development of intellectual skills, noting that "brain development depends on cell division, cell growth, and cell migration, and disturbances in the interrelations of these factors would undoubtedly lead to disturbed brain structure and function" (p.121).

The effect of SCA on brain development may be associated with normal sex-related differences. There is evidence to indicate that in the course of development, the female brain and the male brain are organized somewhat differently (Lake & Bryden, 1976; Lansdell, 1962; McGlone & Kertesz, 1973; Tucker, 1976), and it is believed that these observed differences in cognitive abilities may reflect differences in the rate of maturation of cerebral functions between the sexes (Taylor, 1969; 1971). Certainly, neuropsychologic maturation is a complex process that most probably involves repeated reorganizations of the brain's functions rather than a simple linear development (Waber, 1979), and the rate at which these processes occur may be an important factor. If the sex chromosomes have a role in establishing the rate of brain growth, it follows that the loss or addition of sex chromosomal material could further change the rates of lateralization and hemisphere specialization.

Two researchers (Netley, 1977; Netley & Rovet, 1982; Rovet & Netley, 1983; Netley, 1983) have carefully described the hypothetical process by which brain development and cerebral specialization may be related to SCA. They postulated that the increased cell cycle in individuals with an extra chromosome (47,XXY and 47,XXX) results in slowed brain growth and delay in maturation of left hemisphere sites important to language. In 45,X individuals the shortened cell cycle would increase brain growth rate, allowing inadequate time for right hemisphere development and specialization for nonverbal tasks. Some indirect evidence supporting a relationship between growth and cognitive deficits was also offered. Delayed bone age correlated significantly with verbal impairment in 47,XXY and 47,XXX subjects, which is consistent with the idea that slower brain growth results in poorer verbal ability. In addition, dermal ridge counts, believed to reflect fetal growth rate, were increased in 45,X females and decreased in language-impaired 47,XXY and 47,XXX subjects, again supporting a correlation between growth and cognitive parameters in the predicted directions (Netley & Rovet, 1982). This hypothesis is subject to theoretical and methodologic criticism. First, the statement that "phenotypic females with a 47,XXX complement have essentially identical deficits in verbal ability as phenotypic males with a 47,XXY constitution" (Netley, 1983, p. 184), reflecting the same impeded brain growth, conflicts with other evidence that the two groups have distinctly different profiles of language impairment (Bender et al., 1983a). Data from the Denver study indicate that, aside from spatial ability, the 47,XXX girls are more similar to the 45,X and partial X monosomy girls than to 47,XXY boys in general intellectual and neuromotor impairment, despite the fact that the two groups of girls have opposite chromosome anomalies and supposedly opposite brain growth patterns. Second, the correlational bone age data used by Netley and Rovet (1982) was obtained at 3 to 6 years of age and has no proven relationship to prenatal or neonatal brain growth. In addition, one study utilizing local norms did not confirm delayed bone age in children with SCA except X monosomy girls (Webber et al., 1982). Nevertheless, the brain growth theory of Netley and Rovet (1982) accounts for much of what is known about patterns of cognitive development in specific SCA karyotype groups and has some correlating physiologic evidence.

The presence of abnormal levels of sex hormones in some individuals with SCA has led to the hypothesis that their altered cognitive patterns may have an origin in disturbed endocrine functioning (Money & Erhardt, 1972; Reinisch et al., 1979), most probably occurring during embryonic differentiation (Money, 1973). When the theory is applied to available data, however, some inconsistencies arise. First, 47,XXX girls are the most cognitively impaired of all individuals with SCA yet are usually fertile and have no consistent endocrine dysfunction. Second, 47,XYY boys, who appear to be cognitively similar to 47,XXY boys, do not share their characteristic hypogonadism. Third, 47,XXY boys may enter puberty normally but develop hypergonadotropic hypogonadism by midpuberty, resulting in a leveling off of testosterone levels while those of age peers continue to rise (Salbenblatt et al., 1985). Thus, while their impaired language skills have been identified at preschool age, atypical endocrine functions are not apparent before puberty (Stewart, 1982).

The possiblity of an endocrine-cognition relationship cannot be completely dismissed. Prenatal variations in hormone levels may occur in the presence of fertility and normal hormone levels at puberty. After the third month of embryonic life, the gonial cells of 45,X girls undergo a very high rate of atresia (Polani, 1981), with full gonadal dysgenesis usually developing after early infancy (Illig et al., 1975). Deficient germinal epithelium and hyperplastic Leydig cells have been reported in one 47,XXY fetus (Murken et al., 1974). In addition, abnormally low levels of testosterone have been found in the amniotic fluid (Citoler & Aechter, 1978). Reinisch (1976) presented evidence that the growing brain is affected by hormones. Hence, the presence of atypical hormone levels in prenatal or neonatal individuals with SCA might be of significance. As with the hypothesis of Netley and Rovet (1982), discussed previously, the hormone hypothesis postulates that changes in early neurodevelopment in children with SCA result in alteration of cerebral organization and the lateralization responsible for normal male-female differences. Inadequate data are available to endorse a theory of disturbed brain development based upon abnormal growth or hormone levels in infancy. Both theories stress the importance of investigating disturbances in normal psychophysiologic development for understanding why children with SCA have difficulty processing information and learning in school.

Summary

Most children with sex chromosome abnormalities have learning disorders and, rarely, mental retardation. Mild depression of intellectual, language, and neuromotor skills is characteristic of all nonmosaic groups. Some specific LD patterns are present in the karyotype groups: 47,XXY boys often develop dyslexia associated with their language disorders, and 45,X and partial X monosomy girls frequently have impaired spatial ability. Girls with SCA are more generally impaired than boys. All groups include both severely learning disabled and college-bound children. Two theories involving the effects of changes in hormones and growth rate have been discussed but thus far have not adequately explained the biologic mechanism responsible for cognitive deficits associated with SCA.

References

Alexander, D., & Money, J. (1965). Reading ability, object constancy and Turner's syndrome. *Perceptual and Motor Skills, 20,* 981–984.

Alexander, D., Walker, H.T., & Money, J. (1964). Studies in direction sense: I. Turner's syndrome. *Archives of General Psychiatry, 10,* 337–339.

Annell, A.L., Gustavson, K.H., & Tenstam, J. (1970). Symptomatology in school-boys with positive sex chromatin (the Klinefelter syndrome). *Acta Psychiatrica Scandinavica, 46,* 71–80.

Ayres, J. (1982). *Sensory integration and the child.* Los Angeles: Western Psychological Services.

Barker, T.E., & Black, F.W. (1976). Klinefelter syndrome in a military population: Electroencephalographic, endocrine, and psychiatric status. *Archives of General Psychiatry, 33,* 607–610.

Barlow, P. (1973). The influence of inactive chromosomes on human development. *Humangenetik, 17,* 105–136.

Bender, B., Fry, E., Pennington, B., Puck, M., Salbenblatt, J., & Robinson, A. (1983a). Speech and language development in 41 children with sex chromosome anomalies. *Pediatrics, 71,* 262–267.

Bender, B., Puck, M., Salbenblatt, J., & Robinson, A. (1983b). Hemispheric organization in 47,XXY boys. *Lancet, 1,* 132.

Bender, B., Puck, M., Salbenblatt, J., & Robinson, A. (1984a). Congitive development of unselected girls with complete and partial X monosomy. *Pediatrics, 73,* 175–182.

Bender, B., Puck, M., Salbenblatt, J., & Robinson, A. (1984b). The development of four unselected 47,XYY boys. *Clinical Genetics, 25,* 435–445.

Boles, D.B. (1980). X-linkage of spatial ability: A critical review. *Child Development, 51,* 625–635.

Buckely, F. (1971). Preliminary report on intelligence quotient scores of patients with Turner's syndrome: A replication study. *British Journal of Psychiatry, 119,* 513–514.

Citoler, P., & Aechter, J. (1978). Histology of the testis in XXY fetuses. In J. Murken, S. Stengel-Rulkowski, & E. Schwinger (Eds.), *Prenatal diagnosis: Proceedings of the Third European Conference on Prenatal Diagnosis of Genetic Disorders* (pp. 336–337). Stuttgart: Emke Verlag.

Eller, E., Frankenburg, W., Puck, M., & Robinson, A. (1971). Prognosis in newborn infants with X-chromosomal abnormalities. *Pediatrics, 47,* 681–688.

Funderburk, S.J., & Ferjo, N. (1978). Clinical observations in Klinefelter (47,XXY) syndrome. *Journal of Mental Deficiency Research, 22,* 207–212.

Garron, D.C. (1977). Intelligence among persons with Turner's syndrome. *Behavior Genetics, 7,* 105–127.

Garron, D.C. (1978). Comment on "Spatial and temporal processing in patients with Turner's syndrome" letter. *Behavior Genetics, 8,* 289–295.

Graham, J.M., Bashir, A.S., Walzer, S., Stark, R.E., & Gerald, P.S. (1981). Communication skills among unselected XXY boys. *Pediatric Research, 15,* 562.

Hier, D.B., Atkins, L., & Perlo, V.P. (1980). Learning disorders and sex chromosome aberrations. *Journal of Mental Deficiency Research, 24,* 17–26.

Illig, R., Tolksdorf, M., Murset, G., & Prader, A. (1975). LF and FSH response to synthetic LH-RH in children and adolescents with Turner's and Klinefelter's syndrome. *Helvetica Paediatrica Acta, 30,* 221–231.

Kirk, S.A., McCarthy, J.J., & Kirk, W. (1968). *Illinois Test of Psycholinguistic Abilities.* Urbana, IL: University of Illinois Press.

Lake, D., & Bryden, M.P. (1976). Handedness and sex differences in hemisphere asymmetry. *Brain and Language, 3,* 266–282.

Lansdell, H. (1962). A sex difference in the effect of temporal lobe neurosurgery on design preference. *Nature, 194,* 852–854.

Lyon, M.F. (1962). Sex chromatin and gene action in the mammalian X-chromosome. *American Journal of Human Genetics, 14,* 135.

McGlone, J., & Kertesz, A. (1973). Sex differences in cerebral processing of visuospatial tasks. *Cortex, 9,* 313–320.

Money, J. (1964). Two cytogenetic syndromes: Psychologic comparisons. I. Intelligence and specific-factor quotients. *Journal of Psychiatric Research, 2,* 223–231.

Money, J. (1973). Turner's syndrome and parietal lobe functions. *Cortex, 9,* 385–393.

Money, J., & Alexander, D. (1966). Turner's syndrome: Further demonstration of the presence of specific cognitional deficiencies. *Journal of Medical Genetics, 3,* 47–48.

Money, J., & Ehrhardt, A.A. (1972). *Man and woman, boy and girl.* Baltimore: Johns Hopkins University Press.

Murken, J., Stengel-Rutkowski, S., Walther, J., Westenfelder, S., Remberger, K., & Zimmer, F. (1974). Klinefelter's syndrome in a fetus. *Lancet, 2,* 171.

Netley, C. (1977). Dichotic listening of callosal agenesis and Turner's syndrome patients. In S.J. Segalowitz & F.A. Gruber (Eds.), *Language development and neurological theory* (pp. 133–143). New York: Academic Press.

Netley, C. (1983). Sex chromosome abnormalities and the development of verbal and nonverbal abilities. In C. Ludlow & J. Cooper (Eds.), *Genetic aspects of speech and language disorders* (pp. 179–195). New York: Academic Press.

Netley, C., & Rovet, J. (1982). Handedness in 47,XXY males. *Lancet, 2,* 267.

Nielsen, J. (1969). Klinefelter's syndrome and the 47,XYY syndrome. A genetical, endocrinological and psychiatric-psychological study of thirty-three severely hypogonadal male patients and two patients with karyotype 47,XYY. *Acta Psychiatrica Scandinavica, 45* (Suppl. 209), 353.

Nielsen, J., Nyborg, M., & Dahl, G. (1977). Turner's syndrome. *Acta Jutlandica, 45,* Medicine Series 21.

Nielsen, J., Sillesen, I., Sørensen, A.M., & Sørensen, K. (1979). Follow-up until age 4 to 8 of 25 unselected children with sex chromosome abnormalities compared with sibs and controls. *Birth Defects Original Article Series, 15,* 15–73.

Nielsen, J., Sørensen, A.M., & Sørensen, K. (1982). Follow-up until age 7 to 11 of 25 unselected children with sex chromosome abnormalities. *Birth Defects Original Article Series, 18,* 61–97.

Paradise, J.L. (1981). Otitis media during early life: How hazardous to development? A critical review of the evidence. *Pediatrics, 68,* 869.

Pasqualini, R.Q., Vidal, G., & Bur, G.E. (1957). Psychopathology of Klinefelter's syndrome: Review of 31 cases. *Lancet, 2,* 164–167.

Patten, M.D. (1983). Relationships between self-esteem, anxiety, and achievement in young learning disabled students. *Journal of Learning Disabilities, 16,* 43–45.

Pennington, B., Bender, B., Puck, M., Salbenblatt, J., & Robinson, A. (1982). Learning disabilities in children with sex chromosome anomalies. *Child Development, 53,* 1182–1192.

Pennington, B., Heaton, R., Karzmark, P., & Pendleton, M. (1985). The neuropsychological phenotype in Turner syndrome. *Cortex, 21,* 391–404.

Pennington, B., Puck, M., & Robinson, A. (1980). Language and cognitive development in 47,XXX females followed since birth. *Behavior Genetics, 10,* 31–41.

Polani, P.E. (1960). Chromosomal factors in certain types of educational subnormality. In P.W. Bowman & H.B. Mautner (Eds.), *Mental retardation: Proceedings of the First International Congress* (pp. 421–438). New York: Grune & Stratton.

Polani, P.E. (1977). Abnormal sex chromosomes, behavior and mental disorder. In J.M. Tanner (Ed.), *Developments in psychiatric research* (pp. 89–128). London: Hoddler & Staughton.

Polani, P.E. (1981). Abnormal sex development in man. I. Anomalies of sexdetermining mechanisms. In C.R. Austin & R.G. Edwards (Eds.), *Mechanisms of sex differentiation in animals and man* (pp. 465–547). New York: Academic Press.

Puck, M., Tennes, K., Frankenburg, W., Bryant, K., & Robinson, A. (1975). Early childhood development of four boys with 47,XXY karyotype. *Clinical Genetics, 7,* 8–20.

Pyfer, J., & Carlson, B. (1972). Characteristic motor development of children with learning disabilities. *Perceptual and Motor Skills, 35,* 291–296.

Ratcliffe, S.G., Axworthy, D., & Ginsborg, A. (1979). The Edinburgh study of growth and development in children with sex chromosome abnormalities. *Birth Defects Original Article Series, 15,* 243–260.

Ratcliffe, S.G., Bancroft, J., Axworthy, D., & McLaren, W. (1982a). Klinefelter's syndrome in adolescence. *Archives of Diseases in Childhood, 57,* 6–12.

Ratcliffe, S.G., & Tierney, I. (1982). 47,XXY males and handedness. *Lancet, 2,* 716.

Ratcliffe, S.G., Tierney, I., Nshaho, J., Smith, L., Springbett, A., & Callan, S. (1982b). The Edinburgh study; of growth and development of children with sex chromosomal abnormalities. *Birth Defects Original Article Series, 18,* 41–60.

Reichman, J., & Healey, W. (1983). Learning disabilities and conductive hearing loss involving otitis media. *Journal of Learning Disabilities, 16,* 272–278.

Reinisch, J., Gandelman, R., & Spiegel, F. (1979). Prenatal influences on cognitive abilities. In M. Wittig & A. Petersen (Eds.), *Sex-related differences in cognitive functioning* (pp. 215–239). New York: Academic Press.

Reinisch, J.M. (1976). Effects of prenatal hormone exposure on physical and psychological development in humans and animals: With a note on the state of the field. In E.J. Sachar (Ed.), *Hormones, behavior, and psychopathology* (pp. 69–94). New York: Raven Press.

Robinson, A., Bender, B., Borelli, J., Puck, M., Salbenblatt, J., & Webber, M.L. (1982). Sex chromosomal abnormalities (SCA): A prospective and longitudinal study of newborns identified in an unbiased manner. *Birth Defects Original Article Series, 18,* 7–39.

Robinson, A., Bender, B., Puck, M., & Salbenblatt, J. (1983). Sex chromosomal anomalies: Prospective studies in children. *Behavior Genetics, 13,* 321–329.

Robinson, A., Lubs, H., and Bergsma, D. (Eds.). (1979a). Sex chromosome aneuploidy: Prospective studies on children. *Birth Defects Original Article Series, 15.*

Robinson, A., Lubs, H., Nielsen, J., & Sørensen, K. (1979b). Summary of clinical findings: Profiles of children with 47,XXY, 47,XXX, and 47,XYY karyotypes. *Birth Defects Original Article Series, 15,* 261–266.

Robinson, A., Puck, M., Pennington, B., Borelli, J., & Hudson, M. (1979c). Abnormalities of the sex chromosomes: A prospective study on randomly identified newborns. *Birth Defects Original Article Series, 15,* 203–241.

Robinson, A., & Puck, T. (1967). Studies on chromosomal and nondisjunction in man. II. *American Journal of Human Genetics, 19,* 112–129.

Rovet, J., & Netley, C. (1980). The mental rotation task performance of Turner syndrome subjects. *Behavior Genetics, 10,* 437–443.

Rovet, J., & Netley, C. (1982). Processing deficts in Turner's syndrome. *Developmental Psychology, 18,* 77–94.

Rovet, J., & Netley, C. (1983). The triple X syndrome in childhood: Recent empirical findings. *Child Development, 54,* 831–845.

Salbenblatt, J., Bender, B., Puck, M. Robinson, A., Faiman, C., & Winter, J. Pituitary-gonadal function in Klinefelter syndrome before and during puberty. *Pediatric Research, 19,* 82–86.

Sandberg, S., Rutter, M., & Taylor, E. (1978). Hyperkinetic disorder in psychiatric clinic attenders. *Developmental Medicine and Child Neurology, 20,* 279–299.

Shaffer, J. (1962). A specific cognitive deficit observed in gonadal aplasia (Turner's syndrome). *Journal of Clinical Psychology, 18,* 403–406.

Silbert, A., Wolff, P.H., & Lilienthal, J. (1977). Spatial and temporal processing in patients with Turner's syndrome. *Behavior Genetics, 7,* 11–21.

Smith, S.D., Kimberling, W.J., Pennington, B.F., & Lubs, H.A. (1983). Specific reading disability: Identification of an inherited form through linkage analysis. *Science, 219,* 1345–1347.

Stewart, D.A. (Ed.). (1982). Children with sex chromosome aneuploidy: Follow-up studies. *Birth Defects Original Article Series, 18.*

Stewart, D.A., Bailey, J.D., Netley, C.T., Rovet, J., Park, E., Cripps, M., & Curtis, J.A. (1982). Growth and development of children with X and Y chromosome aneuploidy from infancy to pubertal age: The Toronto study. *Birth Defects Original Article Series, 18,* 99–154.

Stewart, D.A., Netley, C.T., Bailey, J.D., Haka-Ikse, K., Platt, J., Holland, W., & Cripps, M. (1979). Growth and development of children with X and Y chromosome aneuploidy: A prospective study. *Birth Defects Original Article Series, 15,* 75–114.

Taylor, D.C. (1969). Differential rates of cerebral maturation between sexes and between hemispheres. *Lancet, 2,* 140–142.

Taylor, D.C. (1971). Ontogenesis of chronic epileptic psychoses: A reanalysis. *Psychological Medicine, 1,* 247–253.

Taylor, E. (1980). Development of attention. In M. Rutter (Ed.), *Scientific founda-*

tions of developmental psychiatry (pp. 185–197). London: Heinemann Medical Books.

Tennes, K., Puck, M., Bryant, K., Frankenburg, W., & Robinson, A. (1975). A developmental study of girls with trisomy X. *American Journal of Human Genetics, 27,* 71–80.

Tennes, K., Puck, M., Orfanakis, D., & Robinson, A. (1977). The early childhood development of 17 boys with sex chromosome anomalies: A prospective study. *Pediatrics, 59,* 574–583.

Theilgaard, A., Nielsen, J., Sørensen, A., Frøland, A., & Johnsen, S.G. (1971). A psychological-psychiatric study of patients with Klinefelter's syndrome, 47,XXY. *Acta Jutlandica, 43,* 1–148.

Tjio, J.H., & Levan, A. (1956). The chromosome number in man. *Hereditas, 42,* 1–6.

Tucker, D.M. (1976). Sex differences in hemispheric specialization for synthetic visuospatial functions. *Neuropsychologia, 14,* 447–454.

Vale, J. (1980). *Genes, environment, and behavior.* New York: Harper & Row.

Vellutino, F.R. (1979). *Dyslexia: Theory and research.* Cambridge, MA: MIT Press.

Waber, D. (1979). Neuropsychological aspects of Turner's syndrome. *Developmental Medicine and Child Neurology, 21,* 58–70.

Walzer, S., Wolff, P.H., Bowen, D., Silbert, A.R., Bashir, A.S., Gerald, P.S., & Richmond, J.B. (1978). A method for the longitudinal study of behavioral development in infants and children: The early development of XXY children. *Journal of Child Psychology and Psychiatry, 19,* 213–229.

Webber, M.L., Puck, M.H., Maresh, M.M., Goad, W., & Robinson, A. (1982). Skeletal maturation of children with sex chromosome abnormalities. *Pediatric Research, 16,* 343–346.

11
Psychosocial Functioning of Individuals with Sex Chromosome Abnormalities

DANIEL B. BERCH AND ELIZABETH MCCAULEY

This chapter reviews the extant literature pertaining to the psychosocial functioning of individuals who have sex chromosome abnormalities (SCA). Psychosocial research has generally been concerned with three relatively broad domains: gender identity and sexual functioning, psychopathology, and personality characteristics. The most extensive portion of the chapter contains reviews of these topics as they relate to each of the major SCA karyotypes. When appropriate, we have attempted to briefly discuss the early findings emanating primarily from studies of adults found in mental hospitals and penal institutions. Next, we summarize the results of more recent research with children, adolescents, and adults. Where available, we present the latest evidence that has emerged from prospective, longitudinal studies of unselected SCA individuals identified through screening of large newborn populations. This is followed by a discussion of methodological considerations and suggestions for future directions in research.

Summary and Discussion of Specific Karyotypes

45,X

Gender Identity and Sexual Functioning

Results from 10 separate studies conducted at different points during the past 30 years almost uniformly show 45,X girls (Turner syndrome) as (1) having clearly female gender identity and (2) displaying a traditional pattern of female development with strong interests in doll play and child care (Ehrhardt et al., 1970; Hampson et al., 1955; Money & Mittenthal, 1970; Nielsen et al., 1977; Nielsen & Sillesen, 1981; Sabbath et al., 1961; Senzer et al., 1973; Shaffer, 1963; Taipale, 1979; Theilgaard, 1972).

The data on sexual behavior patterns are much less complete, with most studies including only a cursory assessment. While sexual orientation is consistently described as heterosexual, some studies suggest that sexual drive is low (Garron & Vander Stoep, 1969; Hampson et al., 1955), and others describe a tendency to begin dating later than peers (McCauley et al., 1986b; Senzer et al., 1973). Nielsen et al. (1977) found that both Turner syndrome women and short-statured, amenorrheic controls (no chromosome anomalies) were less likely to have moved into an independent life style or to have married than unaffected sisters of the Turner women. This resulted even though the three groups did not differ in either academic or vocational achievement. The Turner syndrome and short-statured women also indicated less frequent and less satisfactory sexual relationships than the unaffected sisters. These findings suggest that there may be a sexual inhibition which characterizes at least some Turner syndrome women, but is also found in other short-statured women. Alternatively, because of their short stature and knowledge of infertility, these women may simply be less socially adept and more self-conscious. The data currently available are not adequate to determine if genetic/hormonal or social/environmental factors are the more significant underlying variables.

Psychopathology

Although there are case reports describing individuals with significant psychiatric impairments including depression and schizophrenia (Raft et al., 1976; Sabbath et al., 1961), studies of groups of women with the Turner syndrome have failed to identify a high incidence or particular pattern of psychiatric disorders. At least eleven cases of Turner syndrome women with anorexia nervosa have been reported (Darby et al., 1981; Halmi & DeBault, 1974; Kron et al., 1977; Walinder & Mellbin, 1977), which raised speculation as to an association between the two disorders. However, anorexia nervosa is still uncommon in the population of females with X chromosome anomalies, and Darby et al. (1981) argue that the co-occurrence of the two disorders is solely due to chance.

At least four studies drawing on adequate sample sizes and utilizing patient and parent interviews have been conducted to assess the psychiatric and psychosocial functioning of Turner syndrome girls and young women (Money & Mittenthal, 1970; Nielsen et al., 1977; Nielsen & Sillesen, 1981; Taipale, 1979). Only one included a control group. These studies consistently reported a very low prevalence of "severe" psychological problems and a clear association between emotional instability and familial factors. Parental rejection and, to a lesser extent, overprotection were the major factors associated with increased emotional difficulty and poor psychosocial adjustment. With a Finnish sample, Taipale (1979) evaluated the impact of timing of hormonal replacement on psychiatric and psychosocial functioning. Four (8%) of the 49 females who participated were felt to have significant emotional problems. These four were among those who did not have hormonal replacement until late teenage years. Based on the data from this study, Taipale and colleagues (Perheentupa et al., 1974; Taipale, 1979) argued vigorously that hormonal replacement should be initiated in a time frame that allows the young woman to go through pubertal changes along with age mates. In Denmark an attempt was made to follow up and evaluate every girl with the Turner syndrome born there between 1955 and 1966 (Nielsen & Sillesen, 1981). One hundred and three girls were identified and 86 of these girls along with their parents participated in a psychiatric interview. While none were determined to have a "severe" emotional disturbance, many (23) were judged as having difficulties with shyness, insecurity, and increased sensitivity, and 58% were rated by parents as below average for their age in terms of social maturity.

The data on psychopathology in the Turner syndrome are consistent across multiple studies and support the conclusion that abnormalities of the second X chromosome are not associated with an increased risk for psychiatric disturbance. Although some individuals in any group may have difficulties, this is not the norm and appears to be closely associated with parents' ability to cope productively with their child's diagnosis.

Personality Style and Social Adjustment

While major psychopathology is seldom documented, difficulties with social adjustment, such as the social immaturity noted earlier, have been reported frequently. Furthermore, early investigations suggested that there was a personality style associated with this syndrome characterized by high compliance, limited emotional reactivity, and "high tolerance for adversity" (Baekgaard et al., 1978; Money & Mittenthal, 1970; Shaffer, 1962). Others have not supported the idea of a common underlying personality pattern. Taipale's (1979) psychiatric evaluations of 49 adolescent girls revealed no uniformity in personality configuration. She described these young women as anxious and upset about their diagnosis, with difficulties in the area of peer relationships. Rather than attributing these problems to personality structure, Taipale and colleagues (Perheentupa et al., 1974; Taipale, 1979) interpreted these problems as behavioral adaptations to short stature and delayed puberty.

While recent data are inconsistent with the notion of an underlying personality style of low reactivity/arousal, they do suggest problems in social and emotional adjustment (Peri & Molinari, 1983; Rovet, 1986; Sonis et al., 1983). Both Sonis and colleagues (1983) and Rovet (1986) collected parental reports of social and behavioral adjustment using the Child Behavior Checklist. In both studies, the girls with Turner syndrome were rated significantly higher than controls on the hyperactivity/immaturity and anxiety dimensions. The younger girls were more often seen as hyperactive and immature with difficulties concentrating, while the older girls were more often seen as socially withdrawn and depressed/anxious. It should be noted that these problems were not at the level considered to be clinically suggestive of major psychopathology, but rather pointed to behavioral or adjustment concerns. On the Middle Childhood Temperament Questionnaire, the parents of 18 Turner syndrome girls described their daughters as having an outgoing but easily distractable and nonpersistent temperamental style (Rovet, 1986).

Rather than supporting the idea of a passive and nonresponsive personality style, these studies suggest something of a different picture: heightened activity in younger girls, and social withdrawal along with more depressed and anxious affect appearing in older girls. Some researchers contend that these problems are behavioral adaptations to being short statured and sexually immature, while others suggest that there is an immaturity in the CNS which is associated with the genetic anomaly.

In an initial effort to investgate the relative contribution of short stature to the behavioral profile identified in girls with Turner syndrome, McCauley et al. (1986a) assessed the psychosocial functioning of a group of Turner syndrome girls in contrast to girls with familial short stature. The two groups were matched for height, weight, full-scale IQ. and family socioeconomic status. The psychosocial assessment included the Child Behavior Checklist (CBCL), the Piers-Harris Children's Self Concept Scale, and child, parent, and teacher ratings of social maturity and peer interactions. The results of this assessment revealed significantly more difficulties among the Turner syndrome girls than girls with familial short stature. On the CBCL, the Turner syndrome girls were depicted as having significantly more difficulties with peer relationships. These difficulties included frequent teasing by other children and behaving more immaturely than peers.

Cognitive and Psychosocial Interactions

As described previously, the more recent studies suggest that girls with Turner syndrome are at increased risk for social and emotional adjustment problems which are not solely due to coping with short stature. Because of the growing body of literature documenting both cognitive and social-behavioral problems, McCauley, Kay, Ito, & Treder (1987) attempted to determine if these two problem areas are related. As mentioned earlier, the girls with Turner syndrome had significantly more social and behavioral difficulties than the short-stature controls. The cognitive assessment also identified group differences, with the Turner syndrome sample showing significant deficits in the areas of spatial problem solving and attentional skills. Furthermore, the Turner syndrome girls also differed significantly from the controls in their ability to interpret facial affect. Additional analyses indicated that the ability to interpret facial affect and visuospatial problem-solving skills were independent of each other, but both may depend on underlying, right cerebral hemispheric functioning.

While further study to replicate and extend these findings is critical, the possibility of a connection between the social and cognitive spheres is thought-provoking. It may be that the kinds of behavioral patterns thought to be due to an underlying personality style might reflect a cognitively based perceptual problem. These kinds of problems may in turn be more easily remedied via socialization training that focuses on the need to attend more carefully to facial cues, thereby compensating for deficits by relying on the stronger verbal mediation skills found in many females with the Turner syndrome.

47,XXY

Gender Identity and Sexual Functioning

Investigations of 47,XXY boys (Klinefelter syndrome) have provided consistent evidence of increased shyness, passivity, and fatigability (Annell et al., 1970; Nielsen & Sillesen, 1976; Stewart et al., 1986), with some impact on gender-related behaviors but little evidence of actual gender identity disturbance. In a matched control study of 12 47,XXY adolescent males identified by newborn screening, Bancroft, Axworthy, and Ratcliffe (1982) administered the Bem Sex Role Inventory. In addition to significantly lower masculinity scores, the 47,XXY boys scored lower than controls on the femininity and neutral subscales, a factor interpreted as reflecting overall self-esteem problems. Bancroft et al. found no differences in involvement in masculine activities.

Several studies of 47,XXY men suggest the presence of gender-related insecurities. Kvale and Fishman (1981) found disturbances in body image in their sample of 12 47,XXY men, with some of them expressing the desire to have a more masculine physique and engaging in body-building activities. In Theilgaard's (1984) study contrasting the psychosocial functioning of 47,XXY, 47,XYY, and unaffected controls, she found evidence on projective testing of insecurities regarding masculine role in the 47,XXY males. This included reports of less involvement in masculine activities (gymnastics and boys' games) in childhood and more difficulties with body image than the control subjects (Schiavi et al., 1988). The significance of these findings is underscored as this sample was identified via population screening rather than referral from a clinical setting.

Taken as a whole these data provide some modest support for the presence of gender-related problems. However, the gender problems are minimal and appear to be closely tied to the more general developmental difficulties identified in 47,XXY boys and men, such as neuromuscular and language developmental delays. The muscular delays lead to problems with fine and gross motor behavior, poor coordination, and diminished strength (Robinson et al., 1986), which would clearly impact on skill and consequently motivation to participate in more active, typically male-dominated sports activities.

With regard to sexual behavior, several studies of adolescent and young adult 47,XXY males have provided evidence of later onset of masturbation, dating, initial attempts at intercourse, and first coital experience, with reports of less interest in these activities than control groups (Nielsen & Sorensen, 1984; Raboch et al., 1979; Ratcliffe et al., 1982). Reports on adult male sexual activity have been more variable, with most describing clinical samples who have presented in part because of concern about sexual functioning. Diminished sexual drive or libido has been documented in most studies (Becker, 1972; Money & Pollitt, 1964; Wakeling, 1972), although Raboch and Starka (1973) found no differences in frequency of coital activity in 47,XXY men seen in a fertility clinic.

In their study of a nonclinical, population-based sample of 14 47,XXY men with 47,XYY and normal controls, Schiavi et al. (1988) found clear evidence of later age of first intercourse, less sexual desire, less frequent sexual activity, and fewer sexual partners in the 47,XXY men. Although low testosterone levels have been implicated in the reduced sexual arousal observed in 47,XXY men, all the men in the Schiavi et al. (1988) study had testosterone values that were within normal limits. Furthermore, studies of the impact of testosterone therapy on sexual behavior in 47,XXY men have yielded mixed results (Luisi & Franchi, 1980; Schiavi et al., 1978; Wu et al., 1982), suggesting that there may be great variability in the sensitivity of these men and/or behaviors to androgen treatment. Finally, although there have been clinical reports of sexual deviancy (pedophilia, sexual aggression, exhibitionism) in 47,XXY men (Crowley, 1965; Hoaken et al., 1964; Mosier et al., 1960), these findings have not been replicated when population-based samples are examined and thus appear to represent isolated incidents rather than a syndrome-associated phenomenon.

Psychopathology

While there have been numerous case reports of individual 47,XXY men with serious psychopathology including schizophrenia (Nielsen, 1969) and manic depressive illness (Caroff, 1978), no one diagnosis appears to characterize these men as a group. There was some initial speculation of an association between the Klinefelter syndrome and schizophrenia, but this was not substantiated when data from multiple samples were pooled (Nielsen, 1969), and no association between the two disorders has been found in more recent studies using population-based samples. The literature also includes approximately 12 case reports of co-occurring manic-depressive illness and the 47,XXY karyotype [see Roy (1985) for a thorough review of psychiatric disorders in the Klinefelter syndrome], which is within the limits of change given the high prevalence rates of each disorder.

An increased prevalence of criminal behavior has been found in some investigations of men with the Klinefelter syndrome (Nielsen, 1970;

Wakeling, 1973), with nonaggressive crimes against property being the most common type of offense. Nielsen and Pelsen (1987) have conducted a 20-year follow-up of 36 Klinefelter men who initially had a higher rate of criminal behavior than a control group of hypogonadal men with no chromosomal anomaly. At follow-up the two groups had few areas of significant difference, with overall good adjustment and no ongoing criminal behavior for the 47,XXY men. Furthermore, there was no evidence of increased criminality in Theilgaard's (1984) population-based sample. Although some acting out has been identified in the recent multicenter newborn screening studies, the subjects are still too young for researchers to gain an accurate sense of their long-term risk for criminal behavior.

In sum, the evidence does not suggest that men with the Klinefelter syndrome are at increased risk for either major psychopathology or criminal behavior. Nielsen and colleagues (Nielsen & Pelsen, 1987; Nielsen & Sorensen, 1985) see early diagnosis and intervention/education for parent and child as critical factors in modulating the potential for individuals with the Klinefelter syndrome to develop serious problems.

Personality Style and Social Adjustment

Early studies of clinical samples of 47,XXY boys characterized them as having difficulties in school, with a shy, passive style and low energy (Annell, 1970; Caldwell & Smith, 1972; Funderburk & Ferjo, 1978). The more recent studies with population-based samples have provided findings consistent with these initial reports. Theilgaard (1984) characterized her sample as "indecisive," "submissive," and "dependent." Ratcliffe and colleagues found that in comparison to matched controls, 47,XXY boys (16 to 18 years old) identified by newborn screening experienced significantly greater difficulties with language development and more frequent learning problems in school due to immaturity in verbal skills (Bancroft et al., 1982; Ratcliffe et al., 1982). These boys were also seen as shy and easily stressed with immaturity in social interactions, marked by distractability and fighting with peers.

In their Toronto sample of 47,XXY boys also identified through newborn screening, Stewart and colleagues (1986) had parents complete three personality/behavior inventories. The boys were seen by parents as having significantly more difficulties as compared with population means on all scales of the Child Behavior Checklist, with the Immature, Hostile/Withdrawn, and Delinquent scales being the highest. It should be noted, however, that no scales were elevated into the range suggestive of psychopathology, but reflected behavioral concerns. On the Personality Inventory for Children the 47,XXY boys were seen as having significantly more development, achievement, and intellectual problems and were rated as significantly below the normative group on the Hyperactive Scale. Findings based on the Middle Childhood Temperament Survey, presented with a word of caution because some of the children were older than the test limits, characterized the 47,XXY boys as being introverted and unassertive, with low levels of social comfort and skill.

In an effort to better evaluate findings across samples of 47,XXY boys identified through newborn screening, Netley (1986) pooled behavioral data provided by a number of study centers from around the world. Of the 65 47,XXY boys represented in this larger pool, 44 had academic problems and 19 were judged as having behavioral problems. The number with academic problems was significantly greater than those in the control group, but there were no differences between the control and index groups in terms of behavior problems. In addition, Netley found that family functioning was a critical variable in psychosocial adjustment, with the 47,XXY boys no more vulnerable to family dysfunction than controls.

The data just described consistently support a temperament or personality style characterized by low energy and poor social skills. These features appear to reflect some of the sequelae of the syndrome such as language and motor skill delays which in turn have a negative impact on 47,XXY boys' self-confidence and ability to interact socially. Some positive changes in behavior and self-esteem have been documented following hormonal therapy, particularly as it impacts energy level and strength. As mentioned previously, Nielsen et al. have found that early diagnosis coupled with parental education, counseling, and hormone treatment can contribute greatly to increasing positive adjustment and reducing the negative social, emotional, and educational impacts.

47,XXX

Gender Identity and Sexual Functioning

Most reports of the psychosexual functioning of 47,XXX females have had next to nothing to say concerning the adequacy of sex role development, gender identity, or sexual functioning. Perhaps the lack of data in this regard constitutes in itself an indication that few if any such problems arise for these girls. Unlike 45,X females, triple X girls do not manifest an identifiable set of physical features other than tallness. Moreover, despite some variation in findings across prospective studies, standard indicators suggest that the onset of puberty is not delayed (Ratcliffe & Paul, 1986).

Psychopathology

In an extensive survey, Barr, Sergovich, Carr, and Shaver (1969) reviewed 143 triple X cases from chromosomal screening of several phenotypically abnormal and normal populations in 17 different countries, along with individuals referred for cytogenetic analysis because of clinical findings. Collating the data from 10 studies in 7 countries concerning the incidence of triple X females identified in hospitals for the mentally ill, Barr et al.

reported a frequency 4.8 times greater than that for newborn babies in the general population. This overrepresentation led Barr et al. to conclude that the 47,XXX karyotype predisposes to mental illness. In contrast, since no behavioral abnormalities were found among triple X women located in nonpsychiatric clinical settings, Barr et al. speculated that some unknown proportion of this population is normal and that the behavioral phenotype is most likely variable. Furthermore, these investigators explicitly pointed out that bias due to the methods of ascertainment precluded an accurate assessment of the magnitude of this predisposition to psychopathology. Finally, they noted that little information was available concerning less severe forms of psychopathology. Their suggestion that progress in this area must come from cytogenetic surveys of unselected newborns has recently come to fruition.

Personality Style and Social Adjustment

Combining the results of nine prospective studies of SCA children from six different countries, Stewart, Netley, and Park (1982) reported that similar proportions of triple X girls and controls exhibited behavior problems (24.1% and 22.6%, respectively). Furthermore, while there was little evidence of atypical personality patterns among the triple X girls, they tended to experience difficulties in establishing good interpersonal relationships. Netley (1986) recently collated and summarized the latest behavioral data from many of these same prospective studies. Eleven of 35 or 31.4% exhibited significant behavioral problems as compared to 11 of 29 or 37.9% of the controls (a nonsignificant difference). Poor family functioning was reliably associated with behavioral problems at comparable levels for the triple X and control children.

Results from the prospective studies described earlier indicate that the overall psychosocial functioning of 47,XXX girls appears to be relatively normal through early adolescence. Nevertheless, as described by Bender et al. (this volume), these girls evidence much greater cognitive impairment (e.g., decreased intelligence, language, learning and short-term memory dysfunctions) than other types of SCA children. Thus, it is possible that psychosocial problems could develop for triple X girls during middle and later adolescence secondary to their cognitive difficulties. Evidence pertaining to this speculation must naturally await reports forthcoming from the prospective, longitudinal studies.

47,XYY

Gender Identity and Sexual Functioning

A recently published report by Schiavi, Theilgaard, Owen, and White (1988) has provided the most detailed information to date concerning the psychosexual characteristics of 47,XYY males. Twelve 47,XYY men were identified in a case-finding study in Denmark of men with sex chromosome abnormalities in the general population, unselected with regard to institutionalization. These investigators found that in comparison to controls, fewer 47,XYY men were married, and those who were experienced greater sexual dissatisfaction with their wives and in general. While not differing from controls in reported frequency of homosexual behavior, 47,XYY men were rated by the interviewer and themselves as less masculine than controls. The 47,XYY men reported that they had fantasized about and engaged in unconventional sexual activities to a greater extent than controls. However, none of these men had been convicted of deviant sexual acts. As Shiavi et al. point out, in contrast to the earlier stereotype of hypermasculinity, their results, along with those of two other surveys of noninstitutionalized 47,XYY men (Nielsen et al., 1973; Zeuthen et al., 1975) suggest that these men "are insecure in their masculine role, lack sexual confidence, and have difficulties developing stable and satisfying relationships with women" (p. 22).

Psychopathology

A case history (Hauschka et al., 1962) of the putatively first observed 47,XYY human revealed essentially normal psychosocial adjustment. Not long afterward, however, Jacobs, Brunton, Melville, Brittain, and McClemont (1965) reported a higher than expected prevalence of 47,XYY males in a maximum security hospital. Subsequent surveys of institutionalized populations, primarily mental hospitals and prisons, also suggested an association between having an extra Y chromosome and aggressive, antisocial behavior. Despite cautions concerning such generalizations from many of the original investigators, a public stereotype of the 47,XYY male emerged—they were seen as "supermales" in possession of a "murder chromosome." This view was seemingly corroborated when two men on trial for murder were discovered to have a 47,XYY karyotype. However, following an extensive review and critique of the early 47,XYY literature (1961–1971), Owen (1972) concluded that there was little objective evidence for this aggressive stereotype. In addition to problems with sampling and selection biases, much of the data was based on clinical impressions influenced by high expectancies of disturbed behavior.

In an initial report of the 12 47,XYY men from the Danish case-finding study described earlier (Schiavi et al., 1988), Witkin et al. (1976) examined penal registers for evidence of violations of the penal code that had resulted in convictions. Although the prevalence of criminal convictions was higher for the 47,XYY group than the control group, this finding was explained in part by other differences between these groups. That is, intelligence test scores were substantially lower for the 47,XXYs, and their educational levels were lower. Once these differences were controlled for, along with parental socioeconomic status, the crime-rate difference was greatly reduced. Moreover, the large majority of the crimes committed by the 47,XYY men were against property, not involving personal violence.

Schiavi, Theilgaard, Owen, and White (1984) have recently presented a more detailed assessment of the aggressive behavioral tendencies of the 47,XYY men from the Danish cohort. While the probands reported exhibiting greater physical aggression toward their wives than the 46,XY controls, there were no differences in other aspects of physical aggression or in verbal aggression. In addition, there was clear evidence of elevated levels of pituitary gonadotropins and plasma testosterone (previously reported by Schiavi et al., 1978). However, the authors note that the hormonal values were within the normal range and that the statistically significant difference in these levels as compared to controls may only have become apparent because 47,XYYs and controls were individually matched. There was no evidence that the elevated testosterone levels were a mediating factor in the criminal behavior of the 47,XYY men.

In sum, the evidence to date suggests that contrary to earlier findings, 47,XYY individuals are not necessarily prone to violent or aggressive behavior in adulthood. Moreover, in the most recent account of the Danish cohort described in the Schiavi et al. (1984) paper, Theilgaard (in press) has noted that the 47,XYY men report the quality of their childhood as significantly poorer in comparison to controls. This finding suggests that studies of 47,XYY children which include measures of family functioning would appear to be most critical for providing a truly complete picture concerning the relative contributions of factors that influence the psychosocial development of SCA individuals. As will be evident in the next section, a number of studies fortunately have provided such data.

Personality Style and Social Adjustment

This section includes a review of data that has emerged from prospective, longitudinal studies of unselected samples of 47,XYY boys identified at birth. Bender, Puck, Salbenblatt, and Robinson (1984) recently reported the psychosocial characteristics (in addition to other kinds of behavioral data) of four 47,XYY boys originally identified through newborn screening. Mild depression was apparent in all four boys, as reflected in sad affect, lethargy, and reduced enthusiasm for age-appropriate activities and interests. While one of them had been diagnosed as suffering from relatively serious emotional disturbance, so were his brother and sister, which the authors ascribed to dysfunctional family dynamics. Finally, Bender et al. noted that in addition to expressions of anxiety from two sets of parents concerning implications of the extra Y chromosome for aggressive behavior, one boy's private pediatrician described him at two years of age as excessively oppositional and unmanageable. Since this characterization conflicted with observations made during multiple clinic visits as well as with parental descriptions, the authors concluded that the physician's judgment may be attributable to biased expectations for the behavioral sequelae associated with possessing an extra Y chromosome.

There have been three attempts to combine the emotional disturbance/ behavior disorder data of 47,XYY boys from various prospective studies conducted around the world. These three have all incorporated data from some of the same samples, but they contain information bearing on different time periods in the developmental histories of these boys, along with results from nonoverlapping samples. Stewart, Netley, and Park (1982) estimated the extent of temperament and behavior disorders based upon data provided by nine study centers in six different countries. They reported an overall incidence of 15.8%, which did not differ significantly from that of controls. Bender et al. (1984) combined the results of eight unselected samples of 47,XYY boys from around the world. This analysis revealed that while emotional disturbance (defined in a variety of ways) was not generally characteristic of these boys, 11 of 42 or 26% had exhibited emotional disturbance.

Finally, Netley (1986) has more recently summarized results regarding behavior problems in 47,XYY boys drawn from data submitted by six study centers in four different countries. Ten of 27 or 37% of the 47,XYY boys were judged to have exhibited behavior problems, as compared with five of nine or 55.5% of the chromosomally normal controls. As Netley points out, not only is the difference between these proportions nonsignificant, but the incidence for 47,XYY boys is comparable to those obtained for the 47,XXY and 47,XXX cases from the same study centers (29.2% and 31.4%, respectively). In addition, poor family functioning was reliably associated with behavior problems at comparable levels for the 47,XYY and control children.

Ratcliffe, Jenkins, and Teague (in press) have provided the most recent data yet concerning psychosocial functioning of 47,XYY boys from a prospective study of the largest extant sample of such children identified at birth. While 18 47,XYY boys were identified, this report was limited to the 12 boys who were at least 13 years of age at the time of data analysis. The controls consisted of 22 unrelated, chromosomally normal boys who were the oldest of a larger group recruited at random from the same newborn population. Behavior at school was assessed by use of the British Social Adjustment Guide (Stott, 1976), a questionnaire completed by the teachers. The 47,XYY boys were reported as more underreactive (withdrawn and depressed) and anxious, but did not differ from controls in overreaction (includes hostility and aggression).

Behavior at home at 12 years of age was assessed by means of the Rutter Parent Questionnaire (Rutter, Tizard, & Whitmore, 1970). In comparison to controls, more than twice as many of the 47,XYY boys were reported as exhibiting tempers, stealing, and having speech problems, as well as being miserable, solitary, and not liked by other children. However, a greater number of controls than 47,XYY boys were reported as having problems with sleeping, eating, worrying, and being fussy. Not one of the 47,XYY boys was reported to have bullied other children.

Additional findings were provided concerning impulsivity, incidence of family pathology, and psychiatric referrals. First , there was no difference in impulsivity between the 47,XYY and control groups [as measured by the matching familiar figures test (Kagan, 1965)]. With regard to family dysfunction and pathology, three times as many of the 47,XYY boys as controls experienced separation/divorce and maternal psychiatric illness, yet there was no difference in paternal psychiatric illness. Finally, the incidence of psychiatric referral was five times greater for the 47,XYY boys. SCA diagnosis had still not been disclosed to parents at the time of referral, but the principal investigator, who was also the pediatrician for these boys, suggests that her knowledge of their chromosomal constitution may have lowered her threshold for psychiatric referral following repeated reports of behavior problems by parents. In any event, a combination of counseling, environmental changes, and the use of pharmacological agents where appropriate resulted in a gradual resolution of the presenting complaints. Indeed, at the time of the writing of the Ratcliffe et al. paper, none of the 47,XYY boys were receiving psychiatric treatment.

In general, the results of prospective, longitudinal studies are relatively consistent in showing that aggressive, antisocial behaviors are no more frequent in 47,XYY males during childhood or early adolescence than in the chromosomally normal population. Similarly, there is little evidence of other types of emotional disturbance or behavior disorders, at least with a frequency high enough to be considered as characteristic of this SCA condition. Furthermore, when such problems do arise, they can frequently be linked to family dysfunction, but no more so than with chromosomally normal children. The collation of data emerging from the major prospective studies over the next decade will provide the most unbiased evidence yet concerning the prevalence of any psychosocial dysfunctions that may arise in 47,XYY males between middle adolescence and young adulthood.

Methodological Considerations

Much of the early research concerning psychosocial dysfunctions in SCA individuals relied on the use of semistructured interviews. Since the relevant data were gathered as part of a generic clinical evaluation process, this work lacked the rigor of more recently developed, structured interview procedures. Thus it is not surprising that findings from studies using more systematic techniques have not always been consistent with prior results.

One of the major methodological concerns in SCA research is the nature of the sampling technique used. As noted earlier, sampling bias was characteristic of many of the initial behavioral studies of SCA individuals, since subjects were primarily selected from mental and/or penal institutions. As Bender and Berch (1987) have pointed out, while such studies established an increased risk for behavioral abnormalities, since the aneuploid individuals in these surveys represented less than 1% of all living persons with SCAs, the results could not provide accurate estimates of the range of phenotypic expression. It should be noted that a sampling or selection bias may also occur when subjects are recruited from medical center records. That is, the ones who are willing to participate in research may not be representative of the general population of SCA individuals.

Another major problem is that of sample size. The primary way in which researchers have attempted to overcome this limitation is to include a relatively wide age range in their sample. For example, a number of studies of Turner syndrome have included females ranging in age from early teens to middle twenties. While this may not be particularly problematic for investigations of cognitive and intellectual dysfunctions, it certainly may be for studies of psychosocial dysfunctions, primarily because of developmental changes in sex role orientation and/or sexual functioning that may occur between early adolescence and early adulthood. Berch and Kirkendall (1986) attempted to handle the problem of sample size in their study of cognitive dysfunctions in Turner syndrome girls by traveling to several different midwestern states, testing children in a mobile behavior laboratory. However, the best solution may be a multicenter collaborative study.

Since sample size is a problem, it is important to at least try to combine the results of independent studies. As noted earlier, several such attempts have been made. Naturally, one of the major problems in accomplishing this is that investigators have used different measurement techniques of psychosocial dysfunction. At times they have used the same label (e.g., emotional disturbance), but have chosen to operationalize it differently. As research design improves and the commonality of measurement devices increases, the need to employ various meta-analytic procedures for combining the results of independent studies will also become critical.

One of the major methodological problems in SCA research involves the formation of appropriate control groups. The ideal approach would be to use a monozygotic twin who does not have an SCA condition (Rovet & Netley, 1982). However, this situation rarely arises. Many investigators (especially those involved in prospective, longitudinal research) have used siblings as controls. While this procedure provides the advantage of controlling for the effects of familial factors on psychological development, it usually has the disadvantage of unequal group sizes, wider age variation among sibs, and/or other differences that could potentially confound interpretation of the results.

A particular problem in the study of psychosocial dysfunctions of SCA individuals is that personality styles and/or specific behavior patterns may emerge in part as adaptations to physiological or somatic sequelae, and thus not be directly attributable to a given sex chromosome anomaly. Appropriate comparison groups must be formed to control for the possible influence of such factors.

Another important methodological consideration is the use of a blind evaluation. While this can be accomplished with 47,XXX, 47,XXY, and 47,XYY individuals, it is generally not possible with 45,X females, because of somatic stigmata (e.g., short stature) that usually make them readily identifiable. Where feasible, some investigators have even attempted to use a double-blind technique to protect against possible effects of disclosure of diagnosis on subsequent psychological development. However, Puck (1981) has provided some evidence that if disclosure is coupled with the continued, long-term availability of supportive guidance, then this so-called "self-fulfilling prophecy" does not necessarily occur.

Future Directions

Until recently, the bulk of the research efforts in the psychosocial arena have been driven by empirical questions concerning the nature and incidence of various types of behavioral dysfunctions associated with the different SCA conditions. However, the time is ripe for drawing on theoretical concepts in allied domains (e.g., neuropsychology and psychoendocrinology) to develop more rigorous, testable models of psychosocial dysfunction. A recent example of this comes from the work of Charles Netley and his colleagues at the Hospital for Sick Children in Toronto, concerning the psychosocial characteristics of XXY males (Netley, in press; Stewart et al., 1986). As described earlier, there exists relatively strong evidence that prior to puberty, 47,XXY boys are less social, assertive, and active than normal 46,XY boys. First, Netley found that the integrity of left-hemisphere functioning was the single best predictor of these temperament qualities. However, he also discovered that during puberty, degree of both left and right hemispheric specialization along with testosterone levels and quality of parenting jointly predicted variations in the extent to which 47,XXY boys were free from a variety of types of maladjustment.

In attempting to explain these findings, Netley chose to draw on theoretical views of Tucker and Williamson (1984). Basically, with regard to social and emotional behaviors, they have suggested that the left hemisphere is concerned wih goal-related or motivational behavior, while the right is primarily responsible for mediating affect. Netley has reasoned that the unresponsive, passive behavior patterns characteristic of prepubertal 47,XXY boys may thus be attributable to impairments in left-hemisphere-based motivational systems. After the onset of puberty, emotional resiliency is enhanced in 47,XXY males who have particularly well-developed right hemisphere mechanism. As he points out, this would permit them to cope better with stresses generated by the onset of puberty, testosterone deficiencies, and poor parenting.

Further evidence is needed, especially from other study centers, before the biopsychosocial relationships reported by Netley's group can be considered reliable enough to warrant further attempts at their explanation. Nevertheless, Netley's theoretical model constitutes a bold step that will at the very least provide a heuristic function for the future study of psychosocial dysfunctions associated with SCA conditions. Moreover, these efforts point up the need for a multidimensional, theoretical approach to understanding the complexities inherent in this research domain.

References

Annell, A.L., Gustavson, K.H., & Tenstam, J. (1970). Symptomatology in schoolboys with positive sex chromatin (the Klinefelter's syndrome). *Acta Psychiatrica Scandinavia, 46,* 71–80.

Baekgaard, W., Nyborg, H., & Nielsen, J. (1978). Neuroticism and extroversion in Turner's syndrome. *Journal of Abnormal Psychology, 87,* 583–586.

Bancroft, J., Axworthy, D., & Ratcliffe, S. (1982). The personality and psychosexual development of boys with 47,XXY chromosome constitution. *Journal of Child Psychology and Psychiatry, 23,* 169–180.

Barr, M.L., Sergovich, F. R., Carr, D.H., Shaver, E.L. (1969). The triplo-X female: An appraisal based on a study of 12 cases and a review of the literature. *The Canadian Medical Association Journal, 101,* 247–258.

Becker, K.L. (1972). Clinical and therapeutic experiences with Klinefelter's syndrome. *Fertility Sterility, 23,* 568–578.

Bender, B.G., & Berch, D.B. (1987). Sex chromosome abnormalities: Studies of genetic influences on behavior. *Integrative Psychiatry, 5,* 171–178.

Bender, B.G., Puck, M.H., Salbenblatt, J.A., & Robinson, A. (1984). The development of four unselected 47,XYY boys. *Clinical Genetics, 25,* 435–445.

Bender, B.G., Puck, M.H., Salbenblatt, J.A., & Robinson, A. (1988). Cognitive development of children with sex chromosome abnormalities. In C.S. Holmes (Ed.), *Psychoneuroendocrinology: Brain, behavior, and hormonal interactions* (pp. 138–163). New York: Springer-Verlag.

Berch, D.B., & Kirkendall, K.L. (May, 1986). *Spatial information processing in 45,X children.* In A. Robinson (Chair), Cognitive and psychosocial dysfunctions associated with sex chromosome abnormalities. Symposium presented at the meeting of the American Association for the Advancement of Science, Philadelphia.

Caldwell, P.D., & Smith, D.W. (1972). The XXY (Klinefelter's) syndrome in childhood: Detection and treatment. *Journal of Pediatrics, 80,* 250–258.

Caroff, S. (1978). Klinefelter's syndrome and bipolar affective illness: A case report. *American Journal of Psychiatry, 135,* 748–749.

Crowley, T.J. (1965). Klinefelter's syndrome and abnormal behavior: A case report. *International Journal of Neuropsychiatry, 1,* 359–363.

Darby, P.L., Garfinkle, P.E., Vale, J.M., Kirwan, P.J., & Brown, G.M. (1981). Anorexia nervosa and 'Turner syndrome': Cause or coincidence? *Psychological Medicine, 11,* 141–145.

Ehrhardt, A.A., Greenberg, N., & Money, J. (1970). Female gender identity and absence of fetal hormones: Turner's syndrome. *Johns Hopkins Medical Journal, 126*, 234–248.

Funderburk, S.J., & Ferjo, N. (1978). Clinical observations in Klinefelter's (47,XXY) syndrome. *Journal of Mental Deficiency Research, 22*, 207–212.

Garron, D.C., & Vander Stoep, L.R., (1969). Personality and intelligence in Turner's syndrome. *Archives of General Psychiatry, 21*, 339–346.

Halmi, K.A., & DeBault, L.E. (1974). Gonosomal aneuploidy in anorexia nervosa. *American Journal of Genetics, 26*, 195–198.

Hampson, J.L., Hampson, J.C., & Money, J. (1955). The syndrome of gonadal agenesis (ovarian agenesis) and male chromosomal pattern in girls and women: Psychologic studies. *Bulletin of Johns Hopkins Hospital, 97*, 207–226.

Hauschka, T.S., Hasson, J.E., Goldstein, M.N., Koepf, G.F., & Sandberg, A.A. (1962). An XYY man with progeny indicating familial tendency to nondisjunction. *American Journal of Human Genetics, 14*, 22–30.

Hoaken, P.C.S., Clark M., & Breslin, M. (1964). Psychopathology in Klinefelter's syndrome. *Psychosomatic Medicine, 26*, 207–223.

Jacobs, P.A., Brunton, M., Melville, M.M., Brittain, R.P., & McClemont, W.F. (1965). Aggressive behavior, mental subnormality and the XYY male. *Nature, 208*, 1351–1352.

Kagan, J. (1965). Reflection-impulsivity and reading ability in primary grade children. *Child Development, 36*, 609–628.

Kay, T. (1983). *Individual differences in children's abilities to discriminate positive and negative affect.* Unpublished doctoral dissertation, Emory University, Atlanta.

Kron, T., Katz, J.L., Gorzynski, G., & Weiner, H. (1977). Anorexia nervosa and gonadal dysgenesis. *Archives of General Psychiatry, 34*, 332–335.

Kvale, J.N., & Fishman, J.R. (1981). The psychosocial aspects of Klinefelter's syndrome. *Journal of the American Medical Association, 193*, 567–572.

Luisi M., & Franchi, F. (1980). Double-blind group comparative study of testosterone undecanoate and mesterolone in hypogonadal male patients. *Journal of Endocrinological Investigations, 3*, 305–308.

McCauley, E., Ito, J., & Kay, T. (1986a). Psychosocial functioning in girls with the Turner syndrome and short stature. *Journal of the American Academy of Child Psychiatry, 25*, 105–112.

McCauley, E., Kay, T., Ito, J., & Treder, R. (1987). The Turner syndrome: Cognitive deficits, affective discrimination and behavior problems. *Child Development, 58*, 464–473.

McCauley E., Sybert, V., & Ehrhardt, A.A. (1986b). Psychosocial adjustment of adult women with Turner syndrome. *Clinical Genetics, 29*, 284–290.

Money, J., & Mittenthal, S. (1970). Lack of personality pathology in Turner syndrome: Relations to cytogenetics, hormones and physique. *Behavior Genetics, 1*, 43–56.

Money, J., & Pollitt, E. (1964). Cytogenetic and psychosexual ambiguity: Klinefelter's syndrome and transvestism compared. *Archives of General Psychiatry, 11*, 589–595.

Mosier, H.D., Scott, L.W., & Dingman, H.F. (1960). Sexually deviant behavior in Klinefelter's syndrome. *Journal of Pediatrics, 57*, 479–483.

Netley, C. (1986). Summary overview of behavioural development in individuals with neonatally identified X and Y aneuploidy. In S. G. Ratcliffe & N. Paul (Eds.), *Prospective studies on children with sex chromosome aneuploidy* (pp. 293–306). Birth defects: Original article series, Vol. 22, No. 3. New York: Alan R. Liss.

Netley, C. (in press). Behavior and extra X aneuploid states. In D.B. Berch & B.G. Bender (Eds.), *Sex chromosome abnormalities and behavior: Psychological studies*. Boulder: Westview Press.

Nielsen, J. (1970). Criminality among patients with Klinefelter's syndrome and the XXY syndrome. *British Journal of Psychiatry, 117*, 365–369.

Nielsen, J., Christensen, A.L., Schultz-Larsen, J., & Yde, H. (1973). A psychiatric-psycholgoical study of patients with the XYY syndrome found outside of institutions. *Acta Psychiatrica Scandinavia, 49*, 159–168.

Nielsen, J., Nyborg, H., & Dahl, G. (1977). Turner's syndrome: A psychiatric-psychological study of 45 women with Turner's syndrome, compared with their sisters and women with normal karyotypes, growth retardation and primary amenorrhea. *Acta Jutlandica, XLV, Medicine Series 21*, Arhus.

Nielsen, J., & Pelsen, B. (1987). Follow-up 20 years later of 34 Klinefelter males with karyotype 47,XXY and 16 hypogonadal males with karyotype 46,XY. *Human Genetics, 77*, 188–192.

Nielsen, J., & Sillesen, I. (1981). Turner's syndrome in 115 Danish girls born between 1955 and 1966. *Acta Jutlandica LIV, Medicine Series 22*, Arhus.

Nielsen, J., & Sorensen, K. (1984). The importance of early diagnosis of Klinefelter's syndrome. In H.J. Bandman & R. Breit (Eds.), *Klinefelter's syndrome* (pp. 170–187). NewYork: Springer-Verlag.

Owen, D.R. (1972). The 47,XXY male: A review. *Psychological Bulletin, 78*, 209–233.

Perheentupa, J., Lenko, H.L., Nevalainen, I., Nittymaki, M., Soderholm, A., & Taipale, V. (1974). Hormone therapy in Turner's syndrome: Growth and psychological aspects. Pediatric XIV. *Growth and Developmental Endocrinology, 5*, 121–127,

Peri, G., & Molinari, E. (1983). Psychological aspects in gonadal dysgenesis. *Acta Medical Auxology, 15*, 75–84.

Puck, M.H. (1981). Some considerations bearing on the doctrine of self-fulfilling prophecy in sex chromosome aneuploidy. *American Journal of Medical Genetics, 9*, 129–137.

Raboch, J., & Starka, L. (1973). Reported coital activity of men and levels of plasma testosterone. *Archives of Sexual Behavior, 2*, 309–316.

Raft, D., Spencer, R. F., & Toey, T.C. (1976). Ambiguity of gender identity fantasies and aspects of normal and abnormal pathology in hypopituitary dwarfism and Turner syndrome: Three cases. *Journal of Sex Research, 12*, 161–172.

Ratcliffe, S., Bancroft, J., Axworthy, D., & McLaren, W. (1982). Klinefelter's syndrome in adolescence. *Archives of Diseases in Childhood, 57*, 6–12.

Ratcliffe, S.G., Jenkins, J., & Teague, P. (in press). Cognitive and behavioral development of the 47,XYY child. In D.B. Berch & B.G. Bender (Eds.), *Sex chromosome abnormalities and behavior: Psychological studies*. Boulder: Westview Press.

Ratcliffe, S.G., & Paul, N. (Eds.), (1986). *Prospective studies on children with sex chromosome aneuploidy*. Birth defects: Original article series, Vol. 22, No. 3. New York: Alan R. Liss.

Robinson, A., Bender, B.G., Borelli, J.B., Puck, M.H., Salbenblatt, J.A., & Winter, J.S.D. (1986). Sex chromosome aneuploidy: Prospective and longitudinal studies. In S.G. Ratcliffe & N. Paul (Eds)., *Prospective studies on children with sex chromosome aneuploidy* (pp. 23–73). Birth defects: Original article series, Vol. 22, No. 3. New York: Alan R. Liss.

Rovet, J. (May, 1986). *Processing deficits in 45, X females*. In A. Robinson (Chair), Cognitive and psychosocial dysfunctions asociated with sex chromosome abnormalities. Symposium presented at the meeting of the American Association for the Advancement of Science, Philadelphia.

Rovet, J., & Netley, C. (1982). Processing deficits in Turner's syndrome. *Developmental Psychology, 18,* 77–94.

Roy, A. (1984). Psychiatric disorders in relation to Klinefelter's syndrome. In H.J. Bandman & R. Briet (Eds.), *Klinefelter's syndrome* (pp. 192–201). New York: Springer-Verlag.

Rutter, M., Tizard, J., & Whitmore, K. (1970). *Education, health, and behaviour*. London: Longmans Green.

Sabbath, J.C., Morris, T.A., Menzer-Benaron, D., & Strugist, S.H. (1961). Psychiatric observation in adolescent girls lacking ovarian function. *Psychosomatic Medicine, 23,* 224–231.

Schiavi, R.C., Owen, D., Fogel, M., White, D., & Szechter, R. (1978). Pituitary-gonadal function in XYY and XXY men identified in a population survey. *Clinical Endocrinology, 9,* 223–239.

Schiavi, R.C., Theilgaard, A., Owen, D.R., & White, D. (1984). Sex chromosome anomalies, hormones, and aggressivity. *Archives of General Psychiatry, 41,* 93–99.

Schiavi, R.C., Theilgaard, A., Owen, D.R., & White, D. (1988). Sex chromosome anomalies, hormones, and sexuality. *Archives of General Psychiatry, 45,* 19–24.

Senzer, N., Aceto, T., Cohen, M.M., Ehrhardt, A. A., Abbassi, V., & Capraro, V.J. (1973). Isochromosome X. *American Journal of Diseases of Children, 126,* 312–316.

Shaffer, J.W. (1963). Masculinity-femininity and other personality traits in gonadal aplasia (Turner's syndrome). In H.C. Beigel (Ed.), *Advances in Sex Research* (pp. 219–232). New York: P.P. Hoeber & Sons, Inc.

Sonis, W.A., Levine-Ross, J., Blue, J., Cutler, G.B., Loriaux, P.L., & Klein, R.P. (October, 1983). *Hyperactivity of Turner's syndrome*. Paper presented at the meeting of the American Academy of Child Psychiatry, San Francisco.

Stewart, D.A., Bailey, J.D., Netley, C.T., Rovet, J. & Park, E. (1986). Growth and development from early to midadolescence of children with X and Y chromosome aneuploidy: The Toronto study. In S.G. Ratcliffe & N. Paul (Eds.), *Prospective studies on children with sex chromosome aneuploidy* (pp. 119–182). Birth defects: Original article series, Vol. 22, No. 3. New York: Alan R. Liss.

Stewart, D.A., Netley, C.T., & Park, E. (1982). Summary of clinical findings of children with 47,XXY, 47,XYY, and 47,XXX karyotypes. In D.A. Stewart (Ed.), *Children with sex chromosome aneuploidy: Follow-up studies* (pp. 1–5). Birth defects: Original article series, Vol. 18, No. 4. New York: Alan R. Liss.

Stott, D.H. (1976). *Bristol social adjustment guides (5th ed.)*. London: Hodder & Stoughton.

Taipale, V. (1979). *Adolescence in Turner's syndrome*. Monograph from Children's Hospital, University of Helsinki, Finland.

Theilgaard, A. (1984). A psychological study of the personalities of XYY- and XXY-men. *Acta Psychiatrica Scandinavica, 69*(315).

Theilgaard, A. (in press). Men with sex chromosome aberrations—as subjects and human beings. In D.B. Berch & B.G. Bender (Eds.), *Sex Chromosome Abnormalities and Behavior: Psychological studies*. Boulder: Westview Press.

Tucker, D.M., & Williamson, P.A. (1984). Asymmetric neural control systems in human self-regulation. *Psychological Review, 91*, 185–215.

Wakeling, A. (1972). Comparative study of psychiatric patients with Klinefelter's syndrome and hypogonadism. *Psychosomatic Medicine, 2*, 139–154.

Walinder, J., & Melbin, G. (1977). Karyotyping of women with anorexia nervosa. *British Journal of Psychiatry, 13*, 48–49.

Witkin, H.A., Mednick, S.A., Schulsinger, F., Bakkestrom, E., Christiansen, K.O., Goodenough, D.R., Hirschhorn, K., Lundsteen, C., Owen, D.R., Philip, J., Rubin, D.B., & Stocking, M. (1976). Criminality in XYY and XXY men. *Science, 193*, 547–555.

Wu, F.C.W., Bancroft, J., Davidson, D.W., & Nicol, K. (1982). The behavioural effects of testosterone undecanoate in adult men with Klinefelter's syndrome: A controlled study. *Clinical Endocrinology, 16*, 489–497.

Zeuthen, E., Hansen, M., Christensen, A.L., & Nielsen, J. (1975). A psychiatric-psychological study of XYY males found in a general male population. *Acta Psychiatrica Scandinavia, 51*, 3–18.

12
Disorders of the Sex Hormones: Medical Overview

GILBERT P. AUGUST

During fetal development, anatomical and physiological development of the neuroendocrine system occurs concomitantly. Neuroendocrine maturation continues through infancy, childhood, and puberty. It is important to understand this process and its role in the disorders of puberty and of the sex hormones.

Embryology of the Hypothalamus and Anterior Pituitary Gland

The pituitary gland has two separate embryologic origins. The anterior pituitary arises from an ectodermal outgrowth of the oral cavity termed Rathke's pouch. By 11 weeks of fetal development it meets a downgrowth of the diencephalon portion of the brain which will develop into the posterior lobe of the pituitary gland.

A capillary network surrounds the nerve endings of the hypothalamic region and drains into the anterior pituitary gland. This pituitary portal system transfers hypothalamic hormones to the anterior pituitary gland. Its development begins by $8\frac{1}{2}$ weeks of gestation and is completed by 19 to 21 weeks.

Immunofluorescence techniques permit the identification of adrenocorticotropin hormone (ACTH) secreting cells at 7 weeks of gestation, growth hormone containing cells at 10 weeks, luteinizing hormone (LH) at 10 weeks, prolactin at 12 weeks, and thyroid stimulating hormone (TSH) at 13 weeks.

Embryology of Gonadal Development

The fetal gonads, internal sex ducts, and external genitalia are all at first bipotential and the fetus can follow either a male or female developmental tract.

The key factor(s) involved in testicular development may be testicular-determining factor genes on the short arm of the Y-chromosome, or HY antigen, coded for by genes near the centromere of the Y-chromosome, or the gonad differentiation locus on the short arm of the Y-chromosome, or a combination of these and other structural and regulatory genes on the X-chromosome and autosomes. Under the influence of testicular organizing factors, the primitive gonad begins to undergo testicular differentiation from days 48 to 60. Leydig cells (which secrete testosterone) appear around 60 days.

Concomitant with testicular differentiation, the ducts which in the female would have become the oviducts and uterus (the mullerian ducts) regress between days 56 to 70. This regression is brought about by the local action of a protein hormone, antimullerian hormone, secreted by Sertoli cells in the tubules of the testes. Masculinization of the external genitalia and growth of the spermatic (Wolffian) duct system are dependent upon androgen production. The external genitalia begin to virilize about 65 days and virilization is complete by 75 days. There appears to be a narrow window of opportunity for full virilization because after the twelfth week, when the vagina has separated from the urogenital sinus, androgen stimulation will not result in labioscrotal fusion but only in clitoral enlargement.

In the absence of testicular development, the fetus will express a female phenotype. However, ovarian development generally requires the presence of two functional X chromosomes.

Abnormalities of the External Genitalia

Ambiguous genitalia is commonly defined as the absence of either completely developed male external genitalia or the presence of any evidence of virilization of the female external genitalia. This very stringent definition is adhered to in order to avoid missing a life-threatening medical emergency or of facing the virilization of an adolescent female or feminization of an adolescent male.

Most children with ambiguous genitalia are divided into two broad subdivisions based upon their genotype—female pseudohermaphrodites and male pseudohermaphrodites. True hermaphroditism, the presence of both well-defined ovarian and testicular tissue, is rare.

Other than for chromosomal disorders which are discussed elsewhere, the most common etiology of pseudohermaphroditism is some form of congenital adrenal hyperplasia, an enzyme defect in the pathway of cortisol and aldosterone biosynthesis, or, additionally, in males a defect in testosterone synthesis or action (see Figure 12.1).

Congenital Adrenal Hyperplasia

The most common type is the 21-hydroxylase deficiency. About a third of these patients have a severe enough deficiency to impair aldosterone production as well. The phenotypic appearance of the external genitalia of females can vary from minimal clitoral enlargement to that of a fully

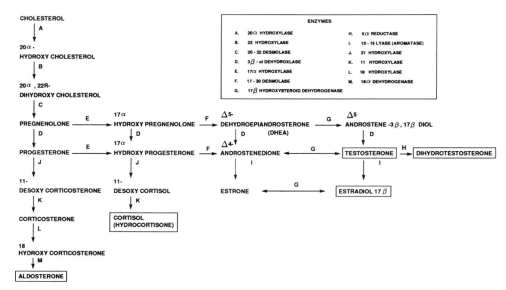

FIGURE 12.1. Enzymes in the etiology of abnormalities of the external genitalia.

virilized cryptorchid "male." Generally, patients with salt-losing adrenal hyperplasia are more virilized. This is a relatively common autosomal recessive trait with a gene frequency of 1/150 in the United States. There is close linkage between the gene locus for 21-hydroxylase deficiency and the HLA-B complex on chromosome 6. The steroids which build up to abnormally high concentrations behind the enzymatic block are shunted into other pathways of biosynthesis leading to the formation of testosterone. It is the abnormally high level of testosterone in the fetal circulation which leads to labio-scrotal fusion during the eighth through twelve week of fetal development and then to clitoral enlargement during the third trimester of pregnancy.

Girls with this condition (boys with a 21-hydroxylase deficiency are phenotypically normal) must undergo a series of surgical procedures to correct the external genitalia. The number and extent of these surgeries vary depending upon the extent of virilization. Surgery can vary from simple clitoral recession, not resection if it can be avoided, to an extensive multistage repair of the external genitalia and of the urogenital ducts and vagina. Although the clitoral recession is carried out in infancy, girls with marked virilization face additional surgery in childhood and vaginal reconstruction in adolescence.

Both males and females with congenital adrenal hyperplasia must undergo medical replacement therapy with hydrocortisone given three times a day in order to prevent virilization, provide for adequate amounts of glucocorticoids, and to maintain normal metabolic homeostasis. These children need to be seen two to four times a year to monitor the dose of hydrocortisone and growth rates in order to prevent undertreatment as well as overtreatment. In the salt-losing form of this disorder, the aldosterone deficiency must also be treated with the mineralocorticoid 9-alpha fluorocortisol. During illness or any sort of medical stress, the dose of hydrocortisone needs to be increased and these children are at increased risk during any illness associated with vomiting.

The other forms of congenital adrenal hyperplasia are less common and are not linked to the HLA-B loci, although they are also autosomal recessive disorders. Some of the rarer forms are also associated with inadequate virilization of the male fetus.

Sex Assignment

It is of utmost importance to assign the sex of rearing of children born with ambiguous genitalia as quickly as possible. We do not recommend naming the child until an unequivocal decision has been made as to the sex of rearing.

The decision is made jointly with the pediatric endocrinologist, geneticist, surgeon, and/or urologist and the parents. Although we are influenced by the genetic sex of the child and potential fertility, the major consideration is the functional anatomy of the patient as an adult. Accordingly, it is unusual for girls with congenital adrenal hyperplasia not to be raised as females. They are fertile, the internal organs are normal, and the external genitalia are surgically correctable (albeit some require extensive and repeated operations). The most difficult decisions we face are in males with ambiguous genitalia. We must determine the etiology of the problem if possible in order to draw upon past experience with similar patients. And we must be assured that the penis (or phallus) will virilize and grow during puberty; to establish this, the patient must often be given a series of testosterone injections over a period of 2 to 3 months. Only if we are reasonably assured that the penis will normally respond to androgen stimulation and that the external genitalia are surgically correctable should we recommend male sex assignment. Long-term results in these patients are not always as good as anticipated in infancy.

Normal Pubertal Development

During the first half of gestation the fetal serum concentrations of LH (luteinizing hormone) and FSH (follicle-stimulating hormone) are high, comparable to adult castrate concentrations but they decline toward term, presumably due to maturation of feedback inhibition by gonadal steroids on the fetal hypothalamus. Shortly after birth the serum concentrations of both LH and FSH rise again in both males and females. The rise in pituitary gonadotropin is followed by a rise in gonadal steroids which is transient in the female. In males, the rise of testosterone lasts from 4 to 9 months and the serum concentration of testosterone is consistent with early to midpubertal development in the male adolescent.

It is not until 2 to 4 years of age, earlier in the male than the female, that the hypothalamus is exquisitely sensitive to feedback inhibition by the very small concentration of circulating gonadal steroids in the typically prepubertal child. As long as the hypothalamus remains inhibited by the small quantities of circulating gonadal steroids, the pituitary gland does not receive a sufficient number of pulses of LH-releasing factor (LRF) from the hypothalamus to release LH and FSH. What triggers puberty is a gradual loss of the sensitivity of the hypothalamus to feedback inhibition by gonadal steroids. It is not known why the "gonadostat" in the hypothalamus loses its sensitivity but the first sign of decreased sensitivity is an increasing number of nocturnal pulses of hypothalamic LRF which stimulates the pituitary release of nocturnal pulses of LH and FSH. As puberty progresses, daytime pulses of LRF become more frequent. In the male, FSH stimulation of the testes is required for stimulation of spermatogenesis and LH stimulation of the testicular Leydig cells is reponsible for testosterone production. Both FSH and testosterone are eventually required for fully developed spermatogenesis. In the male, gonadal steroids are primarily responsible for feedback inhibition of LH while feedback inhibition of FSH is accomplished primarily by a protein hormone— inhibin, a product of the Sertoli cells of the seminiferous tubules. In females inhibin is produced by the ovarian follicles and is responsible for inhibition of FSH release. In females, a certain degree of FSH stimulation of the ovarian follicle is required before LH receptors appear and ovarian steroidogenesis can proceed.

The first visible sign of pubertal development in girls is usually breast development, although pubic hair can appear first in as many as 25% of girls. Fully 98% of North American girls start breast development between 8 and 13 years of age. It generally takes 2 to 2.5 years from the start of breast development to menarche. In boys, pubic hair is considered the first visible sign of puberty, although testicular enlargement can be noted about six months earlier. In 98% of boys, pubic hair starts between the ages of 9 and 14 years of age.

Delayed Puberty

The most common cause of delayed puberty (lack of breast development in girls by age 13 years or lack of pubic hair and testicular enlargement in boys by 14 years) is constitutional delay. Constitutional delay of puberty is often familial and we can elicit a history of late menarche in the mother or late development in the father. Children with constitutional delay do not require any medical treatment but do need to understand that they are normal and will develop and grow to become normal men and women, but will take longer to finish their growth and development—somewhat akin to the story of the tortoise and the hare. Many children with constitutional delay of puberty have emotional problems which may be an indication for the use of sex steroid therapy to induce breast development in girls and pubic hair in boys. This will fulfill their need to look more like their peers. Therapy with testosterone can be short term, just six months, to induce pubic hair and then treatment can be discontinued. Only a low dose of testosterone enanthate, 50 mg of intramuscularly monthly, is needed. In girls it is important not to use too high a dose of estrogen to induce breast development. Excessive amounts of estrogen will inhibit linear growth. Small doses are adequate for full sexual development and a normal growth spurt in pubertal girls. Currently, we use just 4 to 10 μg of ethinyl estradiol daily, starting with the lower dose.

The diagnosis of constitutional delay of puberty is one of exclusion. Organic causes must be excluded. Hypergonadotropic hypogonadism is due to a defect in the gonads. Chromosomal causes are discussed elsewhere in this volume. Agonadism can occur in nonchromosomal disorders. In the vanishing testis syndrome the testes are present during the critical period required for masculinization of the male genitalia but are lost due to torsion and vascular occlusion during testicular descent at term. In girls, functional agonadal states can be due to an autoimmune oophoritis, resistant ovary syndrome (inability to respond to LH and FSH with the presence of primordial follicles) and pure gonadal dysgenesis, a 46,XX syndrome that may be autosomal recessive. In all of these hypogonadal conditions, full replacement with sex steroids is necessary and must be continued in adult life as well.

Hypogonadotropic hypogonadism is a difficult diagnostic problem because constitutional delay of puberty is a natural hypogonadotropic state. If, through the use of dynamic testing of the hypothalamic-pituitary system, we can demonstrate an appropriate LH rise in response to LRF, then we can anticipate spontaneous pubertal development later. On the other hand, a lack of an LH response may be indicative of the normal prepubertal condition. Only time can differentiate these conditions and if puberty is sufficiently delayed, sex steroid therapy is indicated. But it is important under these circumstances to look for organic causes of hypopituitarism and X-rays of the sella turcica are needed to look for a craniopharyngioma. In addition, serum prolactin levels are indicated to look for hyperprolactinemia which can be caused by either a macroadenoma or microadenoma of the pituitary gland. This condition is treatable either medically with bromocriptine or surgically. Kallmann's syndrome is the association of hypogonadotropic hypogonadism with a diminished or absent sense of smell, congenital deafness, and midline facial defects. It is inherited as an autosomal dominant trait with incomplete expressivity.

Precocious Puberty

Incomplete Precocious Puberty

Premature adrenarche is the appearance of pubic and/or axillary hair in boys under nine and girls under eight years of age. It is caused by a premature rise in adrenal androgens without maturation of the hypothalamic gonadostat and is distinct from precocious puberty. It is a diagnosis of exclusion and we must rule out virilizing conditions such as late-appearing congenital adrenal hyperplasia and virilizing adrenal or ovarian tumors. The latter two conditions are rare but late-appearing congenital adrenal hyperplasia may be more common than previously appreciated.

Premature thelarche is the appearance of breast development prior to eight years of age in girls. Prepubertal gynecomastia is uncommon in boys. In premature thelarche, the breasts develop to a certain point over a few months but do not develop further and pubic hair does not appear. The breasts regress in size over a variable time course, sometimes taking two years. Premature thelarche is probably due to a transient ovarian cyst. If the serum estradiol, LH, and FSH levels are low, then observation is still needed over several months to be certain that there is no progression of breast development. If progression occurs then an LRF test is needed to confirm the diagnosis of precocious puberty. High serum estradiol coupled with low LH and FSH concentrations is highly suggestive of an autonomous source of estradiol and a pelvic sonogram is needed to look for an ovarian cyst. Estrogen-producing adrenal or ovarian tumors are rare and would cause more rapid progression of sexual development but can be seen on ultrasound examination.

Complete Precocious Puberty

Pituitary Gonadotropin-Dependent Precocious Puberty

True precocious puberty is defined as the early activation of the hypothalamic-pituitary-gonadal axis. It is associated with normal testicular enlargement in boys and with ovarian enlargement and follicular development as seen by ultrasound in girls. About 70% of boys and 95% of girls with this condition do not have an organic cause and have idiopathic true precocious puberty. The others may have some central nervous system abnormality such as tumor, cyst, hydrocephalus, or hamartoma. Therefore, children with precocious puberty should have either a CT scan or an MRI scan of the brain. There are two problems caused by precocious puberty. First is advancement of bone age and cessation of linear growth at an earlier age but with an adult height less than expected from the genetic potential of the child. Another problem is that the child is fully developed from a sexual viewpoint but not emotionally mature enough to handle it. The younger the child with precocious puberty, the more likely that treatment is indicated to slow or abort puberty. Current treatment in the United States involves the use of medroxyprogesterone acetate or the experimental use of an LRF-agonist which acts by down-regulating the LRF receptors on the pituitary gonadotrophs. The latter agent appears to be more effective and does not have the undesirable side effects of medroxyprogesterone acetate such as weight gain.

Pituitary Gonadotropin-Independent Precocious Puberty

This group comprises a wide variety of disorders. Some, such as chorionic gonadotropin-secreting tumors, can mimic true precocious puberty in boys. They are diagnosed by the very high levels of serum LH in clinical LH immunoassays which also measure chorionic gonadotropin. Because prior stimulation by FSH is required for the appearance of LH receptors on ovarian follicular cells and estradiol production, chorionic gonadotropin-secreting tumors are not associated with precocious puberty in girls.

McCune-Albright syndrome is associated with characteristic cafe-au-lait spots and skeletal lesions. Precocious puberty may precede the other findings. Conversely, precocious puberty need not occur in this disorder. Most likely, precocious puberty in McCune-Albright syndrome is due to autonomous ovarian cysts producing estradiol. Other such autonomous endocrine gland hyperfunctioning occurs in this syndrome as well—Cushing's disease, growth hormone excess, and hyperthyroidism. Although medroxyprogesterone acetate and LRF-agonists have been used in this condition, if they are not successful one can try to use either testolactone, an aromatase inhibitor, or ketoconazole, a 17-20 lyase inhibitor, as experimental agents reported to be successful in controlling pubertal advance in this condition. Testotoxicosis is a familial disorder in which the Leydig cells produce testosterone independent of LH stimulation. Ketoconazole has been reported to work in this syndrome to suppress serum testosterone levels.

Tumors can produce either isosexual precocious puberty when they secrete hormones appropriate to the sex of the patient or heterosexual precocious puberty when they secrete hormones inappropriate to the patient's sex. An example of the latter condition is a virilizing tumor in a girl. An example of the former is a granulosa cell tumor of the ovary in a female. Isosexual precocious puberty caused by granulosa cell tumors or by ovarian cysts are also associated with pubic hair because the ovaries secrete androgens as well as estrogens. These conditions cause pseudoprecocious puberty because they do not function through the hypothalamic-pituitary-axis. However, all conditions causing pseudoprecocious puberty, whether isosexual or heterosexual, can eventually trigger true precocious puberty as the bone age advances. This circumstance is relatively common in inadequately treated congenital adrenal hyperplasia.

References

Boepple, P.A., Mansfield, M.J., Wierman, M.E., Rudlin, C.R., Bode, H.H., Crigler, Jr., J.F., Crawford, J.D., & Crowley, Jr., W.F. (1986). Use of a potent, long acting agonist of gonadotropin-releasing hormone in the treatment of precocious puberty. *Endocrine Reviews, 7,* 24–33.

German, J. (1987). Gonadal dimorphism explained as a dosage effect of a locus on the sex chromosomes; The gonad differentiation locus (GDF). *American Journal of Human Genetics, 42,* 414–421.

Griffin, J.E., & Wilson, J.D. (1985). Disorders of the testes and male reproductive tract. In J.D. Wilson & D.W. Foster (Ed.), *Textbook of Endocrinology* (pp. 259–311). Philadelphia, PA: W.B. Saunders Co.

Grumbach, M.M., & Conte, F. A. (1985). Disorders of sexual differentiation. In J.D. Wilson & D.W. Foster (Ed.), *Textbook of Endocrinology* (pp. 312–401). Philadelphia, PA: W.B. Saunders Co.

Josso, N. (1986). AntiMullerian hormone: New perspectives for a sexist molecule. *Endocrine Reviews, 7,* 421–433.

McLachlan, R.I., Robertson, D.M., De Kretser, D.M., & Burger, H.G. (1988). Advances in the physiology of inhibin and inhibin-related peptides. *Clinical Endocrinology, 29,* 77–114.

Mulchahey, J.J., DiBlasio, A.M., Martin, M.C., Blumenfeld, Z., & Jaffe, R.B. (1987). Hormone production and peptide regulation of the human fetal pituitary gland. *Endocrine Reviews, 8,* 406–425.

New, M.I., & Speiser, P.W. (1986). Genetics of adrenal steroid 21-hydroxylase deficiency. *Endocrine Reviews, 7,* 331–349.

Ohno, S. (1985). The Y-linked testis determining gene and H-Y plasma membrane antigen gene: Are they one and the same? *Endocrine Reviews, 6,* 421–431.

Ojeda, S.R., Andrews, W.W., Advis, J.P., & White, S.S. (1980). Recent advances in the endocrinology of puberty. *Endocrine Reviews, 1,* 228–257.

Pescovitz, O.H., Hench, K.D., Barnes, K.M., Loriaux, D.L., & Cutler, Jr., G.B. (1988). Premature thelarche and central precocious puberty: The relationship between clinical presentation and the gonadotropin response to luteinizing hormone-releasing hormone. *Journal of Clinical Endocrinology and Metabolism, 67,* 474–479.

Rosenfield, R.L. (1986). Low-dose testosterone effect on somatic growth. *Pediatrics, 77,* 853–857.

Van Dop, C., Burstein, S., Conte, F.A., & Grumbach, M.M. (1987). Isolated gonadotropin deficiency in boys: Clinical characteristics and growth. *Journal of Pediatrics, 111,* 684–692.

13
Time of Puberty Onset and Intellectual and Neuropsychological Functioning

Jennifer A. Karlsson

During puberty, secondary sexual characteristics emerge and reproductive potential is attained as a result of hormonal changes dictated by the hypothalamic-pituitary-gonadal axis (Finklestein, 1980). Although pubertal development typically begins during early adolescence and certain variability in pubertal onset is considered within normal limits, puberty may also be significantly accelerated (i.e, precocious) or significantly delayed.

Interest in the psychological implications of the timing of puberty is longstanding and has yielded a large, diverse body of research. Initially, research focused on the psychosexual development of children with clinically accelerated (e.g., Hampson & Money, 1955; Money & Hampson, 1955) or delayed (e.g., Ehrhardt & Meyer-Bahlburg, 1975) puberty. Later, various investigators studied the psychosocial status of children with non-clinical variations in rate of physical maturation (e.g., Gross & Duke, 1980; Jones, 1957; Jones & Bayley, 1950; Jones & Mussen, 1958; Mussen & Jones, 1957; Peskin, 1973; Weatherly, 1964). More recently, intriguing questions have arisen as to whether individuals with age-discrepant physical development also experience significant accelerations or lags in their cognitive development.

This chapter will focus on a review of the research emerging from this inquiry. First, research on the intellectual and neuropsychological abilities of individuals with either clinically precocious puberty or nonclinical variations in pubertal onset will be reviewed. Then, mechanisms theorized as responsible for these findings will be discussed, methodological issues considered, and areas for future research outlined.

Precocious Puberty

Intellectual Abilities

John Money's work represents some of the earliest, systematic investigations of the intellectual abilities of children with idiopathic precocious puberty (IPP). In an early study (Money & Meredith, 1967), children with

194   J.A. Karlsson

IPP were found to possess average intellectual abilities overall (IQ $\overline{X}$ = 108.4, SD = 22.7). However, when Wechsler verbal IQ (VIQ) and nonverbal IQ (PIQ) scores were examined separately, a number of findings emerged. First, the average score on verbal subtests was higher than the average score on performance subtests. Second, the VIQ-PIQ discrepancy was most striking in individuals with full scale IQs (FSIQ) of 110 or higher. When males' and females' IQ scores were examined separately, males with IPP appeared to have higher VIQs and PIQs than females, as well as larger VIQ-PIQ discrepancies. Serial IQ testing with three individuals with IPP revealed they continued to display a PIQ < VIQ profile over a 12-year period. Although duration of IPP did not appear to affect intellectual ability, onset of puberty prior to 5 years of age did seem to be associated with lower IQ.

A second study conducted by Money (Money & Walker, 1971) obtained similar results: girls with IPP had high average overall intelligence (FSIQ $\overline{X}$ = 113.2, SD = 12.5), and a VIQ-PIQ discrepancy favoring stronger verbal skills emerged (VIQ $\overline{X}$ = 116, SD = 15.7; PIQ $\overline{X}$ = 107.7, SD = 8.7).

Meyer-Bahlburg matched adolescent girls with IPP with normal controls on the basis of SES and menarcheal status. Girls with IPP appeared to have average overall intelligence (Wechsler FSIQ $\overline{X}$ = 104.6, SD = 16.4). Comparison of IPP and control subjects' IQ scores did not yield significant group differences. Unlike Money and his colleagues' findings, however, Meyer-Bahlburg did not find a significant VIQ-PIQ discrepancy (Meyer-Bahlburg et al., 1985).

Rovet's research also utilized controls (matched on the basis of age and IQ), and included boys as well as girls. Similar to previously cited studies, youngsters with IPP appeared to have average intelligence overall (boys FSIQ $\overline{X}$ = 100.9, SD = 12.9; girls FSIQ $\overline{X}$ = 109.2, SD = 12.8). In contrast to Money's findings (Money & Meredith, 1967; Money & Walker, 1971), girls with IPP did not show a pattern of VIQ > PIQ. Boys with IPP exhibited a larger VIQ-PIQ discrepancy than did girls with IPP, but this discrepancy favored their nonverbal abilities. Furthermore, males with IPP earned significantly lower VIQs than did their matched controls ($\overline{X}$ VIQ 97 vs 103).

Neuropsychological Findings

A number of researchers have conducted more extensive evaluations of the cognitive abilities of individuals with precocious puberty. Rovet (1983), for example, utilized tasks thought to tap verbal (Sentence Verification, Wechsler Vocabulary) and spatial (Mental Rotation) processing as well as an auditory task thought to localize cerbral organization for language (dichotic listening). Girls with IPP earned significantly lower Vocabulary scores but scored significantly higher on the Mental Rotation task than did controls. There was a trend for boys with IPP to do less well on both Sentence Verification and Mental Rotation tasks than controls. Similarly, a trend was obtained on the dichotic listening task, suggesting that male controls and girls with IPP had stronger lateral asymmetries than did female controls or boys with IPP.

Meyer-Bahlburg administered the Wechsler intelligence test, the Reasoning and Spatial Relations subtests of Thurstone's Primary Mental Abilities (PMA), and tests of hemispheric lateralization (monotic word discrimination, dichotic consonant-vowel discrimination, the Staggered Sporadic Word test) to adolescent girls with IPP and normal controls. The two groups' performance on Wechsler verbal subtests and the PMA Reasoning and Spatial Relations subtests did not vary significantly. However, girls with IPP scored significantly lower on Wechsler Block Design and Object Assembly subtests than did controls. Computation of a Cohen difference factor suggested girls with IPP performed less ably on spatial than on verbal tasks. Lateralization task performance was comparable between the two groups, with the exception of the Staggered Sporadic Word test: here, girls with IPP showed a larger disparity between competing and noncompeting modes than did control subjects, a finding that was more pronounced for the left than the right ear (Meyer-Bahlburg et al., 1985).

In a subsequent investigation (Bruder et al., 1987), adolescent females with IPP and matched controls competed dichotic listening tasks in which consonant-vowel pairs were presented at varying interaural delays. Similar to the Meyer-Bahlburg et al. (1985) findings, no significant group differences in dichotic listening was obtained: both groups had a right ear advantage, had similar accuracy, and improved performance as interaural delays increased.

Summary

Overall, individuals with IPP appear to have average (Meyer-Bahlburg et al., 1985; Money & Meredith, 1967; Rovet, 1983) to high average (Money & Walker, 1971) intelligence. Variable VIQ-PIQ patterns have been obtained, with some investigators finding a pattern of VIQ > PIQ (e.g., Money & Meredith, 1967; Money & Walker, 1971), and others finding roughly equivalent PIQ and VIQ scores (Meyer-Bahlburg et al., 1985; Rovet, 1983, females only). Results of neuropsychological tests have also yielded somewhat variable results. For example, results of lateralization tasks have suggested the lateral asymmetries of girls with IPP are stronger (e.g., Rovet, 1983) or equivalent (e.g., Bruder et al., 1987; Meyer-Bahlburg et al., 1985) to various comparison groups. A fairly consistent finding, however, is that females with IPP have stronger verbal than spatial skills (Meyer-Bahlburg et al., 1985). (See Table 13.1).

TABLE 13.1. Summary of precocious puberty studies.

Author	Main findings
Money & Meredith, 1867	Average IQ; VIQ > PIQ, most striking if male or IQ > 110; males' VIQ & PIQ > females' VIQ & PIQ; IQ lower if puberty before 5 years of age
Money & Walker, 1971	High average IQ; VIQ > PIQ
Meyer-Bahlburg et al., 1985	Average IQ; VIQ ~ PIQ; girls WISC Block Design and Object Assembly ↓, spatial < verbal skills, more lateralized
Bruder et al., 1987	Lateralization N.S. differences
Rovet, 1983	Average I.Q.; VIQ ~ PIQ; boys: VIQ < PIQ; girls: ↓ Vocabulary, ↑ Mental Rotation; trends on other tasks

Nonclinical Variations in Pubertal Onset

Intellectual Abilities

The intellectual abilities of individuals whose pubertal development is somewhat (rather than clinically) accelerated or delayed compared to agemates and/or standardized norms have been examined in a number of investigations. In an early study, Waber (1977) utilized norms for the Tanner staging criteria for sexual development to categorize youngsters as early or late maturers. Although the IQs of older (eighth grade) girls differing in degree of sexual maturation did not differ significantly, early-maturing younger (fifth grade) girls earned significantly lower IQ scores than did their later-maturing agemates (early $\overline{X}$ IQ = 109.7; late $\overline{X}$ IQ = 120). The IQs of early and late maturing boys did not differ significantly, however.

Duke selected adolescent boys who had more (early) or fewer (late) secondary sexual characteristics than agemates. Twelve-year-old boys varying in rate of sexual maturation did not appear to vary significantly in Wechsler IQ test scores. However, fourteen-year-old boys with somewhat delayed pubertal onset had slightly lower IQ test scores ($\overline{X}$ = 99) than did boys whose pubertal onset was average ($\overline{X}$ = 102) or early ($\overline{X}$ = 105). Differences in IQ test scores as a function of maturation rate appeared to be even greater in the sixteen year olds examined. Teacher evaluations of subjects' intellectual abilities and academic achievement parallelled these IQ score patterns (Duke et al., 1982).

Neuropsychological Functioning

The neuropsychological abilities of individuals with nonclinical variations in pubertal onset have been the focus of numerous investigations—most notably those of Deborah Waber and her colleagues.

In an early study, Waber (1976) hypothesized that the sex differences in cognitive abilities obtained in numerous investigations (e.g., Maccoby & Jacklin, 1974) might be less a function of gender and more a function of maturation rate, since males and females typically reach puberty at different ages. Boys and girls with more (early maturers) or fewer (late maturers) secondary sexual characteristics were selected and tests thought to measure verbal abilities (WISC Digit Symbol, Stroop Color Naming, PMA Word Fluency), spatial abilities (WISC Block Design, Embedded Figures, PMA Spatial Abilities), and lateralization (dichotic listening involving phoneme identification) were administered. Regardless of gender, early maturers scored better on verbal than on spatial tasks, while the reverse was true for later maturers (i.e., spatial > verbal). On dichotic listening, older (i.e., 13-year-old females, 16-year-old males) later maturers demonstrated larger ear advantages than older early maturers. Younger subjects failed to exhibit significant dichotic listening performance discrepancies, however.

In Waber's (1977) subsequent investigation, younger (i.e., fifth-grade girls, eighth-grade boys) and older (eighth-grade girls, tenth-grade boys) adolescents were classified as late or early maturers on the basis of Tanner staging criteria, and tests of verbal (WISC Digit symbol, Stroop Color Naming, PMA Word Fluency), spatial (WISC Block Design, PMA Spatial Abilities, Embedded Figures), and dichotic listening abilities administered. Late maturers performed significantly better than early maturers on tasks involving spatial abilities. Analyses for verbal tasks did not yield a significant maturation effect initially. However, when intraindividual verbal-spatial difference scores were analyzed, a significant maturation effect emerged, with early maturers scoring better on verbal than spatial tasks and late maturers scoring better on spatial than verbal tasks. The earliest-maturing group (early-maturing girls) demonstrated the largest verbal-spatial discrepancy (V > S), while the latest-maturing group (late-maturing boys) exhibited the largest spatial-verbal discrepancy (S > V). Analyses of dichotic listening performance suggested older early maturers were significantly less lateralized than younger early maturers, while older late maturers were significantly more lateralized than younger late maturers.

A third study (Waber et al., 1981) utilized a slightly different test battery (Stroop Word and Color Naming, Rhythmic finger Tapping, Locomotor Mazes) administered to adolescent boys from varying socioeconomic backgrounds. Results indicated that while maturational status and neuropsychological test performance were associated, this relationship varied as a function of socioeconomic class: In upper-middle-class boys, physical maturity was negatively associated with performance on Locomotor Mazes, while physical maturity was positively correlated with stability of Rhythmic Finger Tapping in boys from lower-middle-class backgrounds. No association between physical maturity and neuropsychological test performance was detected in boys from blue collar backgrounds, however.

A fourth investigation (Waber et al., 1985) questioned whether the relationship between cognitive performance and maturation rate could be detected prior to puberty. An unselected group of prepubertal third-grade girls and fifth-grade boys from two communities completed verbal (WISC Digit Symbol, Stroop Color Naming, PMA Word Fluency), spatial (WISC Block Design, Embedded Figures, PMA Spatial Abilities), and lateralization (dichotic listening) tasks. Two years later, this test battery was expanded and readministered to subjects, and physical examinations were conducted in order to identify early and late maturers. Only when data were grouped by hometown did the predicted maturation effects emerge from the original test battery data. Maturation-related differences were found, however, in some of the supplementary test data: Late maturers produced better organized Rey-Osterrieth drawings, their accuracy on the Mental Rotation task increased when cues were provided, and they exhibited a recency effect on a selective attention task; early maturers were more accurate on the dichotic listening task.

Other investigators have failed to detect the hypothesized relationship between maturational rate and cognitive abilities. Strauss and Kinsbourne (1981), for example, failed to detect a significant relationship between age at menarche and adult subjects' performance on either a verbal fluency test or a modified version of Piaget's water level task. Rierdan and Koff (1984) also failed to obtain findings consistent with Waber's in a study using subjects and measures similar to that of Strauss and Kinsbourne's (1981).

Summary

Examination of IQ data from individuals with nonclinically delayed or accelerated puberty yields results that appear to vary by gender: Waber (1977) found early-maturing fifth-grade girls to have lower IQs than later-maturing fifth-grade girls, while Duke et al. (1982) found late-maturing boys to have lower IQs than their earlier-maturing peers. Early neuropsychological investigations yielded group (e.g., Waber, 1976) or intraindividual (Waber, 1977) differences, suggesting that early maturers' verbal skills were stronger than their spatial skills, and late maturers' spatial skills were stronger than their verbal skills. Performance on dichotic listening tasks appeared to vary as a function of both age and maturational status: older late-maturing individuals appeared to be more lateralized than younger late maturers, while older early maturers were less lateralized than were younger early-maturing individuals (Waber, 1976, 1977). In subsequent investigations maturation effects on cognitive functioning have varied as a function of social class (Waber et al., 1981), by town (Waber et al., 1985), and by subject selection criteria (Diamond et al., 1983). Studies conducted with adult subjects (e.g., Rierdan & Koff, 1984; Strauss & Kinsbourne, 1981) have failed to detect a relationship between cognitive skills and timing of pubertal onset, however. (See Table 13.2).

TABLE 13.2. Summary of nonclinical variation in pubertal onset findings.

Author	Main findings
Waber, 1976	Early maturers V > S; later maturers S > V; older later maturers more lateralized; younger later maturers N.S.
Waber, 1977	Younger early-maturing girls lower I.Q. than younger later-maturing girls; late maturers spatial > early maturers; V-S scores: early V > S, later S > V; lateralization: older earlys less lateralized than younger earlys, older lates more lateralized than younger lates
Waber et al., 1981	SES effects
Waber et al., 1985	Town effects; late: ↑ spatial, recency on selective attention; early: more accurate on dichotic
Duke et al., 1982	Trend for later-maturing boys' IQ < average or early maturers
Diamond et al., 1983	Face encoding plateau at puberty; Embedded Figures: equivocal findings
Strauss & Kinsbourne, 1981	Adults: N.S.
Rierdan & Koff, 1984	Adults: N.S.

Possible Mechanisms

Major theories regarding possible mechanisms underlying pubertal effects on cognitive functioning fall into three major categories: socialization, cerebral organization, and neurochemical.

Socialization theories maintain that a child's appearance encourages or discourages certain social and intellectual opportunities, and this differential experience is responsible for the cognitive patterns obtained. For example, the older appearance of youngsters with IPP may result in their "skipping" grades for social reasons, providing exposure to more sophisticated verbal concepts and abstract thinking, as well as shifting them to a more socially advanced peer group (Ehrhardt & Meyer-Bahlburg, 1975; Money & Meredith, 1967; Rovet, 1983). Although this theory is commonly cited and intuitively appealing, thus far research has provided little support for the theory that timing of puberty affects cognitive ability via effects on personality and interests (Newcombe & Bandura, 1983).

A second possible mechanism—cerebral organization—received early support from Money and Meredith (1967), who argued that the factors responsible for the etiology of accelerated somatic maturation may also be responsible for differential organization of the central nervous system. Meyer-Bahlburg et al. (1985) theorized that a slight inferiority in relative hemispheric function may underlie the cognitive effects obtained. Waber (1976) has also speculated that slower maturation may allow for the stronger interhemispheric differentiation thought necessary for the development of good spatial skills.

The third major theory—neurochemical—argues that hormones affect central nervous system functioning directly and/or indirectly via their

effects on neurotransmitters. For example, Gordon and Lee's (1986) finding that adult men with high levels of follicle stimulating hormone (FSH)
did poorly on spatial tasks was supportive of their argument that cognitive
abilities may be affected by exposure to specific hormones rather than
maturation per se. This study may also speak to the debate regarding
timing versus rate of pubertal maturation and the inconsistency in results
obtained with prepubertal or pubertal samples (i.e., Waber's work) versus
that obtained with postpubertal samples (i.e., Rierdan & Koff, 1984,
Strauss & Kinsbourne, 1981). As Gordon has recently pointed out, neither
time nor rate of pubertal maturation predicts levels of hormones such as
FSH in postpubertal years, thus explaining why cognitive functioning may
not be as clearly related to onset of puberty in postpubertal subjects
(Gordon, Lee, & Tamres, 1988). Hier and Crowley's (1982) research
supports and expands on the neurochemical view: adult men with androgen deficiency (idiopathic hypogonadotrophic hypogonadism) demonstrated below-average spatial abilities which were not corrected via hormone treatments. Of note was the finding that men who acquired this
disorder postpubertally did not exhibit spatial defects—suggesting that
hormones may have permanent organizing effects on the structure of the
central nervous system. In contrast, Gordon has emphasized neurochemical over neuroanatomical mechanisms, based on diverse research
suggesting the link between hormones and cognitive functioning may be
more fluid in nature (Gordon et al., 1988). Although intriguing, acceptance
of the neurochemical mechanisms must be tempered by the limitations of
this research, as the hormonal assays employed often are incomplete or
confounded by recent medical treatments.

Methodological Issues

Research conducted on the effects of pubertal timing on cognitive function
is vulnerable to a number of methodological difficulties—some unique to
this type of research, and others commonly encountered in other types of
investigations.

Perhaps the most commonly cited methodological issue is the potential
confounding of maturation rate (i.e., age of onset of puberty) with maturational status (i.e., pubertal development when evaluated). For example,
when relatively young subjects with IPP are compared with control subjects of comparable age (e.g., Rovet, 1983), subjects and controls differ in
terms of both maturational rate and status. Conclusions that may be drawn
from such a design are, therefore, limited.

A second major methodological issue involves the selection of subjects.
A number of studies utilize an extreme groups design in which subjects are
chosen to be maximally disparate in terms of their pubertal status (e.g.,

Bruder et al., 1987; Waber, 1976). As Diamond et al. (1983) noted in their series of studies, significant maturation effects appear to emerge more reliably when an extreme groups design is utilized and may be difficult to detect when criteria for subject selection are not as stringent. A number of investigators have questioned the meaningfulness of effects that emerge primarily when an extreme groups design is used.

Results of different investigations are difficult to compare and interpret due to a number of other subject selection factors. Various studies utilize subjects with either varying or unknown reasons for clinically delayed or advanced puberty. How puberty is defined—via retrospective report (e.g., Rierdan & Koff, 1984; Strauss & Kinsbourne, 1981), indirect measures (e.g., Diamond et al., 1983), or physical examinations (e.g., Waber, 1976) —varies widely between studies and may have a significant impact on the interpretation of results. The age of subjects also varies a great deal between studies, with some using adults rather than children or adolescents (e.g., Rierdan & Koff, 1984; Strauss & Kinsbourne, 1981). Age range may also vary considerably within a study (e.g., Rovet, 1983), potentially complicating interpretation of results. Gender effects, which have fairly well-documented effects on certain cognitive functions (e.g., Maccoby & Jackson, 1974; Netley & Rovet, 1983), cannot be investigated when studies use single gender samples (e.g., Waber et al., 1982). Small sample sizes are common in this type of research, which limits the statistical power available to detect significant group differences. Finally, a myriad of seemingly innocuous variables, such as left- versus right-handedness (e.g., Meyer-Bahlburg et al., 1981), geographical location (e.g., Waber et al., 1985), and ethnic group (Meyer-Bahlburg et al., 1985) may have a significant, often confounding, effect on the interpretation of results.

A third major methodological issue concerns the measures utilized in studies of the effects of timing of puberty on cognitive functioning. Measures vary greatly between studies, with a number of studies utilizing as few as two measures when making conclusions about complex verbal and spatial processing capabilities (e.g., Rierdan & Koff, 1984; Strauss & Kinsbourne, 1981). Although dichotic listening tasks are commonly used in this research, these measures are fairly controversial; broad conclusions regarding cerebral organization that are based on a single measure of lateralization are naturally suspect. Similarly, broad statements regarding hormonal effects are questionable when such conclusions are based on single measures (Rubin et al., 1981). Some cognitive measures appear to have been selected on the basis of their usefulness in earlier sex-differences research (e.g., Waber, 1976), while the selection criteria for other tasks is less apparent (e.g., Rierdan & Koff's (1984) use of WAIS Digit Symbol as a measure of verbal fluency). Even when seemingly adequate measures of various cognitive functions are selected, sample characteristics, such as a wide range of ages, may compromise interpretation of findings.

Future Research

Reflection on the numerous, often conflicting empirical findings, thorny methodological issues and somewhat weak theorization that has emerged from research on the effects of pubertal timing on cognitive abilities is apt to leave one somewhat disconcerted. Admittedly, this is formidable research: both clinical and nonclinical subjects are difficult to access; it is perplexing to identify and control numerous intervening and/or confounding variables; complex psychological and physiological processes are difficult to thoroughly assess; and results are rarely clear-cut.

Future research may be able to address a number of these weaknesses. Longitudinal evaluations beginning when children are infants or toddlers have been advocated (Newcombe & Bandura, 1983) as a way to determine when maturation effects on cognitive functioning first emerge. Multicenter collaborative investigations would serve a number of purposes: adequate sample sizes of unusual conditions would be more readily accumulated, and protocols incorporating a number of lines of research would yield more comprehensive appraisals. In addition to using multiple measures of cognitive abilities, tasks need to be chosen that are both sensitive to variations in brain function and related to more complex problem-solving skills. Assessments also need to be linked more closely to putative mechanisms, such as including observational methods to assess social behavior (Newcombe & Bandura, 1983), incorporating more sophisticated hormonal and neurotransmitter assays (Hines & Gorski, 1985), and applying more advanced technology (e.g., PET scanning) to assess brain function. Finally, noncognitive factors that may moderate the relationship between maturation and cognitive functioning warrant further exploration.

References

Bruder, G.E., Meyer-Bahlburg, H.F.L., Squire, J.M., Ehrhardt, A.A., & Bell, J.J. (1987). Dichotic listening following idiopathic precocious puberty: Speech processing capacity and temporal efficiency. *Brain and Language, 31,* 267–275.

Diamond, R., Carey, S., & Back, K.J. (1983). Genetic influences on the development of spatial skills during adolescence. *Cognition, 13,* 167–185.

Duke, P.M., Carlsmith, J.M., Jennings, D., Martin, J.A., Dornbusch, S.M., Gross, R.T., & Siegel-Gorelick, B. (1982). Educational correlates of early and late sexual maturation in adolescence. *The Journal of Pediatrics, 100,* 633–637.

Ehrhardt, A.A., & Meyer-Bahlburg, H.F.L. (1975). Psychological correlates of abnormal pubertal development. *Clinics in Endocrinology and Metabolism, 4,* 207–222.

Ehrhardt, A.A., Meyer-Bahlburg, H.F.L., Bell, J.J., Cohen, S.F., Healey, J.M., Steil, R., Feldman, J.F., Morishima, A., & New, M.I. (1984). Idiopathic precocious puberty in girls: Psychiatric follow-up in adolescence. *Journal of the American Academy of Child Psychiatry, 23,* 23–33.

Ferguson, L.R., & Maccoby, E.E. (1966). Interpersonal correlates of differential abilities. *Child Development, 37,* 549–571.

Finklestein, J.W. (1980). The endocrinology of adolescence. In I. F. Litt (Ed.), *The Pediatric Clinics of North America* (pp. 53–69). Philadelphia: W.B. Saunders Co.

Gordon, H.W., & Lee, P.A. (1986). A relationship between gonadotropins and visuospatial function. *Neuropsychologia, 24,* 563–576.

Gordon, H.W., Lee, P.A., & Tamres, L.K. (1988) The pituitary axis: behavioral correlates. In R.E. Tarter, D.H. Van Thiel, & K.L. Edwards (Eds.), *Medical Neuropsychology: The Impact of Disease on Behavior* (pp. 159–196), New York: Plenum Press.

Gross, R.T., & Duke, P.M. (1980). The effect of early vs. late physical maturation on adolescent behavior. In I.F. Litt (Ed.), *The Pediatric Clinics of North America* (pp. 71–77). Philadelphia: W.B. Saunders Co.

Hampson, J.G., & Money, J. (1955). Idiopathic sexual precocity in the female. *Psychosomatic Medicine, 18,* 16–35.

Hier, D. B., & Crowley, W. F. (1982). Spatial ability in androgen deficient men. *The New England Journal of Medicine, 306,* 1202–1205.

Hines, M., & Gorski, R. A. (1985). Hormonal influences on the development of neural asymmetries. In D.F. Benson & E. Zaidel (Eds.), *The Dual Brain* (pp. 75–96). New York: Guilford Press.

Jones, M.C. (1957). The later careers of boys who were early or late-maturing. *Child Development, 28,* 113–128.

Jones, M.C., & Bayley, N. (1950). Physical maturity among boys as related to behavior. *The Journal of Educational Psychology, 41,* 129–148.

Jones, M.C., & Mussen, P.H. (1958). Self-conceptions, motivations, and interpersonal attitudes of early and late-maturing girls. *Child Development, 29,* 491–501.

Maccoby, E.E., & Jacklin, C.N. (1974). *The psychology of sex differences.* Stanford, CA: Stanford University Press.

Meyer-Bahlburg, H.F.L., Bruder, G.E., Feldman, J.F., Ehrhardt, A.A., Healey, J.M., & Bell, J. (1985). Cognitive abilities and hemispheric lateralization in females following idiopathic precocious puberty. *Developmental Psychology, 21,* 878–887.

Money, J., & Hampson, J.G. (1955). Idiopathic precocity in the male. *Psychosomatic Medicine, 18,* 1–15.

Money, J., & Meredith, T. (1967). Elevated verbal I.Q. and idiopathic precocious sexual maturation. *Pediatric Research, 1,* 59–65.

Money, J., & Walker, P.A. (1971). Psychosexual development, maternalism, non-promiscuity and body image in 15 females with precocious puberty. *Archives of Sexual Behavior, 1,* 45–60.

Mussen, P.H., & Jones, M.C. (1957). Self-conceptions, motivations and interpersonal attitudes of late and early maturing boys. *Child Development, 28,* 243–256.

Netley, C., & Rovet, J. (1983). Relationships among brain organization, maturation rate, and the development of verbal and non-verbal ability. In S.J. Segalowitz (Ed.), *Language functions and brain organization* (pp. 245–266). New York: Academic Press.

Newcombe, N., & Bandura, M.M. (1983). Effect of age at puberty on spatial ability in girls: A question of mechanism. *Developmental Psychology, 19,* 215–224.

Peskin, H. (1973). Influence of the developmental schedule of puberty on learning and ego function. *Journal of Youth and Adolescence, 2,* 273–290.

Rierdan, J., & Koff, E. (1984). Age of menarche and cognitive functioning. *Bulletin of the Psychonomic Society, 22,* 174–176.

Rovet, J. (1983). Cognitive and neuropsychological test performance of persons with abnormalities of adolescent development: A test of Waber's hypothesis. *Child Development, 54,* 941–950.

Rubin, R. T., Reinisch, J.M., & Haskett, R.F. (1981). Postnatal gonadal steroid effects on human behavior. *Science, 211,* 1318–1324.

Strauss, E., & Kinsbourne, M. (1981). Does age at menarche affect the ultimate level of verbal and spatial skills? *Cortex, 17,* 323–326.

Waber, D.P. (1976). Sex differences in cognition: A function of maturation rate? *Science, 192,* 572–574.

Waber, D.P. (1977). Sex differences in mental abilities, hemispheric lateralization, and rate of physical growth in adolescence. *Developmental Psychology, 13,* 29–38.

Waber, D.P., Bauermeister, M., Cohen, C., Ferber, R., & Wolff, P.H. (1981). Behavioral correlates of physical and neuromotor maturation in adolescents from differing environments. *Developmental Psychobiology, 14,* 513–522.

Waber, D.P., Mann, M.B., Merola, J., & Moylan, P.M. (1985). Physical maturation rate and cognitive performance in early adolescence: A longitudinal examination. *Developmental Psychology, 21,* 666–681.

Weatherly, D. (1964). Self-perceived rate of physical maturation and personality in late adolescence. *Child Development, 35,* 1197–1210.

14
Associations Between Pubertal Hormones and Behavioral and Affective Expression

ROBERTA L. PAIKOFF AND J. BROOKS-GUNN

It is commonly believed that the biological changes of puberty exert a strong influence upon adolescent adaptation. Interest in associations between biological change and psychological functioning has always existed within the field of early adolescent developmental psychology (Freud, 1948; Freud, 1905; Hall, 1904). Early psychoanalytic theorists hypothesized that hormonal changes cause libidinal (or sexual instinctual) transformations at puberty (Freud, 1905) as well as accounting for a host of psychological defense mechanisms (Freud, 1948) used by young adolescents to combat these overwhelming libidinal drives. Hall (1904) viewed puberty as inherently stressful and conflictual, describing adolescence as a period of "storm und drang," or storm and stress.

More relevant to current assumptions about adolescents are beliefs and evidence concerning: (a) hormonal effects upon behavior in nonhuman primates and in human adults; (b) hormonal explanations for gender-linked divergencies in adult behavior; and (c) behavioral profiles of individuals who have had "mistakes of nature" occur, with the end result being a difference in fetal hormone environment and, sometimes, in pubertal hormonal functioning. All of this literature focuses, to a greater or lesser extent, upon five behavioral domains: sexual behavior and arousal, parental responsiveness and behavior directed toward infants, aggression and violence, depression and moodiness, and so-called masculine and feminine personality characteristics. All but the parental responsiveness domain will be reviewed in this chapter (see Fleming, in press, and Krasnegor et al., 1987, for current reviews of this literature).

Questions in this area include: (a) which biological changes contribute to adolescent adaptation; (b) which domains of adolescent adaptation are influenced by biological change and which are not; (c) whether gender-related differences in adolescent adaptation may be explained, at least in part, by biological changes; (d) whether biological gender effects when demonstrated are due to internal hormonal changes, and/or to external gender-salient physical changes that are the result of physiological change; and (e) what size are biological effects when found vis-a-vis the size of socially mediated effects. In this chapter, we explore all five issues. First, however, we need to briefly consider the studies available and the methodological issues involved in investigating such effects.

Description of the Studies

Table 14.1 presents a brief description of the studies included in this review. A synopsis of each study follows, as well.

NIMH/NIH Research Group (Susman et al., 1987)

In the NIMH/NIH research program, 56 boys and 52 girls, aged 9 to 14 years and representing the full range of pubertal development, were followed longitudinally. Children were seen three times, at six month intervals, over the course of one year. Hormonal assays were conducted through blood samples, taken 3 times over the course of an hour, during the morning hours (8:00 to 10:00 am). Children and parents filled out questionnaires assessing the child's emotional dispositions and personality attributes. In addition, videotaped interactional data and interview data were collected.

Early Adolescence Project (Eccles et al., 1988)

The Early Adolescence Project investigates hormone behavior relations in a sample of approximately 100 9 to 12 year olds. Data have thus far been reported on a subsample of 21 boys and 24 girls. Hormonal assays on this sample use salivary and urine samples, and subjects were sampled 3 times a week for 1 month. Subjects also reported on moods, aggressive behavior, and energy level three times a week for a month.

Stockholm Research Group (Olweus et al., 1980, 1988)

The Stockholm Research group investigated relations between hormonal status and aggressive behavior in a sample of 58 15- to 17-year-old males. Two morning samples of blood were collected, one month apart, from each boy, and boys filled out questionnaires about aggression level and related behavioral aspects (e.g., impatience, frustration) as well as on personality and psychopathology.

Carolina Population Center (Udry, 1987, 1988)

The relation between pubertal hormones and sexual behavior is being investigated by Udry and his colleagues. Several samples of adolescents aged 12 to 16 have been seen. Data have been reported on 102 males and 78 postmenarcheal females who were asked about sexual behavior, motiva-

TABLE 14.1. Studies examining associations between pubertal hormones and affective or behavioral expression.

Study title	N	Sample ages	Gender	Hormones assessed	Hormonal variables method of assessment	Timing of assessment	Psychological variables	Psychological variables method of assessment	Physical status effects
NIH research group (Susman et al., 1987; Inoff-Germain et al., 1988)	52 56	9-14 10-14	females males	LH FSH T E DHEA DHEAS Δ4-A TeBG T/E C	bloods	am (3× w/in hr.)	Emotional dispositions; Aggressive Behavior	Self and parent report Self and parent report; behavioral observation interview data	Controlled
Early Adolescence Project (Eccles et al., 1988)	24 21	9-10 11-12	females males	LH FSH Androgen E* P*	Saliva Urine	am (3×/ week for 4 weeks)	Moods Aggressive Behavior, related aspects (impatience, frustration)	Self-report (3x/week, for 4 weeks)	Examined for Mood Analyses; not in Aggression Analyses
Stockholm Research group (Olweus et al., 1980, 1988)	58	15-17	males	T	bloods	am (2×, one month apart)	Aggressive Behavior, related aspects (impatience, frustration, psychopathology)	Self-report	Examined

TABLE 14.1. *Continued*

Study title	N	Sample ages	Gender	Hormones assessed	Hormonal variables method of assessment	Timing of assessment	Psychological variables	Psychological variables method of assessment	Physical status effects
Carolina Population Center (Udry, 1987)	102 78	12-16 12-16	males females	LH T FTI DHEA DHEAS Δ4-A TeBG EST* P*	bloods	pm (3× w/in 30 min.)	Sexual Behavior (Coital and noncoital; Sexual and nonsexual motivation; Personality Factors; Deviant Behavior	Self-report	Examined
Adolescent Study Program (Brooks-Gunn & Warren, 1989)	103	10-14	females	LH FSH T E DHEAS	bloods	pm	Affective Expression (Depressive & Aggressive; Life Events; Adjustment	Self-report	Examined
New York Precocious Puberty Study (Meyer-Bahlburg et al., 1985)	16 16	13-20 13-20	IPP females Control females				Psychosexual development	Interview	Reported

208

| NICHD Intervention for IPP (Cutler et al., 1983) | 5 | 2-8 | IPP females | LH FSH E | bloods | am and pm/ every 20 min. for 4 hrs. | Reported |

Key—Hormone Codes

LH	= luteinizing hormone
FSH	= follicle stimulating hormone
T	= testosterone
E	= Estradiol
DHEA	= Dehydroepiandrosterone
DHEAS	= Dehydroepiandrosterone sulfate
Δ4-A	= androstenedione
TeBG	= testosterone-estradiol binding globulin
T/E	= testosterone to estradiol ratio
C	= cortisol
P	= Progesterone
FTI	= free testosterone index
Est	= total estrogens
*	= females only

tion, deviant behavior, and personality factors. Hormones are assessed through afternoon blood samples, with three samples taken over the course of 30 minutes.

The Adolescent Study Program (Brooks-Gunn & Warren, 1989; Warren & Brooks-Gunn, in press)

About 100 girls aged 10 to 14 were followed longitudinally for 3 years in the Adolescent Study Program; all girls participated in yearly hormone assessments collected through afternoon blood samples and 80 girls also gave overnight (morning) and afternoon urine samples. Girls and their mothers reported on a variety of psychological factors, including depressive and aggressive affect, deviant behavior, and life events.

New York Precocious Puberty Study (Meyer-Bahlburg et al., 1985)

The Precocious Puberty Study is a clinical examination of 16 girls aged 13 to 20 diagnosed with idiopathic precocious puberty, and 16 control girls. Current hormone levels were not assessed in these studies. Girls were interviewed with regard to their sexual experiences and sex drive.

NICHD Intervention for Idiopathic Precocious Puberty (Cutler et al., 1983)

The NICHD Intervention for Idiopathic Precocious Puberty (IPP) reports on a potential hormonal treatment for idiopathic precocious puberty. Five girls with IPP were treated daily with a luteinizing-hormone-releasing hormone analog for 2 months. LH and FSH were collected from bloods every 20 minutes over a 4-hour period during the morning and afternoon prior to treatment. Posttreatment LH, FSH, and estradiol levels were measured thrice at 20-minute intervals in 4 of the 5 cases.

Methodological Issues

As is evident from the description of investigations currently being conducted in the area of hormone-behavior links during adolescence, the studies differ in a number of respects: age and pubertal range, gender, number of adolescents seen, and measurement of hormonal and psychological variables. Hormone-behavior associations cannot necessarily be expected to remain constant across the full range of pubertal development nor to be the same for boys and girls. These variations make it difficult to compare findings on hormone-behavior links across studies.

Along with the problem of comparison across studies, researchers examining the effects of pubertal hormones on adolescents face a series of methodological concerns and decisions. One set of concerns involves the timing and nature of endocrine assessment. Hormonal measurements may differ substantially depending on time of day, as well as the nature of measurement (plasma, urinary, and saliva collections have been reported). While plasma assays analyze the levels of bound and unbound ("free") hormones, salivary assays only assess unbound hormone levels (Riad-Fahmy, Read, Joyce & Walker, 1981). Theoretical arguments might be constructed about the relative importance of measuring bound and unbound versus only unbound hormones (perhaps "free" hormones are more important when behavior is considered while overall hormone levels are important when affect is of interest); in any case, the two types of measurement are not the same and may or may not show the same patterns of association with behavior and affect.

Another set of concerns involves the cyclicity in young women. Prior to menstrual cycles, estradiol levels fluctuate on the basis of individual and daily circadian patterns; however, menarche is accompanied by a monthly rhythmicity in estrogen, progesterone, LH, and FSH. Menstrual cycles often do not regulate for several years after onset with great individual variation in the timing and frequency of menstrual cycles (Katchadourian, 1977; Winter & Faiman, 1973). Such cycle inconsistencies in early to middle adolescence render reports of cycle phase less reliable, making it more difficult to assess hormonal functioning at the same cycle phase across girls than in an adult population. Selecting a sample of premenarcheal girls, however, restricts age range, range of maturational timing, and range of hormonal functioning. While no clear-cut solution to this problem is available it is advisable for researchers to analyze data on pubertal females for potential effects of menarche, to attempt to control for cycle phase in postmenarcheal girls, or to use hormonal categories to describe hormone-behavioral associations (see Warren & Brooks-Gunn, 1989).

Directionality of effects must be considered when interpreting hormone behavior associations. Some investigators have gone so far as to suggest that the biological and social changes of adolescence should be thought of as co-occurring processes, without implications of causality (Coe et al., 1988). For example, the hypothalamic-pituitary gonadal system's sensitivity to environmental factors can result in endocrine changes. When substantial weight loss occurs levels of gonadotropic secretions are suppressed in women, with the most obvious manifestation being amenorrhea and anovulatory cycles (Vigersky & Loriaux, 1977; Warren, 1985). In some adolescents with anorexia nervosa, a reversion to the prepubertal pattern of low LH secretion as well as lower amplitude pulsatile secretion, and nocturnal spiking occur (Boyer et al., 1972, 1974). All such changes are reversible with weight gain. Therefore, at the very least, models of hormone-behavior relations must consider the possibility of bidirectional effects.

Other methodological problems involve the interpretation of hormone-behavior associations. First, the endocrine system is an integrated whole, consisting of many hormones and their binding factors. The subsystem of pubertal hormones, while operating on different receptors and different time tables, are often associated with one another. Correlations between 0.30 and 0.60 for estrogens and female gonadotropins, as well as between the androgens have been reported (Brooks-Gunn & Warren, 1989; Susman, et al., 1987). Such high intercorrelations make the teasing apart of particular hormone-behavior links difficult.

Second, that initial hormonal changes precede physical change and that different pathways for various hormonal effects exist suggests that one-to-one correspondence between hormonal and physical changes does not occur. Additionally, individual differences in hormone-secondary sexual characteristic links (see Figure 14.1), as well as in receptor sensitivity to hormones precludes the possibility of perfect correspondence between hormonal and pubertal change over time (Faiman & Winter, 1974), making it difficult to posit specific pathways for hormonal effects as opposed to secondary sex characteristic effects upon early adolescent behavior. We now turn to a review of the current evidence for hormone-behavior associations in several domains of behavioral and affective expression.

Aggressive-Delinquent Behavior

Associations between pubertal hormones (especially testosterone), and aggressive or delinquent behavior have been hypothesized from early psychoanalytic writing (Deutsch, 1944; Freud, 1905) up to the present day

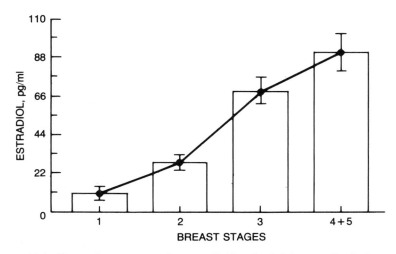

FIGURE 14.1. Tanner breast stages by estradiol levels, Adolescent Study Program.

(Inoff-Germain et al., 1988; Olweus et al., 1980). Contributing to the continued interest in the aggression-testosterone relationship are: (a) the large literature on gender differences in aggressive and delinquent behavior (Cantor, 1982; Maccoby & Jacklin, 1980; Parke & Slaby, 1983); (b) the large increase in delinquency seen around 13 to 15 years of age in boys (Cairns et. al, 1989); (c) the possible links between violent behavior and an extra Y chromosome;[1] and (d) the relatively strong associations between androgens and dominance and aggressive behavior found in nonhuman primates (Joslyn, 1973; Rose et al., 1971). One could hypothesize, then, that androgens would be associated with aggressive feelings and behavior as well as with delinquency in adolescence, and that these links would be stronger for boys than for girls. Few studies, however, have investigated this association at early adolescence. Findings in this area will be discussed separately for adolescent males and females.

The first work investigating links between pubertal hormones and aggressive behavior in males is that of Olweus and his colleagues (Olweus et al., 1980, 1988). While they found no association between testosterone and aggression or antisocial behavior in their sample of 58 adolescent boys, they did report a positive relation between testosterone concentration and physical and verbal aggression in self-reports and peer reports of responses to provocation (Olweus et al., 1980). In a recent reanalysis, testosterone effects on provoked aggressive behavior, as well as indirect effects (via increased impatience and irritability) on aggressive-destructive behavior were found (Olweus et al., 1988). In other words, higher levels of circulating testosterone were associated with increases in impatience and irritability which in turn increased the tendency towards aggressive-destructive behavior. This causal model accounted for a very large percent of the variance in provoked aggressive behavior (48%; Olweus et al., 1988). At the same time, no effects of pubertal status on aggressive behavior or responses to provocation were found.[2]

Estradiol levels were negatively and androstenedione levels positively associated with parent reports of aggression and delinquency in boys (measured by Child Behavior Checklist) in the NIMH/NIH sample (Inoff-Germain, 1986). Pubertal status and age were entered as control variables in the regression equations, and no independent pubertal status effects were found. In another analysis, lower quantities of testosterone-estradiol binding globulin and of estradiol, as well as higher levels of adrenal androgens, especially androstenedione, were related to aggressive attributes in boys (Susman et al., 1987). Verbal aggression was coded for a subsample of these adolescents (30 boys and 30 girls) observed interacting

---

[1] Such claims have yet to be substantiated, however (Witkin et al., 1976).

[2] It should be noted, however, that the boys studied ranged in age from 15 to 17, and tended to fall in the late pubertal status categories; 95% in Tanner 4 and 5. Therefore, the full range of pubertal status effects could not be examined.

with their parents in structured problem-solving situations, as well as unstructured "breaks." (Inoff-Germain et al., 1988) No hormone-behavior associations for boys were found.

Several studies have examined effects of pubertal hormones on the related (but not equivalent) construct of delinquent behavior. Testosterone has been positively and testosterone-estradiol binding globulin negatively associated with boys' norm-violating behavior. Pubertal status effects on boys' norm-violating behavior were mediated by testosterone level (Udry, 1987). Susman et al. (1987) report a strong negative association with estradiol and positive association with andostenedione and delinquent behavior in boys.

In the NIMH/NIH Study and the Adolescent Study Program, DHEA and DHEAS were negatively associated with reports of aggressive behavior in girls (Brooks-Gunn & Warren, 1989; Susman et al., 1987). In other analyses of the feelings and interactions between adolescents and parents, androstenedione and estradiol were positively associated with verbal aggression in girls as measured by observer ratings of dominance, defiance, and anger at parents (Inoff-Germain et al., 1988).[3]

In our own sample of 103 adolescent girls, estradiol categories were created in order to assess nonlinear as well as linear endocrine effects. Brooks-Gunn and Warren (1989) explain our use of estradiol categories as follows:

"In order to test the premise that depressive and aggressive affect will increase during the most rapid endocrine system changes, hormonal categories were derived. Estradiol was used (a) because it is the principal gonadal hormone in girls; (b) because it exhibits the most dramatic increases during puberty (serum estradiol increases approximately 8 to 10 fold, testosterone 5 fold, and DHEAS, LH, and FSH 2 to 4 fold; Apter, 1980; Apter & Vihko, 1977; Faiman & Winter, 1974; Grumbach, 1980); (c) because the rises occur principally between Tanner Stage 1 to 4, which corresponds to our sample's status; and (d) because estradiol is associated with many of the physical changes of puberty. The four groupings were 0 to 25, 26 to 50, 51 to 74, and 75 and higher of estradiol (pg/ml). Levels of less than 25 pg/ml are generally known to have minimal affect. Levels of 25 to 50 pg/ml generally have early visible physiological effects such as secondary sexual development and effects on the vagina. Estrogens at this level are generally not significant enough to cause proliferation of the endometrium and withdrawal bleeding with a progesterone challenge. Levels of 50–75 pg/ml typically are commensurate with mid or late puberty and early follicular levels in menstruating girls. These levels have significant effects on endometrial growth and other organs such as the breast. Girls with levels in this range would generally experience withdrawal bleeding to progesterone and/or have spontaneous periods. Levels greater than 75 pg/ml are associated with cyclicity in women (Gold & Iosimovich, 1980; Grumbach & Sizonenco, 1986).

---

[3] The possibility of outlier effects was examined and ruled out in these analyses (Inoff-Germain, personal communication). Additionally, no effects of pubertal status were reported.

Using these groupings also took into account LH levels, as almost no overlap in LH levels across the estradiol groupings was found.''

Estradiol category was related to both aggressive behavior ($p < 0.10$) and to impulse control ($p < 0.05$) in a nonlinear fashion. Increases in aggressive behavior and decreases in impulse control were reported at the time of initial hormonal change between categories 1, 2, and 3, but not thereafter. The estradiol categories created capture as well as changes in LH, FSH, testosterone, and DHEAS (Warren & Brooks-Gunn, 1989); therefore these results may be indications of general hormonal changes rather than the effects of specific hormones.

In the Michigan Early Adolescence Project high progesterone and low testosterone were significantly associated with reports of impatience in girls ($p < 0.05$; Eccles et al., 1988). These analyses did not control for pubertal status. Udry and Talbert (1988) also found a positive association between progesterone and aggression in their sample of 78 12- to 16-year-old females ($p < 0.01$). No significant associations have been reported between pubertal hormones and delinquency in girls (Paikoff et al., in press; Susman et al., 1987; Udry, 1987).

In summary, a fair amount of evidence exists for activational effects of hormones on adolescent aggressive feelings and behaviors. Many, but not all, of these associations involve the androgens. As expected, larger effect sizes have been reported for boys than for girls. It seems likely that impatience and impulse control will be important mediational factors in these associations, since several investigators report associations between hormonal levels and impatience or impulse control (Susman et al., 1987; Warren & Brooks-Gunn, 1989) and at least one study reports a causal trajectory for effects of pubertal hormones on aggressive behavior via increased impatience and irritability (Olweus et al., 1988). In addition, gender differences in the nature of associations between pubertal hormones and aggressive behavior are likely. These differences may impact upon behavioral differences in aggression commonly reported between males and females (Maccoby & Jacklin, 1980; Parke & Slaby, 1983), due to thresholds for androgen response. In other words, it could be the case that a certain level of androgens, bound or unbound, may be a necessary, but not sufficient, condition for aggressive behavior at adolescence. If such a threshold were relatively high, or close to the modal androgen value for adolescent males, differences in androgen level would have a more profound impact on males than females, and higher associations would be reported for males than for females. Additionally, such a notion would help explain reported gender differences in aggressive behavior, where males tend to exhibit more aggressive behavior than females (Maccoby & Jacklin, 1980; Parke & Slaby, 1983). The notion of androgen thresholds has been invoked to explain gender differences in the correlates of sexual behavior as well (see following); however the concept has not been subjected to empirical analyses, and thus remains a speculative notion.

Sexual Behavior

Sexual arousal and behavior are associated with hormones both in nonhuman primates (Levine, 1971; Rose et al., 1972) and human adults (Bancroft & Shakkebak, 1978; Rose, 1972). A certain level of circulating testosterone may be necessary (but not sufficient) for sexual arousal in both male and female adult humans. While it is commonly believed that testosterone levels impact directly upon male adult sexual arousal, it may in fact be the case that women are more responsive to variations in circulating testosterone levels (Persky, 1974). Female sexual arousal does not decline significantly postmenopause when estrogen levels drop but androgens remain constant (Masters & Johnson, 1970). A certain level of androgen may be needed for normal sexual functioning, but once this level is attained (as is true for all males), additional amounts may have little influence (Rose, 1972). Such a threshold might have more important implications for female sexual arousability, since testosterone levels are significantly lower in females than in males.

The endocrinological changes that occur at puberty affect the fertility and the reproductive maturity of both sexes. These changes also may contribute to increased sexual behavior and coital activity indirectly, through the increased sexual attractiveness of physically mature males and females, and thus increased opportunities for sexual behavior (Udry, 1987). Do pubertal hormones influence sexual motivations, behavior, and coital activity directly as well? Udry and his colleagues at the Carolina Population Center (Udry et al., 1985; Udry et al., 1986; Udry & Billy, 1987; Udry, 1987) present the most direct evidence pertaining to this issue. Again, findings will be discussed separately by gender.

Level of "free" testosterone was associated with a variety of sexual behavioral and motivational factors for boys. Free testosterone predicted all sexual behaviors (including coital activity) and arousal, with effect sizes ranging from 3.06 (intentions to have intercourse) to 0.34 (e.g., any sexual outlet in the past month). Once free testosterone was included in a regression equation, none of the other pubertal hormones assessed added significantly to predicting sexual behavior. Pubertal status and age also did not change these findings.

When girls were examined, however, the adrenal androgens [free testosterone, dehydroepiandrosterone (DHEA) and its sulfate (DHEAS), and androstenedione] were the most frequent hormonal predictors of noncoital sexual behavior and arousal (e.g. masturbation, prior sexual experience, and thinking about sex) and sexual motivation (future intentions). Masturbation (but not coital activity) was predicted both by pubertal status and hormone level.

In summary, the Udry data present a picture of male sexual behavior as more directly controlled by testosterone level than female sexual behavior

(Udry, 1987, 1988). Female sexual motivation and noncoital behavior are directly influenced by a variety of adrenal androgens, similar to findings for males. Unlike males, however, no direct hormone effect for coital activity in girls were found. Pubertal development was a significant predictor of coital activity in girls, along with a variety of social factors, such as age, physical attractiveness, and friends' intercourse behavior (Udry & Billy, 1987).

Negative Affect and Depression

Evidence concerning pubertal hormonal associations with negative or depressive affect is more ambiguous than in previously discussed behavioral domains. However, estrogen level may be associated with positive affect and feelings of well-being in adult women (Klaiber et al., 1979; Melges & Hamburg, 1977). The NIMH/NIH research group (Susman et al., 1987), and our own research group (Brooks-Gunn & Warren, 1989; Paikoff et al., in press) have addressed this issue for adolescents.

The NIMH/NIH group (Susman et al., 1987) examined a wide range of pubertal hormones' effects on various feeling states, and found lower concentrations of hormones generally were related to more positive affect in boys (Susman et al., 1987).

For females, however, the NIMH/NIH research group found no significant associations between pubertal hormones and affective states (Susman et al., 1987). In our own work, estradiol has been associated with girls' reports of depressive affect (Brooks-Gunn & Warren, 1989), with an increase in depressive affect occurring between what we have termed estradiol levels 1 and 3 but not at level 4. When these same estradiol levels are used to predict depressive affect a year later, higher levels of estradiol at Time 1 are associated with higher levels of depressive affect as reported by girls (Paikoff et al., in press). This finding is maintained when prior depressive affect is entered into the regression models, though prior depressive affect accounts for more of the variation in the girls' responses than do hormonal levels. Pubertal status and age, as well as mothers' reports of their daughters' depressive affect, were entered into our regression equations with no significant effects.

Mood Lability

Perhaps the most popularly hypothesized characteristic of adolescence is rapid mood swings, or mood lability. Little empirical data exist regarding this belief. The data that are available, however, suggest that mood variability itself varies substantially across adolescents with most exhibiting fairly stable and predictable mood changes over time while a minority experience wildly fluctuating, unpredictable mood swings (Cskiszent-mihalyi & Larson, 1984; Savin-Williams & Demo, 1984). We do not know how much of the variation in mood lability can be explained by pubertal hormonal changes.[4]

Gender-Typed Behavior/Gender Identity

Gender-typed sexual and aggressive behavior is one area where significant amounts of research have concentrated on hormonal associations in non-human as well as human primates. Both sets of studies will be reviewed here.

In the area of nonhuman primates, Coe and his colleagues (Coe & Levine, 1983; Coe et al., 1988) and Goy (Goy, 1968; Goy & McEwen, 1980; Thornton & Goy, 1986) have examined organizational effects of prenatal pubertal hormones on sex-typed behavior, in squirrel monkeys and rhesus monkeys, respectively.

Coe and Levine (Coe et al., 1988) studied social relationships between intact and gonadectomized (castrated) males with each other and with intact and ovariectomized females throughout the pubertal period. No significant differences were found between intact and gonadectomized males in terms of social proximity and play patterns with one another. Gonadectomized males tended to be smaller and to engage in dominance interactions less frequently than intact males (Coe et al., 1988). In a similar vein, castration of monkey infants at birth did not affect manifestation of sex-typical behavior (Goy & McEwen, 1980).

Findings for female squirrel monkeys (gonadectomized and intact) mirrored those of males in that both ovariectomized and intact females showed species-appropriate levels of play behavior and species-appropriate play preferences (Coe et al., 1988). Dominance behavior was lower than males in both the female groups, and ovariectomized females were smaller than intact females, though they showed similar growth patterns. Female rhesus monkeys exposed to androgen treatment in utero tended to exhibit more masculine rough and tumble play and to emulate the mounting patterns of males in their sex play more than their counterparts

---

[4] To date, a member of the Michigan Early Adolescence Project has directly examined effects of hormonal status on mood shifts at early adolescence in a doctoral thesis as yet unpublished. The effects of hormone concentration and variability upon females' ($N = 24$) mood intensity (relative to sample) and mood variability was examined for a day, as well as over a month (Buchanan, 1989). High FSH variability, but not progesterone or testosterone, was associated with moodiness within a day, $r^2 = 0.23$. Additionally, high FSH concentration was associated with moodiness over one month, $r^2 = .32$. No perceived pubertal status effects were found for moodiness within a day or over a month; however, pubertal status was related to mood intensity, and to mood variability over a one-month period.

who were not exposed to androgen treatment in utero. (Phoenix et al., 1967; Goy, 1970). Thus, the literature on nonhuman primates strongly suggests organizational hormone effects on prepubertal and pubertal sex-typed play behavior.

Organizational effects of pubertal hormones on later gender-typed behavior have been studied in humans as well, in terms of gender identity (affiliation with one's gender) as well as gender role behavior. These studies have relied upon populations with abnormal prenatal exposure to hormones. Since this work by and large involves the prenatal androgenization of females, only females will be discussed.

The major investigators in this area have been Money, Ehrhardt, and colleagues (Ehrhardt & Money, 1967; Ehrhardt et al., 1968; Ehrhardt, 1967; Money & Ehrhardt, 1972). As a result of several studies of females exposed prenatally to excessive androgens (either due to mothers' prescription drug use, or to genetic abnormalities), Ehrhardt (1967) concludes that certain aspects of gender-related behavior, both prepuberty and postpuberty, are influenced by prenatal hormones. The influence of these hormones is primarily on "tomboyishness" or overly energetic and playful behavior. Interest in maternalism may be affected, but many androgenized women who are able to have children do become mothers. Differences in gender identity and sexual orientation have not been found, although androgenized females showed a delay in exhibiting normative teenage sexual experience and motivation, possibly because of the fact that these girls look "different," are less likely to develop the female rounded contours, and may be masculinized if not treated by estrogens. Timing of treatment and surgical correction (if necessary) may strongly influence gender-typed behavior in later life (Ehrhardt, 1967); thus effects may be socially as well as biologically mediated.

In another set of clinical studies, the psychosexual development of girls with idiopathic puberty (IPP) has been examined. In a preliminary intervention with IPP girls, a luteinizing-hormone releasing hormone (LHRH) agonist was found to dampen premature hormonal (and thus physical) effects in 5 girls (Cutler et al., 1983). In another study examining psychosexual development in IPP girls, as well as in matched controls, the IPP girls began masturbation and reported first boy friends and steady relationships at earlier ages (Meyer-Bahlburg et al., 1985). No significant difference was found between the groups in total number of intercourse instances, nor in sexual orientation.

Thus, while clinical studies with humans suggest organizational effects of excessive androgens on some behavior (specifically opposite-sex relationships and activity level), they do not appear to influence sexual behavior, sexual orientation, or many other gender-typed behaviors in females. Finally, prenatal organizational effects of pubertal hormones are often treatable with good medical, and perhaps psychological, results.

Relative Influence of Hormones and Secondary Sex Characteristics

As can be noted from our review thus far, when secondary sexual development and hormonal status are both examined, the former is almost never associated with behavior or affect, while the latter is. Thus, the effects that are reported, if replicable, are probably due to biological change, not to the adolescents' responses to external changes nor to others' responses to these changes. The literature on psychological effects of physical change at puberty suggests that there are psychological effects associated with pubertal change, but that they occur in different behavioral domains than those discussed here.[5]

Relative Influence of Biological and Social Change

Little research has been conducted comparing the relative effects of biological and social change upon adolescent behavior. The one study that examined such effects (Brooks-Gunn & Warren, 1989), found that biological factors (i.e., hormonal levels) accounted for only 4% of the variance in girls' negative affect while social factors accounted for 8% and 18% of the variance in girls' depressive and aggressive affect, respectively.

Conclusions

It should be clear that the study of associations between pubertal hormones and behavior and affect during adolescence is just emerging. The "oldest" hormone-behavior study of a "normal" adolescent sample was published in 1980. Thus, the conclusions we may draw are limited by the small number of studies and the need for replication. When results from the adolescent studies are placed in the context of adult human and nonhuman primate literature, however, some general conclusions regarding associations between pubertal hormones and behavior may be drawn.

Both direct and indirect effects of pubertal hormones on male sexual and aggressive behavior have been reported (Olweus et al., 1980, 1988; Udry, 1987, 1988). For females, the picture is less clear. When estrogen and other hormones (LH and FSH) increase, a small rise in reports of aggressive affect is found in one but not another study (Brooks-Gunn & Warren, 1989; Susman et al., 1987). However, both studies find a negative association

---

[5] For reviews of this literature the reader is referred to: Adams, Montemayor, & Gullotta, 1989; Brooks-Gunn, Petersen, & Eichorn, 1985; Gunnar & Collins, 1988; Lerner & Foch, 1987.

between DHEA and self-reported aggressive feelings and behavior. Females' noncoital sexual behavior and sexual motivations have been associated with pubertal hormone levels, however social and pubertal developmental factors also play a role in this relation (Brooks-Gunn & Furstenberg, 1989). As in human adult models of aggression (Mazur, 1983) and sexual behavior (Bancroft & Shakkebak, 1978), testosterone and the adrenal androgens appear most frequently to be implicated in these associations. Because these effects are reported concurrently, causality cannot be determined; however, these concurrent associations do suggest that the hormonal effects found are activational, not organizational, in nature.

No evidence exists for an association between hormonal status and depressive affect in males. Data for females conflict, although the association we found has not been tested in the NIMH/NIH sample. In adult women, high levels of estrogens have been associated with positive affect, or feelings of well-being. Our findings suggest that initial rises of estradiol, LH, and FSH in pubertal girls may have the opposite effect. Our findings may be compatible with the adult literature if one considers them an "adjustment effect" (Buchanan & Eccles, 1989). The early adolescent surge in hormones may necessitate psychological as well as physical adaptation. If this is the case, then our effect should not persist; we are in the process of longitudinal analyses examining this issue.

Direct organizational effects of pubertal hormones appear to be implicated in the control of certain social behaviors of nonhuman primates. In human females, limited evidence for organizational effects exists for activity level and certain personality characteristics. Organizational effects in female humans, however, appear to be mediated by activational changes, as well as by treatment and surgical conditions in clinical samples. Prenatal androgen levels do not seem to affect female gender identity or sexual orientation, at least in the clinical samples reported upon.

In conclusion, pubertal hormones may exert activational effects on adolescent sexual behavior and arousal, and on aggressive feelings, with the effects being stronger in boys than girls. Whether they have an influence on negative affect and mood lability awaits further investigation. The research projects mentioned in this chapter all continue to flourish, and a wealth of information providing more definitive conclusions on these issues will be forthcoming in the next decade. To facilitate this process, we make the following final suggestions:

(a) *Examine data for nonlinear as well as linear effects.* Our own data examining hormones and affective expression suggest that nonlinear effects may be found in relations between estradiol and affect (Brooks-Gunn & Warren, 1989; Paikoff et al., in press; Warren & Brooks-Gunn, in press). By limiting analyses to linear regression functions, other investigators may miss possible threshold effects for adjustment problems at puberty as a function of endocrinological status especially those occurring during the period of initial hormonal change.

(b) *Examine individual differences in reactivity.* While little evidence exists addressing issues of individual differences and hormone-behavior/ affect links during puberty, we regard this as a most fruitful area to concentrate future research efforts. Studies of relations between physiological and psychological reactivity in human infants (Gunnar et al., 1988) and subhuman primates (Suomi, 1987) have elucidated the importance of physiological reactivity in explaining later psychological reactivity; and, it may well be that reactivity to certain pubertal hormones, or to the more general endocrine changes of puberty, will exert important effects on the behavioral and affective expression of adolescents.

Acknowledgment. This chapter was written while the first author was a Post-Doctoral Fellow at Educational Testing Service and the second author was a Visiting Scholar at the Russell Sage Foundation. The research support of the National Institutes of Health (NICHD) and the W. T. Grant Foundation are greatly appreciated. We wish to thank Dr. Michelle Warren for her helpful comments, and Robin Roth, Rosemary Deibler, and Rhoda Harrison for their assistance in manuscript preparation.

References

Adams, G.R., Montemayor, R., and Gullotta, T.R. (1989). *Advances in adolescent development,* Volume 1. Beverly Hills, CA: Sage Publications.

Apter, D. (1980). Serum steroids and pituitary hormone in female puberty: A partly longitudinal study. *Clinical Endocrinology, 12,* 107–120.

Apter, D., & Vihko, R. (1977). Serum pregnenolone, progesterone, 17-hydroxyprogesterone, testosterone and 5a-dihydrotestosterone during female puberty. *Journal of Clinical Endocrinology and Metabolism, 45*(5), 1039–1048.

Bancroft, J., & Shakkebak, N. (1978). Androgens and human sexual behavior. In *Ciba Foundation Symposium 62 (new series)* (pp. 209–226). Amsterdam: Exerpta Medica.

Boyar, R.M., Finklestein, J., Roffwarg, H., Kapan, S., Wertman, E., & Hellman, L. (1972). Synchronization of augmented luteinizing hormone secretion with sleep during puberty. *New England Journal of Medicine, 287,* 582–586.

Boyar, R.M., Katz, J., Finkelstein, J.W., Kapen, S., Weiner, H., Weitzman, E.D., & Hellman, L. (1974). Anorexia nervosa: Immaturity of the 24-hour luteinizing hormone secretory pattern. *New England Journal of Medicine, 291*(17), 861–865.

Brooks-Gunn, J., & Furstenberg, F.F. Jr. (1989). Adolescent sexual behavior. *American Psychologist, 44*(2).

Brooks-Gunn, J., Petersen, A.C., & Eichorn, D. (1985). The study of maturational timing effects in adolescence. *Journal of Youth and Adolescence, 14*(3), 149–161.

Brooks-Gunn, J., & Warren, M.P. (1989). Biological contributions to affective expression in young adolescent girls. *Child Development, 60,* 372–385.

Buchanan, C.M., & Eccles, J.S. (1988). *Evidence for relationships between hormones and behavior at early adolescence*. University of Michigan, unpublished manuscript.

Buchanan, C.M. (April, 1989) Hormone concentrations and variability: Associations with self-reported moods and energy in early adolescent girls. Paper presented at the biennial meetings of the Society for Research in Child Development, Kansas City, MO.

Cairns, R.B., Cairns, B.D., Neckerman, H.J., Ferguson, L.L., & Gariepy, J.L. (1989). Growth and Aggression: I. Childhood to Early Adolescence. *Developmental Psychology, 25*(2), 320–330.

Cantor, R.J. (1982). *Family correlates of male and female delinquency (Revised)*. Project report of the Institute of Behavioral Science, Boulder, Colorado.

Coe, C.L., Hayashi, K.T., & Levine, S. (1988). Hormones and behavior at puberty. Activation or concatenation? In M. Gunnar & W.A. Collins (Ed.), *Minnesota Symposia on Child Development, Vol. 21* (pp. 17–41). Hillsdale, NJ: Erlbaum.

Coe, C.L., & Levine, S. (1983). Biology of aggression. *Bulletin of the American Academy of Psychiatry and the Law, 11*(2), 131–148.

Csikszentmihalyi, M., & Larson, R. (1984). *Being adolescent: Conflict and growth in the teenage years*. New York: Basic Books.

Cutler, G.B. Jr., Comite, F., Rivier, J., Vale, W.W., Loriaux, D.L., & Crowley, W.F. Jr. (1983). Pituitary desensitization with a long-acting luteinizing-hormone-releasing hormone analog: A potential new treatment for idiopathic precocious puberty. In J. Brooks-Gunn & A.C. Petersen (Eds.), *Girls at puberty: Biological and psychosocial perspectives* (pp. 89–102). New York: Plenum Press.

Deutsch, H. (1944). *The psychology of women, Vol. 1*. New York: Grune & Stratton.

Eccles, J.S., Miller, C., Tucker, M.L., Becker, J., Schramm, W., Midgley, R., Holmes, W., Pasch, L., & Miller, M. (March, 1988). *Hormones and affect at early adolescence*. Paper presented at the Biannual Meeting of the Society for Research on Adolescence, Alexandria, VA.

Ehrhardt, A.A. (1967). Psychosexual adjustment in adolescence in patients with congenital abnormalities of their sex organs. In M. Lewis & L. Rosenbaum (Eds.), *The exceptional infant*. New York: Plenum Press.

Ehrhardt, A.A., Epstein, R., & Money, J. (1968). Fetal androgens and female gender identity in the early-treated adrenogenital syndrome. *Johns Hopkins Medical Journal, 122*, 160–167.

Ehrhardt, A.A., & Money, J. (1967). Progestin-induced hermaphroditism: IQ and psychosexual identity in a study of ten girls. *Journal of Sex Research, 3*, 83–100.

Faiman, C., & Winter, J.S. (1974). Gonadotropins and sex hormone patterns in puberty: Clinical data. In M.M. Brumbach, G.D. Grave, & F.E. Mayer (Eds.), *Control of the onset of puberty*. New York: John Wiley & Sons.

Fleming, A. (in press). Hormonal and experiential correlates of maternal responsiveness in human mothers. In N. Krasnegor et al. (Ed.), *Mammalian parenting*. London: Oxford University Press.

Freud, A. (1948). *The ego an the mechanisms of defense*. New York: International Universities Press.

Freud, S. (1905/1953). *A general introduction to psychoanalysis* (translated by J. Riviere). New York: Permabooks.

Gold, J.J., & Iosimovich, J. (1980). *Gynecologic Endocrinology*. New York: Harper & Row.

Goy, R.W. (1968). Organizing effects of androgen on the behavior of rhesus monkeys. In R.O. Micheal (Ed.), *Endocrinology and human behavior* (pp. 12–31). London: Oxford University Press.

Goy, R.W. (1970). Experimental control of psychosexuality. In G.W. Harris & R.G. Edwards (Eds.), *A discussion on the determination of sex* (pp. 149–162). London: Philosophical Transactions of the Royal Society.

Goy, R.W., & McEwen, B.S. (1980). *Sexual differentiation of the brain*. Cambridge, MA: MIT Press.

Grumbach, M.M. (1980). The neuroendocrinology of puberty. In D.T. Krieger & J.C. Hughs (Eds.), *Neuroendocrinology*. Sunderland, MA: Sinauer.

Grumbach, M.M., & Sizonenko, P.C. (Eds.). (1986). *Control of the onset of puberty II*. New York: Academic Press.

Gunnar, M.R., & Collins, W.A. (Eds.). (1988). *Development during the transition to adolescence: The Minnesota Symposium on Child Psychology, Volume 21*. Hillsdale, NJ: Erlbaum.

Gunnar, M.R., Mangelsdorf, S.C., Kestenbaum, R., Larson, M., & Andreas, D. (April, 1988). *The relevance of temperament to other constructs: Temperament and attachment*. Paper presented in a symposium entitled The Relevance of Temperament to other Constructs. International Conference on Infant Studies, Washington, DC.

Inoff-Germain, G. (1986). Hormones and aggression in early adolescence. Paper presented at the first biennial meeting of the Society for Research on Adolescence, Madison, Wisconsin.

Inoff-Germain, G., Arnold, G.S., Nottlemann, E.D., Susman, E.J., Cutler, G.B., & Chrousos, G.P. (1988). Relations between hormone levels and observational measures of aggressive behavior of young adolescents in family interactions. *Developmental Psychology, 24*(1), 129–139.

Joselyn, W.D. (1973). Androgen-induced social dominance in infant female rhesus monkeys. *Journal of Child Psychology and Psychiatry, 14*, 137–145.

Katchadourian, H. (1977). *The biology of adolescence*. San Francisco, CA: W.H. Freeman.

Klaiber, E.L., Broverman, D.M., Vogel, W., & Kobayashi, Y. (1979). Estrogen therapy for severe persistent depressions in women. *Archives of General Psychiatry, 36*, 550–554.

Krasnegor, N., Elliott, E.M., Blass, Hofer, M.A., & Smutherman, W.P. (Ed.). (1987). *Perinatal development: A psychobiological perspective*. New York: Academic Press.

Lerner, R.M., & Foch, T.T. (1987). *Biological psychosocial interactions in early adolescence*. Hillsdale, NJ: Erlbaum.

Levine, S. (1971). Sexual differentiation: The development of maleness and femaleness. *California Medicine, 114*, 12–17.

Maccoby, E.E., & Jacklin C.N. (1980). Sex differences in aggression: A rejoinder and reprise. *Child Development, 51*, 961–980.

Masters, W.H., & Johnson, V.E. (1970). *Human sexual inadequacies*. Boston: Little Brown.

Mazur, A. (1983). Hormones, aggression, and dominance in humans. In B. Svare (Ed.), *Hormones and aggressive behavior*. New York: Plenum Press.

Melges, F., & Hamburg, D. (1977). Psychological effects of hormonal changes in women. In F.A. Beach (Ed.), *Human sexuality in four perspectives*. Baltimore: The Johns Hopkins University Press.

Meyer-Bahlburg, H.F.L., Ehrhardt, A.A., Bell, J.J., Cohen, S.F., Healey, J.M., Feldman, J.F., Morishima, A., Baker, S.W., & New, M.I. (1985). Idiopathic precocious puberty in girls: Psychosexual development. *Journal of Youth and Adolescence, 14*(4), 339–354.

Olweus, D., Mattsson, A., Schalling, D., & Low, H. (1980). Testosterone, aggression, physical and personality dimensions in normal adolescent males. *Psychosomatic Medicine, 42*(2), 153–169.

Olweus, D., Mattsson, A., Schalling, D., & Low, H. (1988). Circulating testosterone levels and aggression in adolescent males: A casual analysis. *Psychosomatic Medicine, Vol. L*(3), 261–272.

Paikoff, R.L., Brooks-Gunn, J., & Warren, M.P. (in press). *Effects of girls' hormonal status on affective expression over the course of one year*. Journal of Youth and Adolescence.

Parke, R.D., & Slaby, R.G. (1983). The development of aggression. In P.H. Mussen (Ed.), *Handbook of child psychology, Vol. 4, 4th ed.* (pp. 547–642). New York: Wiley.

Persky, H. (1974). Reproductive hormones, moods, and the menstrual cycle. In R.C. Friedman, R.M. Richard, R.L. Vande, & R. Wiele (Eds.), *Sex differences in behavior* (pp. 455–466). New York: Wiley.

Phoenix, C.H., Goy, R.W., & Young, W.C. (1967). Sexual behavior: General aspects. In L. Martini & W.F. Ganong (Eds.), *Neuroendocrinology, Vol. II*. New York: Academic Press.

Riad-Fahmy, D., Read, G.F., Joyce, B.G., & Walker, R.F. (1981). Steroid immunoassays in endocrinology. In A. Vollar, A. Bartlett, & J.D. Bidwell (Eds.) *Immunoassays for the '80's*. Baltimore, MD: University Park Press.

Rose, R.M. (1972). The psychological effects of androgens and estrogens: A review. In R.I. Shader (Ed.), *Psychiatric complications of medical drugs*. New York: Raven Press.

Rose, R.M., Holaday, J.W., & Berstein, I.S. (1971). Plasma testosterone, dominance rank, and aggressive behavior in male rhesus monkeys. *Nature, 231*, 366–368.

Savin-Williams, R., & Demo, D.H. (1984). Developmental change and stability in adolescent self-concept. *Developmental Psychology, 20*(6), 1100–1110.

Suomi, S.J. (1987). Genetic and maternal contributors to individual differences in rhesus monkey biobehavioral development. In N. Krasnegor, E.M. Elliot, Blass, M.A. Hofer, & W.P. Smutherman (Eds.), *Perinatal development: A psychobiological perspective*. New York: Academic Press.

Susman, E.J., Inoff-Germain, G., Nottlemann, E.D., Loriaux, D.L., Cutler, G.B., & Chrousos, G.P. (1987). Hormones, emotional dispositions, and aggressive attributes in young adolescents. *Child Development, 58*(4), 1114–1134.

Udry, J.R. (1987). *Biosocial models of adolescent problem behaviors*. Unpublished manuscript.

Udry, J.R. (1988). Biological predispositions and social control in adolescent sexual behaviors. *American Sociological Review, 53*, 709–722.

Udry, J.R., & Billy, J.O.G. (1987). Initiation of coitus in early adolescence. *American Sociological Review, 52*, 841–855.

Udry, J.R., Billy J.O.G., Morris, N.M., Groff, T.R., & Raj, M.H. (1985). Serum androgenic hormones motivate sexual behavior in adolescent boys. *Fertility and Sterility, 43*(1), 90–94.

Udry, J.R., & Talbert, L.M. (1988). Sex hormone effects on personality at puberty. *Journal of Personality and Social Psychology, 54*(2), 291–295.

Udry, J.R., Talbert, L.M., & Morris, N.M. (1986). Biosocial foundations for adolescent female sexuality. *Demography, 23*(2), 217–230.

Vigersky, R.A., & Loriaux, D. (1977). Anorexia nervosa as a model of hypothalamic dysfunction. In R.A. Vigersky (Ed.), *Anorexia nervosa* (pp. 109–121). New York: Raven Press.

Warren, M.P. (1985). When weight loss accompanies amenorrhea. *Contemporary Obstetrics and Gynecology, 28*(3), 588–597.

Warren, M.P., & Brooks-Gunn, J. (1989). Mood and behavior at adolescence: Evidence for hormonal factors. *Journal of Clinical Endocrinology and Metabolism, 69*(1), 77–83.

Winter, J.S., & Faiman, C. (1973). Pituitary-gonadal relations in female children and adolescents. *Pediatric Research, 7,* 948–953.

Witkin, H.A., Mednick, S.A., Schulsinger, F., Bakkestrom, E., Christiansen, K.O., Goodenough, D.R., Hirchhorn, K., Lundsteen, C., Owen, D.R., Philip, J., Ruben, D.B., & Stocking, M. (1976). Criminality in XYY and XXY men. *Science, 193,* 547–555.

15
Congenital Adrenal Hyperplasia: Intellectual and Psychosexual Functioning

SHERI A. BERENBAUM

Patients with congenital adrenal hyperplasia represent a unique opportunity to study the effects of prenatal sex hormones on the development of the brain and behavior. They also reveal the amazing capacity of people to adjust to major developmental aberrations.

There are several forms of congenital adrenal hyperplasia, each inherited as an autosomal recessive disorder, and associated with enzyme defects which result in various disruptions in adrenal steroid biosynthesis. In the most common form of CAH, there is an adrenal enzyme deficit in 21-hydroxylase (21-OH), and individuals are unable to produce sufficient quantities of cortisol to inhibit the release of adrenocorticotropic hormone (ACTH) and subsequent adrenal steroid synthesis (New, 1985). The 21-OH form of CAH is characterized by an accumulation of cortisol precursors, resulting in increased production of adrenal androgens in utero. Approximately two thirds of CAH individuals also manifest a disturbance in mineralocorticoid regulation and are salt-losers. The incidence of the disorder is estimated to be approximately 1 in 14,200 live births (Pang et al., 1988), with equal rates for males and females. Because androgen levels are high during embryonic differentiation of the external genitalia, females usually have masculinized genitalia, particularly clitoral enlargement and labial fusion. Internal reproductive structures are normal, because there are no testes to produce Mullerian inhibiting factor, and the level of androgen produced is probably not sufficient to induce Wolffian development (New, 1985). Treatment for the 21-OH deficiency has been available since the 1950s; it consists of corticosteroid and (for salt-losers) mineralocorticoid replacement therapy. In addition, surgical reconstruction of malformed genitalia is usually necessary in affected females, although corticosteroid treatment prevents further virilization. Thus, treated individuals have been exposed to excess quantities of androgens during the prenatal and neonatal periods but have normal levels postnatally with optimal hormone therapy.

Knowledge of the behavioral sequelae of CAH is important for the medical management of patients. Further, CAH provides a unique opportunity to study the effects of elevated prenatal and neonatal androgen on later behavior. Studies in laboratory animals indicate that gonadal hormones play a major role in the development of sex differences in behavior and the brain. For example, a genetic female rat who is given testosterone during specific sensitive periods will develop a pattern of behavior typical of normal male rats. This masculinization or defeminization is not limited to sexual behavior, but includes rough play and maze performance (e.g., Goy & McEwen, 1980; Beatty, 1979). Although these effects have been studied most extensively in rats, hormonal influences on sexually dimorphic behavior have been documented in a wide variety of species, including primates and birds (for reviews, see Arnold and Gorski, 1984, and Goy & McEwen, 1980).

In the past few years, evidence has accumulated that gonadal hormones also influence human behavioral development. Most of this evidence comes from studies of individuals exposed to unusual levels of gonadal hormones prenatally [when much of neural development and sexual differentiation appears to occur (Goy & McEwen, 1980; Hines, 1982)]. This includes individuals with CAH, as well as individuals with Turner's syndrome, androgen insensitivity, and those whose mothers were treated during pregnancy with hormones such as progesterone, synthetic progestins, estradiol, and the synthetic estrogen, diethylstilbestrol (DES). In general, these studies indicate that prenatal gonadal hormones affect sex-typed behavior in human beings, particularly play, aggression, cognitive abilities, and sexual behavior.

In this chapter, I will review in detail studies that have been conducted on individuals with CAH. The findings of these studies are important for two main reasons: (a) understanding the behavioral sequelae of CAH and other conditions that produce elevated hormones and for providing proper medical management of the patients and (b) indicating how hormones can affect human behavior, particularly sexually dimorphic behavior.

In a typical study of behavior in individuals with CAH, patients are recruited from a pediatric endocrine clinic at a university-affiliated hospital, and range in age from early childhood to adulthood. Because treatment was started in the early 1950s, it is possible to study both patients with early treatment and those with late (or no) treatment. In order to assess perinatal (rather than postnatal) hormone influences, it is best to study patients with early effective treatment. The major differences among studies are in the behaviors assessed and the instruments used. A major problem in developing a test battery for use with CAH patients is finding tests that are suitable for subjects of varying ages, and that measure the same trait in all ages. Most studies of CAH patients focus on behaviors that show sex differences, because studies in animals clearly indicate that gonadal hormones only influence behaviors that normally show sex differences. Thus, because these traits show large sex differences in normal samples (e.g., Maccoby & Jacklin, 1974), the most commonly studied traits in hormone-exposed subjects are cognitive abilities and aspects of psychosexual differentiation (gender identity, activities and interests, sex-role self-concept, personal attributes, and sexual behavior).

Review of Literature

Cognitive Abilities

General Intellectual Ability

Money's initial studies (Money & Lewis, 1966; Lewis et al., 1968) on general intellectual ability (IQ) in CAH children and adults spurred much of the work in this area. These studies indicated that the mean IQ of CAH patients exceeded that of the general population; 60% of CAH patients had an IQ above 110, compared with 25% of the general population. Money and Lewis (1966) could find no obvious explanation for the elevated IQs: IQ was not related to sex, disease symptoms, age at treatment, or residential proximity to the hospital. Nonetheless, there was some suggestion in their data that the CAH patients seen represented a biased sample of the population, because a small sample of unaffected siblings of CAH patients also showed elevated IQs.

Subsequent studies with sibling controls allowed a further assessment of this IQ elevation. Baker and Ehrhardt (1974) and McGuire and Omenn (1975) confirmed Money and Lewis's (1966) finding of elevated IQ in CAH patients. Importantly, however, they also found elevated IQs in the siblings of the CAH patients, with no differences between the IQs of patients and unaffected siblings. Although both groups of investigators discussed various hypotheses that could explain the intellectual superiority in both CAH patients and their unaffected siblings, including complicated associations between the genes for CAH and intelligence, the most likely explanation for the results is sampling bias. Families with higher than average IQ might come to the clinics where these studies are conducted, and these studies were done in northern states, where the mean IQ is higher than 100. The bias is probably not due to socioeconomic status, because the families in Baker and Ehrhardt's (1974) study did not have high SES. Thus, it does not appear that CAH itself is associated with elevated IQ. As there are no sex differences in general intelligence, and animal studies suggest that androgen affects only those behaviors that show sex differences, it is not surprising that CAH patients do not have higher IQs than their siblings.

Specific Cognitive Abilities

What about abilities that do show sex differences, such as spatial abilities, verbal fluency, and perceptual speed? If prenatal or neonatal hormones act to organize the brain in humans as they do in rodents and primates, we would expect girls and women with CAH to have patterns of abilities more like normal males than normal females. Thus, CAH females would be hypothesized to have higher spatial ability and lower verbal ability and perceptual speed scores than their unaffected sisters. Although these hypotheses are straightforward, they have turned out to be difficult to examine, for several reasons.

Methodological Considerations

There are four major methodological concerns that affect the sensitivity of studies designed to assess hormonal influences on specific cognitive abilities in patient with CAH. First, different cognitive tests are differentially sensitive to sex differences. Although there is a general sex difference in spatial ability, the difference is not found in all aspects of spatial ability. The biggest sex differences are on tests of spatial orientation, whereas there are very small or no sex differences on tests like Block Design. Tests of abilities that do not show sex differences are not informative for assessing differences between CAH patients and controls. Second, sex differences in most aspects of spatial ability do not appear consistently until puberty (Maccoby & Jacklin, 1974). Thus, studies of CAH children are less likely to be able to detect patient-control differences than are studies of CAH adolescents and adults. Third, studies of CAH patients are generally restricted to small samples because of the relatively low incidence of the disorder, making it difficult to see differences between patient and control groups. Consider, for example, the effect of small samples on the likelihood of finding differences between CAH and control females in spatial orientation ability, which shows a large sex difference (approximately one standard deviation on tests of mental rotation). The difference between CAH and control females is likely to be smaller than the sex difference, say three quarters of a standard deviation. In order to have sufficient statistical power to see such a difference (i.e., to maximize the probability of seeing a significant difference between the two groups if a difference does exist in the populations from which they were drawn), it is necessary to have 23 subjects per group (Cohen, 1977). Given the age distribution of patients in a single pediatric endocrine unit, it is usually not possible to obtain samples of this size in a single institution. Of course, when differences are smaller than that expected for spatial ability, larger samples than this will be necessary. Therefore, failures to observe differences between CAH and control females may be explained simply by lack of statistical power associated with small samples. Fourth, the choice of a control group affects the statistical power to see differences between groups. Siblings represent the best comparison group (Reinisch & Gandelman, 1978), because they are matched to patients on general genetic and environmental background, and sampling characteristics. Although they cannot be matched individually on age and birth order, groups of patients and siblings will generally turn out to be matched on these variables, and any differences between the groups can be adjusted with analyses of covariance (using statistical control instead of experimental control). The increase in power obtained from matched-pairs analyses can be substantial in studies of abilities, because correlations between test scores of siblings are approximately 0.4 to 0.5. In the example given above, where differences between patients and controls are estimated to be three quarters of a standard deviation, the use of sibling pairs reduces the number of subjects required by 44%, to 13 pairs. It usually turns out, given family constellations, that only 50% to 70% of patients have a same-sex sibling control, so analyses are done two ways: matched pairs with a reduced sample, and unmatched pairs comparing all female patients with all female controls (including sisters of male patients) and all male patients with all male controls (including brothers of female patients).

Studies of Specific Cognitive Abilities

There are four published studies examining the effects of early hormones on specific abilities. These studies are summarized in Table 15.1, and discussed herein. Resnick et al. (1986) were the only ones to find evidence of an androgenizing influence on spatial abilities, but they were also the only ones to use sibling controls, adolescent and adult subjects, and tests that show large sex differences in normal samples.

Baker and Ehrhardt (1974) examined hormonal influences on cognitive abilities in a sample of 27 CAH children and 27 siblings, with most subjects in middle childhood and early adolescence. Abilities were measured with the age-appropriate Wechsler tests of general intelligence and the age-appropriate form of the verbal, number, and spatial tests of the Primary Mental Abilities Test (PMA). Although Baker and Ehrhardt had a total sample of reasonable size and used sibling controls, they had difficulty testing hypotheses about hormonal influences on specific cognitive abilities because of the widely varying ages of the subjects and the consequent difficulty in finding tests that tap the abilities that show sex differences. For example, the forms of the spatial subtest of the PMA differ substantially across the different age ranges. The spatial subtest for young children is essentially a picture-completion test, whereas the test for adolescents and adults is a measure of two-dimensional spatial orientation. The latter type of ability shows sex differences, whereas the former does not. Therefore, it is not surprising that there were no differences observed between any patient group and any comparison group on either the spatial or verbal tests. The only noteworthy finding was a relative inferiority in numerical skills in CAH subjects (both males and females). This finding is difficult to interpret, however, because sex differences are not usually found on the PMA number subtests. Baker and Ehrhardt's study represents a pioneering effort in our attempts to understand whether hormones affect specific cognitive abilities in CAH patients, but difficulties in their test battery cause these questions to remain open.

TABLE 15.1. Summary of studies of cognitive abilities in CAH patients.

Study	Subjects		Age range	Cognitive tests	Results
	CAH	Controls			
Baker & Ehrhardt (1974)[a,b]	17F 10M	Siblings	4–26 most children and young adolescents	Wechsler Scales of Intelligence Primary Mental Abilities	F, M CAH<sibs, number ability
Perlman (1973)[b]	10F 10M	Matched Normals	3–15	11 tests to detect learning disabilities	F mixed results M no differences
McGuire & Omenn (1975)[a,b]	16F 15M	Matched Normals	5–30	Embedded Figures Digit Symbol Block Design Stroop	No differences
Resnick et al. (1986)	17F 8M	Siblings/ cousins	11–31	Specific cognitive abilities: verbal, spatial, speed; general ability	F CAH>sib, spatial no differences in other abilities M no differences

[a] Used different tests with different age groups.
[b] Tests do not show sex differences.

232

Perlman (1973) studied specific cognitive abilities in male and female patients with CAH, male "pseudohermaphrodites," and matched normal (nonsibling) controls. The test battery consisted of 11 tests chosen for their ability to distinguish learning-disabled from normal children, because Perlman had hypothesized that hormonally unusual children would show patterns of performance similar to learning-disabled children (the reasons for this are unclear). CAH boys did not differ from controls on any tests. CAH girls scored higher than control girls on some tests, but lower than controls on other tests, in no consistent pattern. The differences are difficult to interpret because many of the subtests used do not show sex differences.

McGuire, Ryan, and Omenn (1975) tested 31 CAH patients, matched to normal controls on the basis of age, height, sex, urban/rural residence, and full-scale IQ. Differential abilities were measured with age-appropriate forms of the Embedded Figures Test, Digit Symbol and Block Design subtests of Wechsler intelligence tests, and Stroop Color Naming Test (which shows a female superiority and measures either speed, fluency, or resistance to interference). The lack of sibling controls and the reduction in sample size caused by the separate analysis of data from children and adults resulted in low statistical power to see differences between patients and controls. Further, the tests used made it difficult for McGuire et al. to assess hormone effects. None of the cognitive tests showed large or consistent sex differences in previous studies, especially in children, and they did not show consistent sex differences in the control subjects in the McGuire et al. sample, again particularly in the children. Therefore, it is not surprising that there were no differences between CAH patients and their controls (either males or females, children or adults). The data of McGuire et al. also leave open the question of hormonal influences on sexually dimorphic cognitive abilities.

Resnick, Berenbaum, Gottesman, and Bouchard (1986) designed their study specifically to correct the difficulties that had prevented previous investigators from seeing differences in abilities between CAH patients and controls. They used sibling and cousin controls, subjects in the age range where cognitive sex differences have been consistently found, and tests that differentiate normal male and female subjects. They tested 25 CAH patients and 27 sibling or cousin controls on a battery of cognitive tests selected to emphasize three cognitive abilities for which sex differences have been reported: spatial ability, verbal fluency, and perceptual speed and accuracy. In addition, tasks that measure general intellectual ability (PMA Vocabulary and Raven's Progressive Matrices) and that do not show sex differences were included to assess the comparability of patient and control groups. CAH females scored significantly higher than unaffected female relatives on three of five tests of spatial ability: Hidden Patterns (a measure of disembedding) and the two measures of spatial orientation, Card Rotations and Mental Rotations. There were no differences between CAH females and controls on tests of verbal ability, perceptual speed and accuracy, or general ability. There were no significant differences between CAH males and their brothers on any test. Resnick et al. failed to confirm the previous reports of a numerical deficit in CAH patients. Thus, studying subjects in the age range in which sex differences can be reliably found, using tests sensitive to sex differences, Resnick et al. were able to document enhanced spatial ability in CAH females, consistent with the hypothesis that exposure to androgens early in development influences later patterns of cognitive abilities.

Summary and Interpretation

Only one of four studies was able to document differences in cognitive ability between CAH and control females that parallel sex differences in normal subjects, that is, Resnick et al. found enhanced spatial ability in CAH adolescent and adult females. It is important to emphasize, however, that this was the only study that specifically addressed the important methodological issues outlined earlier. Clearly, these results need to be replicated and extended. For example, do CAH females show enhanced performance in all aspects of spatial ability? Do they also show enhanced mathematical ability, as Benbow (1988) would suggest? (Note that mathematical ability is different from the simple numerical skills assessed in early studies.) Is the failure to find differences in verbal fluency and perceptual speed and accuracy replicable? Such results would suggest that CAH males are masculinized but not defeminized. It is important to note that findings of enhanced ability in individuals with CAH are remarkable in light of the fact that genetic diseases are often associated with impaired intellectual functioning (e.g., Smith, 1976).

Psychosexual Differentiation

Ehrhardt and Money and colleagues (Ehrhardt et al., 1968a; Ehrhardt et al., 1968b) were the first to recognize that people with clinical syndromes associated with unusual prenatal hormone exposure (including CAH) represented an opportunity to study hormonal influences on human behavior parallel to primate studies then being conducted which showed that female monkeys given neonatal testosterone were behaviorally masculinized. Thus, in the mid-1960s, they embarked upon a series of studies that provide us with valuable information about the relationship between early hormones and subsequent gender identity, gender-role behavior, and sexual preference (see Money & Ehrhardt, 1972), and set the stage for studies by others.

Methodological Considerations

Some of the same methodological issues important in studying hormonal influences on cognitive abilities apply to the study of psychosexual differentiation, and additional issues emerge. First, tests need to be sensitive to sex differences and appropriate for subjects in the particular age range studied. Second, sample size and choice of control group affect statistical power to see differences between groups. Third, the nature of the behavior of interest needs to be clearly defined and related to other aspects of sex-typed behavior. Thus, for example, there are several proposed schemes for classifying sex-typed behavior that might be influenced by early hormones (Ehrhardt & Meyer-Bahlburg, 1979; Huston, 1983). Following convention, I have included under psychosexual differentiation studies of gender identity, gender-role behavior, and sexual behavior. Fourth, gender-related behaviors show developmental change in form as well as amount, so different behaviors must be assessed at different ages. For example, sex-typed play can be assessed only in children, whereas sexual behavior cannot be assessed before adolescence.

Gender Identity

In general, genetic females with CAH reared as girls develop normal (female) gender identity. Interviews with CAH females indicate that they identify themselves as females (Ehrhardt et al., 1968a; Ehrhardt & Baker, 1974), and CAH girls usually draw a girl first on the Draw-A-Person test (McGuire et al., 1975; Perlman, 1973).

There are a few cases of genetic females with CAH reared as males because they were very virilized at birth. In those cases, sex of rearing was male either because the diagnosis was made when the child was too old for a change or the parents refused medical advice. These patients have not been studied systematically, but reports suggest that they develop normal male gender identity.

Gender-Role Behavior

Gender-role behavior refers to aspects of behavior in which males and females differ from each other in our culture at this point in time (Ehrhardt & Meyer-Bahlburg, 1979). Behaviors primarily studied under this rubric are activities, interests, personality, and social attributes. Because the nature of gender-role behaviors changes across the life span, it is important to separate studies examining gender-role behavior in children from those examining different aspects of gender-role behavior in adolescents and adults. The studies of gender-role behavior in CAH patients are summarized in Table 15.2 and discussed herein.

Childhood Interests and Activities

Gender-role behaviors usually studied in children with CAH include sex-typed play activities and interests, because they are the most salient aspects of children's behavior. Some studies have assessed children's plans for the future. There have been no studies of sex-typed personality in CAH children, because these traits are difficult to measure in children.

TABLE 15.2. Summary of studies of gender-role behavior in CAH patients.

Study	Subjects		Age range	Behaviors studies	Results
	CAH	Controls			
Ehrhardt et al. (1968)	15F	Matched Normals	5–16	Information from interviews: Interest in:	
				reproduction/genitals	no diffs
				romance, marriage	F CAH<ctl
				motherhood	no diffs
				"Boy" play/activities	F CAH>ctl
				Tomboyism	F CAH>ctl
Ehrhardt & Baker (1974)	17F 9M	Siblings	4–26 Most child/ Adolescents	Information from interviews:	
				Rough outdoor play	F CAH>sib, M CAH> sib
				Boy playmates	F CAH>sib, M no diff
				Marriage/parenthood	F CAH<sib, M no diff
				Interest in infants	F CAH<sib, M no diff
				Doll play	F CAH<sib, M no diff
				"Boy" play/activities	F CAH>sib, M no diff
				Tomboyism	F CAH>sib
				Feminine appearance	F CAH<sib, M no diff
				Fighting	no diffs
McGuire et al. (1975)[a,b]	16F 15M	Matched Normals	5–30	Child play/games list Lunneborg scales:	no diffs
				masculinity	no diffs
				femininity	F CAH<ctl, M CAH>ctl
				Tomboyism	no diffs

Study	N	Controls	Age	Tests	Results
Berenbaum & Hines (1989)	26F 11M	Siblings/ cousins	2–8	Observed toy preference boys' toys girls' toys neutral toys	F CAH>sib, M no diff F CAH<sib, M no diff no diffs
Hurtig et al. (1983)	9F	None	adolescent	Bem sex-role inventory	within normal range
Resnick (1986)	17F 8M	Siblings/ cousins	11–31	Personal Attributes Q: masculinity femininity masc/fem Differential Pers. Q: aggression stress reaction harmavoidance constraint negative affectivity Personality Research Form: succorance other personality dims	F no diff, M CAH<sib no diffs F CAH>sib, M CAH<sib F CAH>sib, M no diff F CAH<sib, M CAH>sib F no diff, M CAH>sib F CAH<sib, M CAH>sib F no diff, M CAH>sib F CAH<sib, M no diff no diffs
Berenbaum & Hines (unpublished)	11F 14M	Siblings/ cousins	13–35	Reinisch inventory physical aggression	F CAH>sib, M no diff

[a] Used different tests with different age groups.
[b] Tests do not show sex differences.

Ehrhardt, Epstein, and Money (1968) interviewed 15 early-treated CAH girls and 15 matched (nonsibling) controls and their mothers, assessing the child's interest in reproduction and genital morphology, romance, marriage, and maternalism, and cosmetic interests, physical energy, and tomboyism. Results were presented as number of CAH versus control girls who indicated particular responses, but statistical comparisons were generally not made, presumably because of the small sample size. The major differences between CAH and control girls appeared in terms of play: compared to control girls, more CAH girls preferred boys' toys over dolls, and slacks over dresses, showed interest in outdoor activities, were likely to consider themselves tomboys, and were likely to be dissatisfied with a female sex role (the latter two differences were statistically significant with a chi-square test).

Ehrhardt and Baker (1974) replicated these results in a second sample of CAH females. This study included male CAH patients as well, and sibling controls were used for all patients. The subjects were the same ones whose cognitive abilities were reported previously (Baker & Ehrhardt, 1974). Subjects and mothers were interviewed in a semistructured format about the child's developmental history and gender-related behavior (e.g., play, interest in marriage and parenthood, aggression). In addition, 13 mothers were later interviewed about their own developmental histories. Information about specific items apparently was tabulated from the transcribed interviews by two raters and scored on rating scales, which were then dichotomized. Data were again presented in terms of percentages of CAH versus control subjects who displayed specific behaviors, and group comparisons were tested statistically with chi-square or Fisher's exact tests. Again, the major differences between CAH and control girls were in terms of play: significantly more CAH girls than sisters or mothers were reported to have a high level of physical energy expenditure (rough outdoor play), to show more interest in boys' toys than in girls' toys, and to be tomboys (59% of CAH girls were tomboys compared to no sister who was considered a long-term tomboy). Based on extensive interviews with the parents, Ehrhardt and Baker concluded that parental or patient attitude is unlikely to account for the behavioral differences observed between the CAH girls and their unaffected relatives.

With respect to CAH boys, their reported behaviors were remarkably similar to those of their unaffected brothers in most behaviors assessed. The only significant difference was that more CAH than control boys showed a high level of energy expenditure in sports and rough outdoor activities (compared to moderate or periodic interest in sports and physical activities).

McGuire, Ryan, and Omenn (1975) also examined sex-role behavior in CAH and control children, by assessing play activities (self-report) and general sex-role development (by subject interview). The subjects were the same patients and matched (nonsibling) controls who had been studied on cognitive abilities (reported earlier). Although the results indicated few differences between CAH children and controls, problems in study design make it difficult to infer to the population. The measures used showed small and nonsignificant differences between control boys and girls, making them insensitive to differences between CAH patients and controls, and nonsibling controls were used.

In a recently completed study, Berenbaum and Hines examined the sex-typed behavior of CAH girls and boys. One goal was to expand upon the studies of Ehrhardt and colleagues and McGuire et al. by studying behaviors that showed clear sex differences, using observational measures and questionnaires (rather than interviews), and sibling controls. Results from most aspects of the study are still being analyzed, but data on toy preferences are available (Berenbaum & Hines, 1989). Twenty-six girls and 11 boys with CAH between the ages of $2\frac{1}{2}$ and 8 years were tested. Sibling and cousin controls included 15 unaffected female relatives and 18 unaffected male relatives. Children played individually for 10 minutes with a standard set of toys previously shown to differentiate normal boys and girls (e.g., dolls, transportation toys). Two raters who were blind to the patient or control status of the children scored the amount of time each child spent playing with "masculine," "feminine," and "neutral" toys. Play with the toys showed the expected sex differences. As hypothesized, CAH girls spent significantly more time playing with masculine toys than did their controls (a difference of three quarters of a standard deviation), about as much time as the control boys. As in previous studies, not all CAH girls were masculinized, but we have not been able to identify medical factors (e.g., age at diagnosis, degree of virilization) that differentiate girls who played with boys' toys from those who did not. CAH girls also showed reduced play with feminine toys, although this result is somewhat difficult to evaluate because of the nature of the play task: given the limited play time, play with one set of toys generally precludes play with another set of toys. There were no differences between CAH and control boys on masculine or feminine play. These results are consistent with those of Ehrhardt and colleagues, and strengthen previous conclusions about hormonal influences on play behavior, because objective measures of observed behavior were used and ratings were made without knowledge of the patient or control status of the child.

Summary and Interpretation of Studies of Children

Three of four studies of childhood play activities and interests, using both subjective and objective assessment techniques, indicate that CAH girls differ from control girls in having higher rates or amounts of "masculine" preferences. Four important points bear highlighting. First, the "masculinization" of behavior in CAH girls is well within the normal range, and is not indicative of pathology. Second, although a majority of CAH girls show this masculinization (approximately 50-70%, depending upon the specific behaviors assessed), not all CAH girls do so. It would be interesting to explore factors that might differentiate CAH girls who are masculinized from those who are not. Third, although Ehrhardt and colleagues assessed behaviors as present or absent (they dichotomized rating scales), many of these gender-role behaviors can be considered as continuous traits. For example, intense physical expenditure or interest in infants can be measured quantitiatively, and assumed to reflect traits continuously distributed in the population. Thus, CAH girls might be seen as having higher amounts of the trait than their sisters. Indeed, a continuum of sex-typed toy preference was observed in Berenbaum and Hines' study. Such an approach is likely to facilitate inferences about hormonal influences on normal behavior from studies in CAH patients, because gender-role behaviors generally show continuous distributions in normal samples: although there are mean differences between the sexes, there is considerable variability within sex and overlap in the scores of males and females. Fourth, although there is some suggestion that CAH girls are "defeminized" as well as "masculinized," the data are not clear on this issue, in part because masculine and feminine behaviors have usually been measured as opposite ends of a continuum, and not as distinct dimensions, or because it is not possible to engage in both masculine and feminine behaviors simultaneously. For example, a girl who enjoys rough outdoor activities is likely to prefer functional to attractive clothing, because the former are comfortable and appropriate for active play. Given a choice, then, between functional or attractive, "frilly"clothing, such a girl would likely prefer functional clothing. This does not necessarily mean that she is "defeminized." Fifth, in the studies that have examined possible nonhormonal factors that could account for these differences, none proved able to explain the data, although there have been no systematic studies of these factors. Thus, these data are consistent with the hypothesis that prenatal or neonatal androgen operates on the human brain to masculinize aspects of sex-role behavior, much as it does in rodents, songbirds, and primates.

Adolescent and Adult Personality

The assessment of sex-typed behavior in CAH adolescents and adults has focused primarily on sex-role self-concept and other aspects of personality. In this context, it is important to note that tests that claim to assess the sex-typed sex-role concepts of "masculinity"and "femininity" actually measure the personality dimensions of dominance and nurturance/warmth, respectively (see, e.g., Lubinski et al., 1983).

McGuire, Ryan, and Omenn (1975) assessed sex-role behaviors in adolescents and adults, using Lunneborg's scales of "masculine" and "feminine" behaviors, and a structured interview. The subjects were the same patients and matched (nonsibling) controls who had been studied on cognitive abilities (reported earlier). As with the children they studied, McGuire et al. found few differences between adolescent and adult CAH patients and controls on gender-role behaviors. Again, problems with the study design (specifically failure to find sex differences on the tests used and the control group chosen) make it difficult to assess differences between CAH patients and controls.

Hurtig, Radhakrishnan, Reyes, and Rosenthal (1983) studied gender-role behavior in nine adolescent CAH females. The CAH females scored within "one standard deviation of the norms" (p. 889) on a sex-role inventory assessing masculinity and femininity. Hurtig et al. indicated that CAH females preferred masculine toys and active play (based on information from interviews with the subject or parent), although they did not compare CAH girls to a control group.

Resnick (1982) examined sex-role self-concept and personality in the same sample of adolescent and adult CAH patients and their relatives in which cognitive abilities were studied (see earlier). Subjects completed questionnaires which assessed both sex-typed traits and traits which do not typically show sex differences. The scores of CAH females indicated some behavioral "masculinization" and perhaps "defeminization," but good emotional adjustment. The scores of CAH males suggested some "demasculinization" or "femininization," but Resnick interpreted the results to indicate that CAH males have some emotional distress because sex differences are not usually observed on some of the scales on which CAH males scored high and which measure negative affectivity.

Preliminary data from a small sample of adolescents and adults (Berenbaum & Hines, unpublished data) provide additional evidence of "masculinization" in CAH females on aggression. To date, we have tested 25 patients and 10 controls. We measured propensity for aggression with Reinisch's revision of the Leifer-Roberts Response Hierarchy (Reinisch & Sanders, 1986), in which subjects are asked to indicate how they would have responded to a hypothetical conflict situation when they were 10 years old. The test showed sex differences in normative samples studied by Reinisch and Sanders and in our controls: males more often than females reported that they would have used physical aggression in a conflict situation. CAH females also reported significantly higher use of physical aggression than did their unaffected female relatives (sisters or cousins); there were no signficant differences between CAH males and their unaffected relatives (brothers or cousins). These data are consistent with studies in animals indicating that neonatal testosterone increases aggression in genetic females (Beatty, 1979; Quadagno et al., 1977) and with Reinisch's report (1981) of increased propensity for aggression in human females who were exposed to androgenizing progestins in utero. In contrast, Ehrhardt did not find evidence of increased fighting in CAH girls compared to controls (matched or siblings), and Berenbaum and Hines (unpublished data) did not find differences between CAH females and controls in young subjects (ages $2\frac{1}{2}$ to 8). It is possible that it is difficult to measure aggression in children by self- or parent-report or that puberty plays a role in the development of individual differences in human aggression.

Summary and Interpretation of Studies in Adolescents and Adults

Two of four studies assessing sex-typed behavior in CAH adolescents and adults found evidence of "masculinization" in CAH females. The results in adolescents and adults are less compelling than the results in CAH children, in part because of the measures used. Thus, there have been no studies of current activities and interests, behaviors which show large differences between CAH and control females in childhood. On the other hand, given the nature of behavioral development, it may be useful to examine adult behaviors that are developmentally related to, but not exact parallels of, childhood behaviors. For example, early tomboyism may be associated either with later interest in masculine activities or later personality traits that are more common in men than women (e.g., aggression). Longitudinal studies of "masculine" behaviors in CAH patients should provide important information about these issues.

Sexual Behavior and Preference

The authors of two recent review articles discussed the importance of prenatal hormones in the determination of sexual preference (Ellis & Ames, 1987; Money, 1987). They based their conclusions on evidence from various sources, including women with congenital adrenal hyperplasia. Unfortunately, the evidence from CAH women is currently equivocal with regard to sexual behavior. (There are no studies of sexual behavior in men with CAH.)

Methodological Considerations

There are several problems in assessing sexual preferences in CAH patients. First, there are only a few patients who are old enough to have established sexual preference and behavior patterns, especially if the goal is to study only those patients who received early treatment, thereby minimizing the possibility that behavioral changes are secondary to virilization, anomalous genitalia, or postnatal hormones. Second, in late-treated CAH women (who are the subjects of some studies), circulating levels of androgen are high. Therefore, it is not possible to determine if changes in sexual behavior are the result of prenatal hormones acting to organize the brain or current circulating hormones. Third, rates of homosexuality and bisexuality are relatively low so it would take fairly large samples of CAH patients to determine if prenatal or neonatal androgenization is associated with increased rates of homosexuality or bisexuality. Fourth, inferences about sexual behavior in CAH women require that a study include an appropriate internal control group; it is difficult to compare data obtained on CAH women to data reported from surveys conducted at another point in time. Fifth, it is difficult to assess sexual behavior; all studies rely on information obtained from interviews rather than observation. Sixth, subjects who participate in studies of sexual behavior may not be representative of all CAH patients, given the sensitive nature of assessing sexual behavior. It is possible that those patients with atypical sexual preference are especially likely to participate in such studies, because they are looking for an "explanation" for their behavior. Therefore, it is especially important in studies of sexual behavior to know how subjects were selected.

Studies of Sexual Behavior

There are no studies of sexual behavior that meet all methodological criteria; most do not include an internal control group. Nevertheless, these studies (summarized in Table 15.3) provide a starting point for evaluating sexual behavior in CAH women.

In an early report of seven genetic females with CAH, Money and Dalery (1976) reported that sexual orientation was consonant with sex of rearing. Four patients who were surgically corrected and reared as girls developed a female sexual orientation, whereas three corrected and reared as boys developed a male sexual orientation.

Ehrhardt, Evers, and Money (1968) studied 23 women with CAH who had received either late treatment or no treatment for their disorder. Subjects were at least 16 years old at the time of assessment; the mean age was 33 years. The procedure consisted of compilation of material from the patients' files, supplemented with a contemporary interview when possi-

TABLE 15.3. Summary of studies of sexual preference in CAH women.

Study	Subjects		Treatment	Assessment	Results
	CAH	Controls			
Ehrhardt et al. (1968)	23	Kinsey	Late or none	Files, interview	I homosexual fantasies I arousal I sexual freedom
Lev-Ran (1974)	18	None	Late	Interviews	No homosexual fantasies or experience
Money et al. (1984)	30	Endocrine	Early	Self-report; interviews	Homo/bi preference: CAH>controls Homo/bi experience: no diff?
Mulaikal et al. (1987)	80	None	Some early, some late	Questionnaires	Normal rates of homo/bisexuality D sexual activity associated with inadequate vagina

I increased D decreased relative to control group or comparison CAH group differing in symptoms.

ble, and a questionnaire (returned by nine patients). Data are presented by Ehrhardt et al. in tabulated form based on information extracted from transcripts of the interviews, and subsequently compared with data reported by Kinsey, Pomeroy, Martin, and Gebhard (1953). Although Ehrhardt et al. concluded that CAH women showed "behavior more often found in males than in females" (p. 121), the results are difficult to interpret without an internal control group. Further, it is unclear whether this masculinization refers to an increased rate of preference for women or an increased rate of sexual arousal and activity, but predominantly directed towards men. It is interesting to note here Ehrhardt and Baker's (1974) observation that adolescent CAH girls in their sample tended to start dating at a somewhat later age (another "male" pattern?), but they did not show evidence of homosexual preference. It is also important to remember that late-treated women are exposed to high levels of androgen postnatally, so their behavior could be related to elevation in circulating rather than prenatal androgen.

Lev-Ran (1974) studied the sexual orientation of late-treated CAH women in the USSR. Subjects were 18 female patients, for whom treatment began between the ages of 11 and 40. In contrast to patients studied by Ehrhardt et al. (1968), subjects reported neither homosexual fantasies nor experiences. Lev-Ran invoked different sexual mores and attitudes to explain the conflicting findings.

Money, Schwartz, and Lewis (1984) reported on sexual preferences of women based on nonblind interviews and compiled histories. Subjects were 30 women with early-treated CAH compared to two other groups of women with endocrine abnormalities resulting in low levels of androgen (15 women with androgen insensitivity and 12 patients with the Mayer-Rokitansy-Kuster Syndrome). Subjects were at least 17 years of age (no further information about age was provided). Sexual preference was rated in two ways: by the patient herself, whose self-report was based on arousal imagery in fantasy or dream, and erotic experience alone or with a partner; and by the investigators on the basis of information from the patient's history of imagery and practice. Because this study is widely cited to indicate prenatal hormone influences on sexual preference, it is important to examine the results carefully. Results indicated that 37% of CAH women rated themselves as bisexual (20%) or homosexual (17%), compared to 7% and 0%, respectively, in the control groups, a signficant difference. The numbers of women with homosexual experience were considerably smaller, however: 17% of CAH and 4% of controls; statistical testing of these differences was not reported. The differences between the groups can be seen as reflecting high rates of bisexuality or homosexuality in CAH women or low rates of such preference in the control groups. Money et al. indicate that Kinsey et al. (1953) found 10% of women in the general population to have had homosexual experience. Further, rates of homosexuality increase by age 45 (Kinsey et al., 1953), and Money et al. do not provide the age distributions for the CAH and control subjects.

A recent study on fertility rates in CAH women also addresses the issue of sexual preference, as well as suggesting additional factors to consider in interpreting studies directly assessing sexual behavior. Mulaikal, Migeon, and Rock (1987) studied 80 women with CAH (40 with the simple virilizing form and 40 with the salt-losing form), over the age of 18. These women appear to represent a reasonable sample of the 158 CAH females born between 1938 and 1957 who were seen at the institution of the investigators, and represent a larger sample of patients than assessed by Money et al. (1984) from the same institution. It is unclear how many patients received treatment in the neonatal period. Subjects completed questionnaires assessing treatment and compliance, the presence of hirsutism, the adequacy of results of reconstructive genital surgery, regularity of menstrual cycles, marital status, and fertility. Comparisons were made between women who did or did not have specific symptoms (e.g., salt losing vs. simple virilizing form of CAH, adequate vs. inadequate vaginal reconstruction); there was no control group. There were no differences between the salt-losing and simple virilizing groups on treatment compliance, hirsutism, or regularity of menstrual cycles. The responses to the questions, "Is the vaginal opening adequate for sexual intercourse? Do you have pain or discomfort during intercourse?" significantly differentiated the salt losers from those with the simple virilizing form: 48% of the former versus 83% of the latter reported satisfactory introitus and vagina. Adequacy of vaginal introitus was related to sexual behavior: heterosexual behavior was significantly more frequent, and lack of sexual experience significantly less frequent among patients with an adequate vagina than among those with an inadequate vagina. Rates of homosexual behavior were very low in all groups: one woman with an adequate introitus and three with inadequate introitus reported themselves to be bisexual or homosexual. This rate of 5% (4 out of 80) is substantially below that reported by Ehrhardt et al. (1968b) or Money et al. (1984). (Mulaikal et al. noted that all four of these women were included in the Money et al. study.) Further, there were lower rates of pregnancy in salt losers than in simple virilizers, even adjusting for lower rates of vaginal adequacy among the former. Mulaikal et al. suggested that "a greater emphasis on adequate surgical correction earlier in adolescence, as well as stricter medical management, will be needed to improve the sexual experience and fertility of these patients" (p. 182).

In an editorial accompanying the Mulaikal et al. paper, Federman (1987) suggested that the inadequate introitus may "be not so much the cause of the abnormal psychosexual adjustment as a parallel reflection of the prenatal androgen exposure" (p. 210). This explanation of the results would be more likely if vaginal inadequacy was associated with increased rates of homosexuality, rather than reduced rates of sexual activity in general, especially in light of Ehrhardt, Evers, and Money's (1968) report of increased arousal and libido in (late-treated) CAH women. Nonetheless, given societal pressures against homosexuality, it is possible that reduced heterosexual preference results in reduced sexual activity rather than overt homosexual preference. In this respect, it would be interesting to examine the sexual fantasies of the women studied by Mulaikal et al. Further, if prenatal androgen exposure affects adult sexual behavior, then women with the salt-losing form of CAH (apparently the severe form) should have higher rates of unusual sexual behavior than women with the simple-virilizing form; Mulaikal et al. reported that there were no significant differences between these groups in sexual activity.

Summary and Interpretation

The data presented by Ehrhardt, Evers, and Money (1968) on late-treated CAH women and Money et al. (1984) on early-treated CAH women suggest somewhat elevated rates of homosexual or bisexual fantasies and perhaps behavior in CAH women. These results should be considered tenuous, given low rates of homosexual behavior in Mulaikal et al. (1987) patients, difficulties in separating prenatal from postnatal hormone influences in late-treated women, and methodological problems (nonblind interviews, subjective assessments, possible sampling bias, and inadequate control groups). An open question relates to how prenatal or neonatal hormones might affect this sexual preference. For example, are the differences due to hormonal effects specifically on arousal or on specific behaviors that would lead to preference for members of the opposite sex? Physiological studies of sexual response in CAH women would provide valuable information in this regard.

It is interesting to speculate on the association between "masculinized" toy preferences in CAH girls and possibly increased rates of bisexuality or homosexuality. Several investigators have shown relationships in nonendocrine samples between "cross-sex" child play behavior and later bisexual or homosexual behavior. Grellert, Newcomb, and Bentler (1982) asked homosexual men and women to provide retrospective reports of their childhood toy preferences using a questionnaire composed of toys and games that have been shown to differentiate normal boys and girls. There were large and significant differences between heterosexual males and females, male homosexuals and heterosexuals, and female homosexuals and heterosexuals. A majority of homosexuals reported cross-sex childhood play. Green (1987) recently reported on the results of his longitudinal study of "effeminate" boys, in which he followed prospectively boys with both "cross-sex" and "male-typical" child play patterns. Boys who had preferred girls' toys and activities as children showed higher rates of bisexual and homosexual behavior as adults than did boys who had shown male-typical behavior as children. With respect to the issues raised by Mulaikal et al. and Federman, it would be interesting to see if CAH girls with unsuccessful vaginal surgery would be more likely than those with successful surgery to show masculinized play as children (before the significance of the surgery becomes apparent). Such a result would support Federman's argument that both the vaginal inadequacy and the behavior reflect parallel effects of prenatal androgens.

Emotional Adjustment

There have been no large-scale studies that focused specifically on emotional adjustment in CAH patients. Because there are no sex differences in overall emotional adjustment in the general population, differences between patients and controls would likely reflect consequences of the illness, or effects of the disease not due to gonadal hormones. On the other hand, patient-control differences in specific psychiatric diagnoses that show sex differences in the population (e.g., depression, attention deficit disorder) might be related to the influence of prenatal androgen on the brain. Given the low rates of psychiatric illness, however, such studies would be difficult to do.

There are two studies with small samples that specifically addressed emotional adjustment in CAH girls, and found it to be normal. Gordon, Lee, Dulcan, and Finegold (1986) assessed behavior problems and self-perception in 16 girls with CAH aged 6 to 16, and age-matched controls (endocrine and community samples). Parents completed the Child Behavior Checklist, a measure of behavior problems and social competency, and children younger than age 14 completed a self-perception questionnaire. The results indicated that girls with CAH were generally not different from controls. Hurtig and Rosenthal (1987) assessed psychological adjustment in nine adolescent girls with CAH and six medical controls, using two projective measures, the Rorschach and Thematic Apperception Tests. They found the CAH and control groups to be comparable on measures of personality adjustment. Although these two studies suggest that CAH girls do not have emotional problems, conclusions must remain tentative given the small samples, lack of sibling controls, and limited assessment procedures.

Some studies of CAH patients that addressed sex-typed behaviors also included some assessment of emotional functioning, usually an interview or standardized questionnaires completed by patients and their parents. In general, these studies also indicate that CAH patients show remarkably good emotional adjustment, given the variety of physical anomalies they manifest and the fact that they must adhere to a lifelong regimen of medication. Thus, for example, Resnick (1982) found CAH and control females to score similarly and within the normal range on personality scales that reflect emotional adjustment.

There is some suggestion that males with CAH may have some emotional difficulty. Thus, Resnick (1982) found that, compared to their unaffected brothers, CAH males had higher scores on scales that assess negative affectivity. Further, both McGuire et al. (1975) and Resnick (1982) interpreted "feminized" or "demasculinized" scores of CAH males to indicate emotional insecurity because the scales include items about worrying, and interpersonal anxiety. This emotional distress in CAH males has been viewed as a consequence of their short stature.

The findings of good adjustment in CAH females stand in contrast to physician reports suggesting pathology in CAH females. It is unclear whether physicians are overly sensitive to difficulties in CAH patients because of expectations, or whether systematic studies are insensitive to these emotional problems. Studies may fail to find differences between patients and controls because of the insensitivity of the tests used or because patients who volunteer for studies are not representative of the population of CAH patients, specifically volunteers may have better adjustment than nonparticipants (although the refusal rate is usually low). Clearly, further work is needed in this area.

What Do These Studies Mean?

As summarized in Table 15.4, the results of studies on CAH patients suggest few behavioral differences between CAH males and controls, but a sizable number of differences between CAH females and controls. The differences between male patients and controls are probably a consequence of the disorder and not the direct result of organizational influences of prenatal androgens. The absence of differences between CAH and control males on sex-typed behaviors is compatible with data from animal studies indicating inconsistent effects of excess androgen on behavior in males (Baum & Schretlen, 1975; Diamond et al., 1973). The most robust findings in CAH females seem to be in the area of gender-role behavior, specifically increased interest in masculine activities and physical energy expenditure, and aspects of personality, especially aggression. There is also some support for differences between CAH females and their affected relatives in spatial abilities. There may also be elevated rates of atypical sexual behavior in CAH women.

It is easy to conclude that differences between CAH females and unaffected control females must be due to the effects of excess androgen on the developing central nervous system, because of (a) the robustness of the effects, (b) the fact that differences are found almost exclusively in females, and only on tests that show sex differences in normals, (c) the good emotional adjustment of the patients, (d) failures to find other factors that could account for the differences, (e) the parallels to findings from animals experimentally treated with androgenizing hormones, and (f) the consistency of the findings in CAH and subjects with different abnormalities of hormone exposure (e.g., maternal ingestion of masculinizing hormones) (Goy & McEwen, 1980; Hines, 1982). It is important to note, however, that there have been no systematic studies of other factors that might account for behavioral differences between CAH and control females, including (a) effects of other hormones that are also elevated in CAH, particularly

TABLE 15.4. Summary of behavioral changes in CAH patients.

	Females	Males
Abilities		
IQ	I	I
	but not compared to siblings sampling/referral bias	
Spatial Ability	I	-
Gender Identity	Normal female	Normal male
	(consistent with sex of rearing)	
Activities and interests		
energy expenditure (outdoor play)	I	I ?
Tomboyism	I	-
Preference for boys' toys	I	-
Preference for girls' toys	D	-
Interest in infants	D	-
Interest in marriage and parenting	D ?	-
Personality		
Aggression	I	-
Sexual Behavior		
Rates of homosexual fantasies and homosexual behavior	I ?	N/A
Pattern of arousal and response	"Male" ?	
Fertility	D (related to inadequacy of vaginal introitus)	

I increased, D decreased, - unchanged compared to controls.
? indicates that findings are equivocal.
N/A: behaviors not assessed.

progesterone and corticoids, (b) side effects of cortisone replacement therapy (Quadagno et al., 1977), and (c) parental attitudes towards CAH girls with masculinized genitalia (Quadagno et al., 1977).

Although progesterone and corticoids have been found to influence some sexually dimorphic behaviors in rats (e.g., Hull et al., 1980; Meaney et al., 1982; Shapiro et al., 1976), these influences are smaller and less consistent than those of androgen, and in fact might be expected to prevent rather than promote masculinization (Hull et al., 1980; Shapiro et al., 1976). It is possible that some behavioral changes may be due to elevated levels of circulating ACTH, but treatment should normalize these levels. There is one study that indirectly assessed the behavioral effects of ACTH in CAH patients. Veith, Sandman, George, and Kendall (1985) examined visual and verbal memory and attention in CAH adults before and after changes in cortisone medication dose. They found a reduction in reaction time on one task associated with elevated ACTH levels, which they interpreted as facilitation of "visual attentional functioning and/or overt motor response capacity, rather than alteration of simple cognitive processing" (p. 33). Further studies, particularly in children, might indicate whether the increased energy expenditure in CAH girls can be attributed to ACTH-associated changes in attention or motor response. These results and similar findings on behavioral effects of ACTH in animals and people (discussed by Veith et al., 1985) would also suggest that differences between CAH and control females are unlikely to be due to treatment effects, because treatment actually suppresses the level of the hormone (ACTH) that is likely to affect behavior.

With respect to parenting practices, Quadagno et al. (1977) proposed that parents treat their CAH daughters in a masculine fashion because the girls look masculine (because of their virilized genitalia), but it is also possible that parents would treat their CAH girls in a *feminine* fashion to counteract their masculinized appearance. Data from Berenbaum and Hines' (1989) study indicate no relationship in CAH girls between time spent playing with boys' or girls' toys and degree of virilization at diagnosis assessed from medical records. Data from parental interviews (Ehrhardt & Baker, 1974) and retrospective questionnaires (Resnick, 1982) also argue against differential parental treatment. Nevertheless, these studies were not specifically designed to assess questions about parental expectations and child-rearing practices.

Possible Neural Substrate

If the differences between CAH and control females are the result of androgen acting to organize the brain during prenatal (and perhaps early postnatal) development, then what is the likely mechanism whereby behavior is "masculinized" and perhaps "defeminized"?

Observations of morphological sex differences in the brains of rodents, songbirds, and primates have suggested that sex differences in brain structure and organization may underlie dimorphisms in behavior and neuroendocrine regulation. For example, the preoptic area of the hypothalamus, which is involved in sexual behavior and endocrine control, shows sex differences in rodents, hamsters, and primates (Ayoub et al., 1983; Gorski et al., 1978; Greenough et al., 1977). Sex differences have also been observed in other parts of the rodent brain (e.g., Matsumoto & Arai, 1981; Mizukami et al., 1983), and in areas of songbird brains (Nottebohm & Arnold, 1976). In rodents and songbirds, steroid hormone influence on these structural differences has also been demonstrated by prenatal and neonatal hormone manipulations (Gorski et al., 1978; Greenough et al., 1977; Gurney & Konsihi, 1980). (For reviews, see Arnold & Gorski, 1984, and Goy & McEwen, 1980.)

Similar sex differences in brain structure have been demonstrated in human beings. There are sex differences in an area of the human brain comparable to the sexually dimorphic nucleus of the preoptic area of the rodent hypothalamus (Swaab & Fliers, 1985). Other sex differences have been noted, but may not be reliable. de Lacoste-Utamsing and Holloway (1982), in a study of postmortem brains, reported that the most posterior region of the corpus callosum, the splenium, was larger and more bulbous in women than in men. Witelson (1985) was unable to replicate this finding, and a study of intact people using magnetic resonance imaging also failed to find sex differences in the splenium or any other aspect of the corpus callosum (Oppenheim et al., 1987). No one has yet investigated differences in brain structure between people with unusual prenatal hormone exposure and unexposed controls, for example, CAH women compared to their unaffected sisters, but this would be an obvious step once we identify regions that show reliable sex differences. This would likely be accomplished in living people using magnetic resonance imaging, because it would be extremely difficult to get a sufficient sample of appropriate brains for postmortem analysis. Such studies would be most powerful if they included behavioral measures, with the goal being to relate behavioral differences between CAH women and their sisters (e.g., in spatial ability) to differences in brain structure.

Studies examining variability in human brain function rather than structure have revealed sex differences in both interhemispheric organization (lateralization) and intrahemispheric organization (regional differences) that may underlie sex differences in behavior. These studies have relied on a variety of techniques to assess individual differences in brain organization, including perceptual asymmetries, evoked potentials, regional cerebral blood flow, and studies of brain-damaged patients (Bryden, 1982). Thus, women appear to be less lateralized than men, at least for normal left-hemisphere skills (Bryden, 1979; McGlone, 1980), and to have expressive language and motor skills represented predominantly in the anterior regions of the left hemisphere, compared to equal representation of these functions in anterior and posterior regions in men (Kimura, 1983; Taylor et al., 1988). Gur and colleagues (1982) have also reported sex differences in cerebral blood flow during rest and cognitive activity, including a greater percentage of fast-clearing tissue, presumably gray matter, in right-handed females compared to right-handed males.

There is some evidence that the sex differences in lateralization are influenced by hormones: compared to control women, women with low levels of prenatal hormones (due to Turner's syndrome) are less lateralized (Gordon & Galatzer, 1980; Netley & Rovet, 1982) and women exposed to high levels of prenatal estrogen (due to maternal ingestion of diethylstilbestrol, DES) are more lateralized (Hines & Shipley, 1984). There are no published studies of lateralization in women with CAH. Berenbaum and Hines are examining lateralization differences between CAH patients and their relatives, and asking whether differences in spatial ability between

CAH and control females can be explained by differences in lateralization. We are also testing subjects on a battery of tests known to be sensitive to damage in specific cortical regions (e.g., spatial tasks disrupted by damage in the right parietal lobe), with the goal of delineating the neural substrate of hormonal influences on cognitive ability.

In an indirect assessment of hormonal influences on brain organization in CAH patients, Nass et al. (1987) examined hand preference in CAH patients and their siblings. Males are more often left-handed than females, and Geschwind and Behan (1981) suggested that this sex difference is due to the effect of testosterone on the development of the cerebral hemispheres. Further, handedness is related to brain organization: in general, left handers are more likely than right handers to show reversed or reduced lateralization for language. Thus, the Nass et al. study provides an indirect test of the effects of hormones on the development of brain organization. A handedness inventory was completed by individuals from 80 families, including 42 CAH females (mean age 11.8 years), 46 sisters (mean age 16.5), 45 CAH males (mean age 11.6), and 36 brothers (mean age 14.1). There were 19 male patient-brother pairs and 18 female patient-sister pairs. In matched-pair t-tests, Nass et al. found that CAH females were significantly more left-biased than their sisters; the difference was about 0.6 standard deviations. Analyses comparing all CAH female patients to all sisters (unmatched-pairs analyses) revealed differences in the same direction, but they were not statistically significant. (This study illustrates the power of matched-sibling pairs.) There were no differences between CAH males and brothers on matched or unmatched analyses. The data of Nass et al. are consistent with the notion that prenatal androgens affect the lateral specialization of the brain. It is interesting to note in this context that Resnick (1982) also examined handedness differences between CAH patients and unaffected relatives. She found no differences, but given her relatively small sample size, and the size of the difference obtained by Nass et al., it is not possible to know whether the lack of differences represents a conflicting finding or merely low statistical power.

In sum, there is some weak, indirect evidence that prenatal androgen might affect behavior by acting directly on the brain, but clearly more studies are needed. Studies that examine interhemispheric and intrahemispheric organization of the brain (using evoked potentials, perceptual asymmetries, and regional cerebral blood flow) in CAH patients would be valuable, but such studies should be preceded by studies that delineate in clear detail the extent and nature of the sex differences in human brain structure and function.

Behavioral Mechanisms

Although hormones might act directly on the development of synapses or receptors in specific regions that subserve behaviors that show sex differences (e.g., spatial ability, sexual behavior), directly facilitating behaviors subserved by those regions, hormonal influences might also be indirect. Thus, hormones might affect other structures (neural or peripheral) that subserve other functions (behavioral or nonbehavioral), which in turn may underlie the behaviors that show sex differences and differences between CAH and control females. For example, it is possible that prenatal androgens facilitate the development of specific types of muscle fibers, which in turn facilitate energy expenditure, resulting in different patterns of play in boys compared to girls, and CAH girls compared to their sisters. Further, although CAH and control females differ on a variety of behaviors, we do not know whether androgen influences all behaviors through a single, common mechanism; whether androgen influences one behavior, which in turn, influences other behaviors; or whether these behaviors are influenced by the actions of androgen on several different systems. Such possibilities argue for a better understanding of the relationships among the behaviors that show sex differences and androgen influences.

The various sexually dimorphic traits are correlated with each other in normals. For example, there are correlations between cognitive abilities, on the one hand, and toy preferences (Serbin & Connor, 1979), sex-typed activities (Connor and Serbin, 1977; Fraser, 1982; Newcombe et al., 1983; Serbin & Connor, 1979) and sex-role self-concept (masculine/feminine) (Signorella & Jamison, 1986), on the other hand. These correlations have been used to argue about the direction of effects: sex differences in spatial ability are thought to be "caused" by sex differences in early sex-typed toy play (Sherman, 1967) or sex-role self-concept (Signorella & Jamison, 1986). It is not possible, however, to infer causation from these studies. The relationship between early toy play and later spatial ability may occur because (a) play with boys' toys (e.g., blocks) directly facilitates the development of spatial ability; (b) children *choose* different types of play materials because of differences in ability; or (c) a common factor, such as hormones or social practices, influences both sets of behavior. In cross-sectional studies of normals, it is difficult to decide among alternatives. Studies of hormone-exposed individuals, however, provide an opportunity to study whether the relationships among these variables are mediated by hormonal or nonhormonal factors, because hormonal "sex" and sex of rearing are separated. Thus, studies of CAH girls can tell us, for example, whether the relationship between toy preference and spatial ability is due to hormones influencing both behaviors directly but separately, or to hormones influencing toy preference, which in turn influences spatial ability.

Resnick et al. (1986) provided the only report of relationships among behaviors that showed differences between CAH and control females. They suggested that hormonal influences on activities and cognition appear to be unrelated. Thus, they obtained retrospective reports about the early childhood activities of the same subjects whose cognitive abilities were tested in adolescence or adulthood (as discussed previously), using a section of the Early Life Experiences Questionnaire (Fraser, 1982).

Resnick et al. confirmed expected sex differences on the Spatial Manipulation and Verbal Expression scales, and observed similar differences between CAH and control females. Although Fraser (1982) reported some associations between these activity scales and targeted cognitive abilities (e.g., Spatial Manipulation and spatial ability), the pattern of results obtained by Resnick et al. suggested that differences between CAH females and controls on activity scales were not related to the differences on specific abilities. Thus, although CAH females differed from control females on both spatial ability (see earlier) and the Spatial Manipulation scale, there was no association between spatial ability and the Spatial Manipulation scale on a within-subjects analysis. Further, there was an association between verbal fluency and the Verbal Expression activities scale, and a difference between CAH and control females on the Verbal Expression scale, but not on verbal fluency itself. Resnick et al. (1986) concluded that "to the extent that differential hormone exposure influences both activity and cognitive variables, its effects across the two domains appear to be unrelated" (p. 196).

In a study of children with CAH that I have just begun, I am examining relationships among measures that differentiate CAH and control females. One goal is to follow-up children already tested on play behavior to see if these CAH girls also have enhanced spatial ability, and, if so, to determine whether hormonal influences on spatial ability are the result of increased play with boys' toys. A second goal is to assess if increased motor activity level accounts for participation in masculine activities. In order to accomplish the goals of examining relationships among hormonally influenced behaviors, it is necessary to have measures sensitive enough to see within-group correlations as well as between-group differences. This usually requires multiple measures of behavior aggregated across situations and occasions (Eaton, 1983; Epstein, 1980). To date, studies of hormone-exposed patients have not obtained such aggregated measures. In my current study of CAH patients, all behaviors are being assessed with multiple measures (several methods and several occasions) to obtain stable, reliable individual differences on the traits of interest. This should facilitate answering the question about whether hormonal influences on various sex-typed behaviors are related. The longitudinal nature of the study will also allow examination of the causal nature of these relationships.

Conclusions

A review of the literature inevitably answers some questions and raises others. With respect to behavioral functioning in patients with congenital adrenal hyperplasia, there is good evidence that, on average, females with CAH are behaviorally "masculinized" in play, activities and interests, aggression, spatial ability, and perhaps sexual behavior, whereas males with CAH do not differ from controls with respect to sex-typed behavior.

There are still many questions that remain with respect to the nature and mechanism of the behavioral changes in CAH females. How do parent expectations and child-rearing practices affect the development of sex-typed behavior? What factors (e.g., medical, social) differentiate females who are "masculinized" from those who are not? Are CAH females "defeminized" as well as "masculinized" (e.g., do they have low scores on measures of nurturance and abilities on which females outperform males, such as verbal fluency)? To what extent are behavioral changes related to *postnatal* levels of androgen or ACTH? What are the specific behaviors that are affected (e.g., do scores on "masculinity" scales really reflect personality, such as aggression)? And perhaps most important, what are the neural and behavioral mechanisms that produce the behavioral changes? How does androgen act to organize the development of the brain and subsequent behavior? What are the behavioral mediators that account for the increase in play with boys' toys, for example?

In terms of emotional adjustment, CAH females seem to have few problems, but additional data are necessary here, particularly as increasing numbers of early-treated CAH females enter adulthood and become concerned with sexuality and reproduction. Some CAH males appear to have some emotional difficulty secondary to the disorder. Emotional problems in both males and females are likely to be prevented if the disease is diagnosed and treatment is begun early in life. Such goals are likely to be met with the advent of newborn screening programs for CAH, which are in place in several states (Alaska, Illinois, Texas, and Washington).

The results of the studies on congenital adrenal hyperplasia reviewed in this chapter provide important information for understanding the behavioral sequelae of CAH and for maintaining appropriate medical support of CAH patients. They also contribute to our understanding of the biological factors that can affect normal behavioral development, particularly the development of sex-typed behavior.

Acknowledgment: I thank Joanne Rovet, Clarissa Holmes, and Kenneth Zucker for helpful comments on an earlier version of this paper. Preparation of this chapter and my own research reported here were supported by National Institutes of Health Grant HD19644.

References

Arnold, A.P., & Gorski, R.A. (1984). Gonadal steroid induction of structural sex differences in the central nervous system. *Annual Review of Neuroscience, 7,* 413–442.
Ayoub, D.M., Greenough, W.T., & Juraska, J. (1983). Sex differences in dendritic

structure in the preoptic area of the juvenile macaque monkey brain. *Science,*
219, 197–198.

Baker, S.W., & Ehrhardt, A.A. (1974). Prenatal androgen, intelligence, and cognitive sex differences. In R.C. Friedman, R.M. Richart, & R.L. Vande Wiele
(Eds.), *Sex differences in behavior.* New York: Wiley.

Baum, M.J., & Schretlen, P. (1975). Neuroendocrine effects of perinatal androgenization in the male ferret. *Progress in Brain Research, 42,* 343–355.

Beatty, W.W. (1979). Gonadal hormones and sex differences in nonreproductive
behaviors in rodents: Organizational and activational influences. *Hormones and*
Behavior, 12, 112–163.

Benbow, C.P. (& others with commentaries). (1988). Sex differences in mathematical reasoning ability in intellectually talented preadolescents: Their nature,
effects, and possible causes. *Behavioral and Brain Sciences, 11,* 169–232.

Berenbaum, S.A., & Hines, M. (April, 1989). *Hormonal influences on sex-typed*
toy preferences. Paper presented at the Meeting of the Society for Research in
Child Development, Kansas City.

Bryden, M.P. (1979). Evidence for sex-related differences in cerebral organization.
In M.A. Wittig & A.C. Petersen (Eds.), *Sex-related differences in cognitive*
functioning. New York: Academic.

Bryden, M.P. (1982). *Laterality: Functional asymmetry in the intact brain.* New
York: Academic.

Cohen, J. (1977). *Statistical power analysis for the behavioral sciences* (rev. ed.).
New York: Academic.

Connor, J.M., & Serbin, L.A. (1977). Behaviorally based masculine- and feminine-
activity-preference scales for preschoolers: Correlates with other classroom
behaviors and cognitive tests. *Child Development, 48,* 1411–1416.

de Lacoste-Utamsing, C., & Holloway, R.L. (1982). Sexual dimorphism in human
corpus callosum. *Science, 216,* 1431–1432.

Diamond, M., Llacuna, A., & Wong, C.L. (1973). Sex behavior after neonatal
progesterone, testosterone, estrogen, or antiandrogens. *Hormones and Behav-*
ior, 4, 73–88.

Eaton, W.O. (1983). Measuring activity level with actometers: Reliability, validity,
and arm length. *Child Development, 54,* 720–726.

Ehrhardt, A.A., & Baker, S.W. (1974). Fetal androgens, human central nervous
system differentiation and behavior sex differences. In R.C. Friedman, R.M.
Richart, & R.L. Vande Wiele (Eds.), *Sex differences in behavior.* New York:
Wiley.

Ehrhardt, A.A., Epstein, R., & Money, J. (1968a). Fetal androgens and female
gender identity in the early-treated adrenogenital syndrome. *Johns Hopkins*
Medical Journal, 122, 160–167.

Ehrhardt, A.A., Evers, K., & Money, J. (1968b). Influence of androgen and some
aspects of sexually dimorphic behavior in women with the late-treated adrenogenital syndrome. *Johns Hopkins Medical Journal, 123,* 115–122.

Ehrhardt, A.A., & Meyer-Bahlburg, H.F.L. (1979). Prenatal sex hormones and the
developing brain: Effects on psychosexual differentiation and cognitive function. *Annual Review of Medicine, 30,* 417–430.

Ellis, L., & Ames, M.A. (1987). Neurohormonal functioning and sexual orientation: A theory of homosexuality-heterosexuality. *Psychological Bulletin, 101,*
233–258.

Epstein, S. (1980). The stability of behavior. II. Implications for psychological research. *American Psychologist, 35,* 790–806.

Federman, D.D. (1987). Psychosexual adjustment in congenital adrenal hyperplasia. *New England Journal of Medicine, 316,* 209–211.

Fraser, M.E. (1982). *The early life experiences questionnaire: A test of its validity in measuring differential antecedents of specific mental abilities.* Doctoral dissertation, University of Minnesota. University Microfilms No. 8308047.

Geschwind, N., & Behan, O. (1981). Laterality, hormones, and immunity. In N. Geschwind & A. Galaburda (Eds.), *Cerebral dominance.* Cambridge, MA: Harvard University Press.

Gordon, A.H., Lee, P.A., Dulcan, M.K., & Finegold, D.N. (1986). Behavioral problems, social competency, and self perception among girls with congenital adrenal hyperplasia. *Child Psychiatry and Human Development, 17,* 129–138.

Gordon, H.W., & Galatzer, A. (1980). Cerebral organization in patients with gonadal dysgenesis. *Psychoneuroendocrinology, 5,* 235–244.

Gorski, R.A., Gordon, J.H., Shryne, J.E., & Southam, A.M. (1978). Evidence for a morphological sex difference within the medial preoptic area of the rat brain. *Brain Research, 148,* 333–346.

Goy, R.W., & McEwen, B.S. (1980). *Sexual differentiation of the brain.* Cambridge, MA: MIT Press.

Green, R. (1987). *The "sissy boy syndrome" and the development of homosexuality.* New Haven, CT: Yale University Press.

Greenough, W.T., Carter, C.S., Steerman, C., & DeVoogd, T. (1977). Sex differences in dendritic patterns in hamster preoptic area. *Brain Research, 126,* 63–72.

Grellert, E.A., Newcomb, M.D., & Bentler, P.M. (1982). Childhood play activities of male and female homosexuals and heterosexuals. *Archives of Sexual Behavior, 11,* 451–478.

Gur, R.C., Gur, R.E., Obrist, W.D., Hungerbuhler, J.P., Younkin, D., Rosen, A.D., Skolnick, B.E., & Reivich, M. (1982). Sex and handedness differences in cerebral blood flow during rest and cognitive activity. *Science, 217,* 659–661.

Gurney, M.E., & Konishi, M. (1980). Hormone-induced sexual differentiation of brain and behavior in zebra finches. *Science, 208,* 1380–1383.

Hines, M. (1982). Prenatal gonadal hormones and sex differences in human behavior. *Psychological Bulletin, 92,* 56–80.

Hines, M., & Shipley, C. (1984). Prenatal exposure to diethylstilbestrol (DES) and the development of sexually dimorphic cognitive abilities and cerebral lateralization. *Developmental Psychology, 20,* 81–94.

Hull, E.M., Franz, J.R., Snyder, A.M., & Nishita, J.K. (1980). Perinatal progesterone and learning, social and reproductive behavior in rats. *Physiology and Behavior, 24,* 251–256.

Hurtig, A.L., Radhakrishnan, J., Reyes, H.M., & Rosenthal, I.M. (1983). Psychological evaluation of treated females with virilizing congenital adrenal hyperplasia. *Journal of Pediatric Surgery, 18,* 887–893.

Hurtig, A.L., & Rosenthal, I.M. (1987). Psychological findings in early treated cases of female pseudohermaphroditism caused by virilizing congenital adrenal hyperplasia. *Archives of Sexual Behavior, 16,* 209–223.

Huston, A.C. (1983). Sex-typing. In P.H. Mussen (Ed.), *Handbook of Child Psychology,* Vol. IV (4th ed.). New York: Wiley.

Kimura, D. (1983). Sex differences in cerebral organization for speech and praxic functions. *Canadian Journal of Psychology, 37,* 19–35.

Kinsey, A.C., Pomeroy, W.B., Martin, C.E., & Gebhard, P.H. (1953). *Sexual behavior in the human female.* Philadelphia: W.B. Saunders Co.

Lev-Ran, A. (1974). Sexuality and educational levels of women with the late-treated adrenogenital syndrome. *Archives of Sexual Behavior, 3,* 27–32.

Lewis, V.G., Money, J., & Epstein, R. (1968). Concordance of verbal and nonverbal ability in the adrenogenital syndrome. *Johns Hopkins Medical Journal, 122,* 192–195.

Lubinski, D., Tellegen, A., & Butcher, J.N. (1983). Masculinity, femininity, and androgyny viewed and assessed as distinct concepts. *Journal of Personality and Social Psychology, 44,* 428–439.

Maccoby, E.E., & Jacklin, C.N. (1974). *The psychology of sex differences.* Stanford, CA: Stanford University Press.

Matsumoto, A., & Arai, Y. (1981). Effect of androgen on sexual differentiation of synaptic organization in the hypothalamic arcuate nucleus: An ontogenetic study. *Neuroendocrinology, 33,* 166–169.

McGlone, J. (& others with commentaries). (1980). Sex differences in human brain asymmetry: A critical survey. *Behavioral and Brain Sciences, 3,* 215–263.

McGuire, L.S., & Omenn, G.S. (1975). Congenital adrenal hyperplasia. I. Family studies of IQ. *Behavior Genetics, 5,* 165–173.

McGuire, L.S., Ryan, K.O., & Omenn, G.S. (1975). Congenital adrenal hyperplasia. II. Cognitive and behavioral studies. *Behavior Genetics, 5,* 175–188.

Meaney, M.J., Stewart, J., & Beatty, W.W. (1982). The influence of glucocorticoids during the neonatal period on the development of play fighting in Norway rat pups. *Hormones and Behavior, 16,* 475–491.

Mizukami, S., Nishizuka, M., & Arai, Y. (1983). Sexual differences in nuclear volume and its ontogeny in the rat amygdala. *Experimental Neurology, 79,* 569–575.

Money, J. (1987). Sin, sickness, or status? Homosexual gender identity and psychoneuroendocrinology. *American Psychologist, 42,* 384–399.

Money, J., & Dalery, J. (1977). Hyperadrenocortical 46XX hermaphroditism with penile urethra: Psychological studies in seven cases, three reared as boys, four as girls. In P.A. Lee, L.P. Plotnick, A.A. Kowarski, & C.J. Migeon (Eds.), *Congenital adrenal hyperplasia.* Baltimore: University Park Press.

Money, J., & Ehrhardt, A.A. (1972). *Man and woman, boy and girl.* Baltimore: Johns Hopkins University Press.

Money, J., & Lewis, V. (1966). IQ, genetics and accelerated growth: Adrenogenital syndrome. *Bulletin of the Johns Hopkins Hospital, 118,* 365–373.

Money, J., Schwartz, M., & Lewis, V.G. (1984). Adult erotosexual status and fetal hormonal masculinization and demasculinization: 46, XX congenital virilizing adrenal hyperplasia and 46, XY androgen-insensitivity syndrome compared. *Psychoneuroendocrinology, 9,* 405–414.

Mulaikal, R.M., Migeon, C.J., & Rock, J.A. (1987). Fertility rates in female patients with congenital adrenal hyperplasia due to 21-hydroxylase deficiency. *New England Journal of Medicine, 316,* 178–182.

Nass, R., Baker, S., Speiser, P., Virdis, R., Balsamo, A., Cacciari, E., Loche, A., Dumic, M., & New, M. (1987). Hormones and handedness: Left-hand bias in female congenital adrenal hyperplasia patients. *Neurology, 37,* 711–715.

Netley, C., & Rovet, J. (1982). Atypical hemispheric lateralization in Turner Syndrome subjects. *Cortex, 18,* 377–384.

New, M.I. (Ed.). (1985). Congenital adrenal hyperplasia. *Annals of the New York Academy of Sciences, 458,* 1–290.

Newcombe, N., Bandura, M.M., & Taylor, D.G. (1983). Sex differences in spatial ability and spatial activities. *Sex Roles, 9,* 377–386.

Nottebohm, F., & Arnold, A.P. (1976). Sexual dimorphism in vocal control areas of the songbird brain. *Science, 194,* 211–213.

Oppenheim, J.S., Lee, B.C.P., Nass, R., & Gazzaniga, M.S. (1987). No sex-related differences in human corpus callosum based on magnetic resonance imagery. *Annals of Neurology, 21,* 604–606.

Pang, S., Wallace, M.A., Hofman, L., Thuline, H.C., Dorche, C., Lyon, I.C.T., Dobbins, R.H., Kling, S., Fujieda, K., & Suwa, S. (1988). Worldwide experience in newborn screening for classical congenital adrenal hyperplasia due to 21-hydroxylase deficiency. *Pediatrics, 81,* 866–874.

Perlman, S.M. (1973). Cognitive abilities of children with hormone abnormalities: Screening by psychoeducational tests. *Journal of Learning Disabilities, 6,* 26–34.

Quadagno, D.M., Briscoe, R., & Quadagno, J.S. (1977). Effects of perinatal gonadal hormones on selected nonsexual behavior patterns: A critical assessment of the nonhuman and human literature. *Psychological Bulletin, 84,* 62–80.

Reinisch, J.M. (1981). Prenatal exposure to synthetic progestins increases potential for aggression in humans. *Science, 211,* 1171–1173.

Reinisch, J.M., & Gandelman, R. (1978). Human research in behavioral endocrinology: Methodological and theoretical considerations. In G. Dorner & M. Kawakami (Eds.), *Hormones and brain development.* Elsevier/North Holland Biomedical Press.

Reinisch, J.M., & Sanders, S.A. (1986). A test of sex differences in aggressive response to hypothetical conflict situations. *Journal of Personality and Social Psychology, 50,* 1045–1049.

Resnick, S.M. (1982). *Psychological functioning in individuals with congenital adrenal hyperplasia: Early hormonal influences on cognition and personality.* Doctoral dissertation, University of Minnesota.

Resnick, S., Berenbaum, S.A., Gottesman, I.I., & Bouchard, T.J. (1986). Early hormonal influences on cognitive functioning in congenital adrenal hyperplasia. *Developmental Psychology, 22,* 191–198.

Serbin, L.A., & Connor, J.M. (1979). Sex-typing of children's play preferences and patterns of cognitive performance. *Journal of Genetic Psychology, 134,* 315–316.

Shapiro, B.H., Goldman, A.S., Bongiovanni, A.M., & Marino, J.M. (1976). Neonatal progesterone and feminine sexual development. *Nature, 264,* 795–796.

Sherman, J. (1967). Problem of sex differences in space perception and aspects of intellectual functioning. *Psychological Review, 74,* 290–299.

Signorella, M.L., & Jamison, W. (1986). Masculinity, femininity, androgyny, and cognitive performance: A meta-analysis. *Psychological Bulletin, 100,* 207–228.

Smith, D.W. (1976). *Recognizable patterns of human malformation.* Philadelphia: W.B. Saunders Co.

Swaab, D.F., & Fliers, E. (1985). A sexually dimorphic nucleus in the human brain. *Science, 228,* 1112–1115.

Taylor, M.J., Smith, M.L., & Iron, K. (1988). Sex differences in the ERPs to verbal and nonverbal memory tasks. *Society for Neuroscience Abstracts, 14,* Part 2, 1013.

Veith, J.L., Sandman, C.A., George, J.M., & Kendall, J.W. (1985). The relationship of endogenous ACTH levels to visual-attentional functioning in patients with congenital adrenal hyperplasia. *Psychoneuroendocrinology, 10,* 33–48.

Witelson, S.F. (1985). The brain connection: The corpus callosum is larger in left handers. *Science, 229,* 665–668.

16
Disorders of the Thyroid: Medical Overview

Thomas P. Foley, Jr.

Disorders of the thyroid gland have been recognized for more than 2,000 years. Endemic goiter and cretinism were evident during the fourth century B.C. by Andean sculptures of goitrous dwarfs and descriptions recorded in Europe during the first century B.C. (Gaitan, 1975; Cranefield, 1962). Thyromegaly associated with the symptoms and signs of hyperthyroidism was first recognized by Parry in 1786 although not published until 1825 (Parry, 1825). The most common cause of hyperthyroidism is named after Robert Graves who reported the disease in 1835 and associated the enlargement of the thyroid with the cause of the clinical disease (Graves, 1835). Not until the second half of the nineteenth century was an absence of the thyroid gland associated with clinical features of cretinism (Curling, 1850) and myxedema (Ord, 1878). Medical therapy for hypothyroidism was discovered approximately 100 years ago, and antithyroid drugs and radioiodine were introduced as treatment for hyperthyroidism approximately 50 years ago.

The most common thyroid disorder is still endemic goiter which in most instances is caused by a low dietary supply of iodine. The greatest prevalence of endemic goiter is found in mountainous regions, such as the Andes, Himalayas, and Alps. However, the disease also occurs in low altitude areas, particularly those subjected to glaciation, and iodine supplementation to salt and animal feed causes an eradication of the disease. Endemic goiter is a major public health problem since there are estimates that the disease afflicts more than 200 million people throughout the world.

In iodine-sufficient regions the most common thyroid disorders are the autoimmune thyroid diseases, Hashimoto's disease, and Graves' disease. The most common cause of hyperthyroidism is Graves' disease, estimated to occur in 0.4% of the population of the United States (Ingbar, 1985). From an epidemiological survey in England the incidence of Graves' disease was estimated to be 1 to 2 cases per thousand per year (Tunbridge et al., 1977). In this survey the prevalence of newly diagnosed hypothyroidism was 3.3 per 1,000 females and in men was less than 1 per thousand (Tunbridge et al., 1977). The annual incidence has been estimated to be from 1 to 2 cases per thousand women. Newborn screening programs for congenital hypothyroidism have reported an incidence of 1 in 4,000 (Fisher et al., 1979). During childhood and adolescence thyromegaly is a common clinical disorder that may occur in as many as 6% of school-aged children (Rallison et al., 1975; Inoue, et al., 1975; Trowbridge et al., 1975). According to criteria established by the World Health Organization thyromegaly in children occurs when the lateral lobes of the gland are larger than the terminal phalanx of the thumb (Stanbury & Hetzel, 1980). In cases of autoimmune thyroid disease there is a female to male predominance at all ages (Ingbar, 1985; Foley, 1982).

Regulation of Thyroid Hormone Secretion

Hypothalamic-Pituitary-Thyroid Control

Normal circulating and tissue concentrations of thyroid hormones are controlled by a negative feedback control system at the level of the pituitary gland. The synthesis and release of the major thyroid hormones, L-thyroxine (T_4) and 3,3',5-L- triiodothyronine (T_3), are stimulated by the pituitary glycoprotein, thyroid stimulating hormone (TSH), or thyrotropin. When the free or unbound concentration of T_4 and T_3 in serum decreases, there is an increase in the pituitary secretion of TSH. TSH interacts with a specific membrane-bound surface receptor on the thyroid follicular cell that stimulates the production of cyclic AMP and the active transport of iodide from the circulation into the cell. After a series of enzymatic steps (Figure 16.1), predominantly T_4 and small amounts of T_3 are synthesized and released by the thyroid gland. When the circulating levels of free T_4 (FT_4) and free T_3 (FT_3) increase, the secretion of TSH decreases. When the concentration of FT_4 in the circulation and within the pituitary gland decreases, there is a reduction in the intrapituitary conversion of T_4 to T_3 which causes a decrease in the binding of T_3 to its DNA receptor and an increase in pituitary synthesis of TSH. The release of TSH from the pituitary gland is stimulated directly by the hypothalamic hormone, thyrotropin releasing hormone (TRH).

Peripheral Metabolism and Transport of Thyroid Hormones

During conditions of normal caloric intake and activity circulating free thyroxine is converted by peripheral tissues predominantly to T_3. During starvation, severe illness, and in the presence of certain drugs, thyroxine is preferentially mono-deiodinated to a stereoisomer of T_3 called reverse T_3 (3,3',5'-L-triiodothyronine). Reverse T_3 has no metabolic activity and is produced when there is a desire, teleologically speaking, to conserve metabolic activity and reduce metabolism.

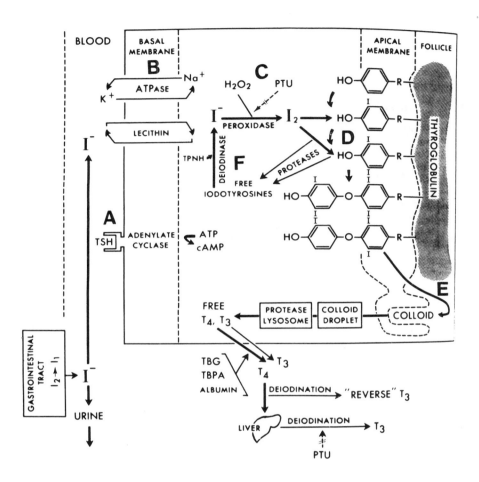

FIGURE 16.1. Biosynthesis of thyroid hormones. The thyroid concentrates iodide from the circulation and immediately oxidizes it to iodine. Tyrosyl residues on the large glycoprotein, thyroglobulin, are iodinated to form mono- and diiodotyrosine. These residues are coupled to form thyronine, either T_4 or T_3. On stimulation and proteolysis the iodothyronines are released into the circulation. The major amount of circulating biologically active thyroid hormone, T_3, is derived from the peripheral monodeiodination of T_4. Both hormones circulate bound to specific binding proteins. Iodide that is not utilized in the formation of T_4 and T_3 becomes available again after the deiodination of tyrosyl residues. (Reprinted with permission of the author, Gerald S. Levey, M.D., and publisher from *The Merck Index,* Figure 89-1, edition 15, Rahway, N.J., Merck, 1987, page 1034.)

In the circulation thyroid hormones are transported bound to specific proteins. The major binding protein for T_4 and T_3 is thyroxine binding globulin (TBG). T_4 and T_3 also bind to thyroxine binding prealbumin, recently named transthyretin, and to albumin. T_4 and T_3 have the greatest binding affinity for TBG and the weakest affinity for albumin (Robbins & Edelhoch, 1986).

Peripheral Action of Thyroid Hormones

After entering the cell and nucleus, T_3 binds to a specific DNA receptor to initiate a cascade of biochemical processes that increase the synthesis of various thyroid-dependent proteins. T_3 also mediates a stimulatory effect on oxygen consumption through a direct effect on oxidative phosphorylation in mitochondria.

Clinical Manifestations of Abnormal Thyroid Hormone Function

Hypothyroidism

Causes

Hypothyroidism may be classified into two major groups, primary thyroid disease (Table 16.1) and diseases of the hypothalamus and pituitary that impair the secretion of TRH and TSH (Table 16.2). Though usually a

TABLE 16.1. Causes of primary hypothyroidism.

1. Defect of embryogenesis
    a. Agenesis, hypoplasia or ectopic thyroid dysgenesis
2. Destruction of the thyroid gland
    a. Autoimmune thyroiditis (Hashimoto disease)
    b. Surgical excision
    c. Irradiation: therapeutic radioiodine
                    external radiation
    d. Postthyroiditis: subacute
    e. Replacement by cancer or other diseases
3. Inhibition of thyroid hormone synthesis
    a. Familial dyshormonogenesis
    b. Iodine deficiency
    c. Iodine excess in susceptible patients
    d. Antithyroid drugs
4. Transient hypothyroidism
    a. After radioiodide or surgical therapy
    b. Postpartum
    c. During the recovery from toxic thyroiditis
5. Peripheral resistance to thyroid hormone action
6. Idiopathic atrophy: possible autoimmune

TABLE 16.2. Causes of hypothalamic and pituitary hypothyroidism.

A. Isolated TRH and TSH deficiency
    1. Familial
    2. Sporadic
        a. Associated with pseudohypoparathyroidism
        b. Idiopathic
B. Deficiencies of TRH/TSH and other pituitary hormones
    1. Familial
        a. Autosomal recessive
        b. X-linked recessive
        c. Inheritance unknown
    2. Sporadic
        a. Congenital
           1) Pituitary aplasia, ectopia and hypoplasia
           2) Dysmorphogenesis: septo-optic dysplasia
                              cleft lip and palate
                              severe forebrain defects
           3) Craniopharyngioma
           4) Perinatal trauma
        b. Acquired
           1) Tumors: hypothalamic
                              pituitary adenomas
                              metastatic tumors
           2) Postirradiation
           3) Trauma
           4) Postinflammation: infectious diseases
                              lymphocytic hypophysitis
           5) Vascular lesions: ischemic necrosis
                              internal carotid aneurysm
           6) Miscellaneous: sarcoidosis
                              histiocytosis
                              hemochromatosis
           7) Idiopathic

permanent impairment of thyroid hormone secretion, primary hypothyroidism can be a transient disorder that may occur at any age. On the other hand, TRH and TSH deficiencies with rare exception are permanent diseases that present as familial or sporadic and congenital or acquired diseases. In acquired primary hypothyroidism the diseases usually are intrinsic to the thyroid gland whereas in acquired TRH and TSH deficiency, the diseases usually are extrinsic and interfere with function by local pressure or invasion of the hypothalamic-pituitary system.

Pathophysiology

Hypothyroidism results from deficient thyroid action at the cellular level either by decreased delivery to the cell of the active, unbound thyroid hormone, T_3, or by an impairment in the cellular response to T_3. In normals, TRH from the hypothalamus is secreted into the portal vasculature to reach the pituitary and stimulate the release of preformed TSH. The release of TSH is directly inhibited when the free T_4 and T_3 concentrations exceed a specific level or set point that differs with each individual within a narrow range. When the production of thyroid hormones decreases because of primary disease in the thyroid gland, there is an increased release of TSH by the pituitary. When there is hypothalamic disease, the circulating levels of free T_4 are low and TSH levels are either low or normal (inappropriately low for the serum free T_4 value); the TSH values increase after the exogenous administration of TRH, and the rise in TSH may be normal or delayed and exaggerated. In pituitary disease TSH concentrations are low or undetectable, and there is no TSH response after TRH administration. An evaluation of thyroid function in hypothyroidism may be summarized as follows:

Tests of thyroid function	Primary hypothyroidism	Hypothalamic hypothyroidism	Pituitary hypothyroidism
TSH	Elevated	Normal or low	Low
Free T_4	Low	Low	Low
TSH response to TRH	Exaggerated	Normal or delayed and exaggerated	Absent

Symptoms and Signs

Primary thyroid diseases usually present with symptoms and signs of thyroid hormone deficiency whereas the clinical features of the primary disease that affects the hypothalamic-pituitary complex or abnormalities of the other pituitary hormones (growth hormone, prolactin, ACTH) usually determine the initial symptoms and signs in patients with TRH and TSH deficiency. The degree of thyroid hormone deficiency in TRH and TSH deficiency usually is very mild, and symptoms and signs may be minimal or absent.

The age at onset and severity of hypothyroidism determine the symptoms and signs of the disease (Table 16.3). The clinician must remember that the symptoms and signs during the early phase of hypothyroidism may be nonspecific and very subtle. In children, growth deceleration may be the only sign of the disease. Although memory impairment and poor school performance are reversible on treatment of the child and adult, permanent mental retardation will occur if hypothyroidism is not diagnosed and treated within a few weeks after birth. The development of hypothyroidism after the second or third year of life is not associated with irreversible mental development.

In the adolescent and adult, behavioral and psychological manifestations initially are not specific and mild. Only as myxedema develops with chronic, advanced hypothyroidism does the patient experience slowing of thought processes, difficulty with calculations and comprehension of complex questions and memory impairment. Late in the disease the patient

TABLE 16.3. The symptoms and signs of hypothyroidism and myxedema in the approximate order of frequency and severity.

Early infancy	Childhood	Adult
Prolonged, indirect hyperbilirubinemia	Growth retardation	Weakness
Edema	Delayed puberty	Dry and coarse skin
Gestation > 42 weeks	Delayed dentition	Lethargy, slow speech
Poor feeding	Constipation	Edema of eyelids, face
Hypothermia	Decreased appetite	Decreased sweating
Large posterior	Cold intolerance	Cold intolerance
fontanelle > 5 mm	Muscle pseudohypertrophy	Thick tongue
Peripheral cyanosis and	Menometrorrhagia	Coarse hair
mottling	Mild obesity	Cold and pale skin
Respiratory distress	Delayed tendon relaxation	Memory impairment
Umbilical hernia	Galactorrhea	Constipation
Failure to gain weight	Precocious Puberty	Weight gain
Decreased stooling < qd	Pale, thick, dry,	Loss of hair
Dry skin	carotenemic skin	Dyspnea
Macroglossia	Impaired school	Peripheral edema
Decreased activity	performance	Hoarseness
Hoarse cry		Anorexia
		Menorrhagia

may develop visual and other hallucinatory distortions with paranoid ideas and unusual behavior. At this stage drowsiness, lethargy and difficulty in arousal occur and progress to stupor and myxedema coma and occasionally convulsions. A careful history should enable the clinician to differentiate hypothyroidism from primary psychiatric disorders.

Evaluation and Management

Once the disease is suspected, confirmation of the diagnosis and method of therapy is easy. Thyroid functions tests (serum TSH, total and free T4) and serologic tests to determine the etiology (thyroid antibodies) will establish the diagnosis of primary hypothyroidism. The presence of thyroid antibodies are indicative of the most common cause of primary hypothyroidism in North America, chronic lymphocytic thyroiditis, or Hashimoto's disease. Further tests may be required to determine the cause of goitrous hypothyroidism if thyroid antibodies are negative. Treatment with sodium L-thyroxine should be initiated promptly. In chronic hypothyroidism, the patient may better tolerate the resumption of the euthyroid state by a gradual increment of dose from 25 μg to a final dose in the adult of 1 to 2 μg/kg/day over a 4- to 6-week interval; some patients who begin a full replacement dose will experience symptoms and signs of hyperthyroidism despite euthyroid or hypothyroid values of thyroid function. Infants and patients with mild hypothyroidism should be started on the full replacement dose that is appropriate for age (Foley, 1985). Once clinical and biochemical euthyroidism is achieved, an annual evaluation of thyroid function will assure compliance with therapy and proper potency of medication.

Thyrotoxicosis

Etiology and Pathogenesis

The clinical response of peripheral tissues to an excess of thyroid hormones, or thyrotoxicosis, may be classified in terms of etiology by the presence or absence of an increase in thyroid hormone synthesis and secretion, or hyperthyroidism (Table 16.4). The abnormal thyroid stimulators cause hyperthyroidism since they mimic the action of TSH to stimulate the TSH receptor. In Graves disease, the thyrotropin receptor antibodies (TRAb) of the immunoglobulin G (IgG) class of antibodies stimulate the thyroid gland to cause thyromegaly and escessive synthesis and secretion of T_4 and T_3. The increased concentrations of thyroid hormones inhibit the pituitary secretion of TSH, but have no effect on the production of TRAb by plasma cells of the immune system. Adults with trophoblastic tumors secrete a glycoprotein that in high concentrations will stimulate the TSH receptor to cause hyperthyroidism. The autonomous hypersecretion of thyroid hormones by tumors and nodules of the thyroid cause hyperthy-

TABLE 16.4. Etiology and pathogenesis of thyrotoxicosis.

Etiology	Pathogenesis
A. Thyrotoxicosis and Hyperthyroidism	
1. Abnormal thyroid stimulator	
a. Graves disease	TSH stimulating receptor antibody
b. Trophoblastic tumor	hCG-like thyroid stimulator
2. Intrinsic thyroid autonomy	
a. Toxic adenoma	Benign tumor
b. Toxic multinodular goiter	Foci of functional autonomy
3. Excessive TSH secretion	
a. Pituitary adenoma	TSH-secreting tumor
b. Pituitary hyperplasia	Pituitary resistance to thyroid hormone
B. Thyrotoxicosis without Hyperthyroidism	
1. Inflammatory disease	
a. Toxic autoimmune thyroiditis	Release of thyroid hormones from damaged follicules
b. Toxic phase of subacute thyroiditis	Release of thyroid hormones from damaged follicules
2. Extrathyroidal source of hormones	
a. Hormone ingestion	Hormones in foods or medications
b. Ectopic thyroid tissue	Functional metastatic thyroid cancer; Struma ovarii

Thyrotoxicosis: the response of peripheral tissues to an excess of thyroid hormone.
Hyperthyroidism: the condition in which there is an increase of thyroid hormone synthesis and secretion.

roidism, inhibit TSH secretion and cause involution of normal thyroid tissue. Diffuse thyromegaly results from autonomous secretion of TSH by a pituitary adenoma and from the excessive release of TSH when the pituitary is resistant to the inhibitory effects of normal to elevated levels of thyroid hormones.

When the thyroid is damaged by autoimmune or other inflammatory agents, preformed thyroid hormones are released by the gland into the circulation to cause thyrotoxicosis. Excessive levels of thyroid hormone also occur from the ingestion of large amounts of thyroid medication or foods containing thyroid hormones. Similarly, other abnormal tissues such as metastatic thyroid cancer may secrete thyroid hormone in excess to cause thyrotoxicosis.

Symptoms and Signs

The clinical presentation of thyrotoxicosis is similar to all ages, though less pronounced with advancing age. There are minor variations in symptoms and signs according to age and diagnosis, particularly with respect to the eye changes (Table 16.5). Certain eye changes are seen with all forms of thyrotoxicosis irrespective of diagnosis. In Graves' disease severe infiltrative ophthalmopathy is rarely seen in children; infiltrative dermopathy, or pretibial myxedema, is not seen in children and occurs in those adults with usually severe infiltrative ophthalmopathy. Many clinical symptoms and signs develop in response to the increased activity of the adrenergic ner-

TABLE 16.5. The symptoms and signs of thyrotoxicosis.

Symptoms	Signs
Nervousness	Tachycardia
Diaphoresis	Thyromegaly
Heat intolerance	Soft, warm, moist skin
Palpitations	Fine tremor
Fatigue	Bruit over thyroid
Weight loss	Eye signs:
Dyspnea	+ retraction of upper eyelid
Weakness	+ lid lag and globe lag
Increased appetite	+ jerky, spasmodic lid movement
Eye complaints	+ tremor of the lid
Swelling of legs	Atril fibrillation
Hyperdefecation without diarrhea	Splenomegaly
Diarrhea	Gynecomastia
Pruritis	Infiltrative ophthalmopathy**
Poor school performance*	conjunctival injection, tearing
Nocturia and enuresis*	exophthalmos, periorbital edema
Insomnia and restless sleep*	extraocular muscle weakness
Mild growth acceleration*	Infiltrative dermopathy

+ Occurs in all forms of thyrotoxicosis.
* Specific for the presentation during childhood and adolescence.
** Occurs in Graves' disease only.

vous system. The mechanism for the augmented adrenergic activity in thyrotoxicosis is poorly understood. Drugs such as propranolol that block beta adrenergic activity may reduce some clinical findings such as tachycardia, tremor, mental symptoms, heat intolerance and sweating, diarrhea, and proximal myopathy. However, those parameters that are not directly influenced by the adrenergic system are not improved by propranolol, such as oxygen consumption, thyromegaly, thyroid bruit, weight loss, exophthalmos, and myocardial contractility. Only with reduction in thyroid hormone concentrations will these latter symptoms and signs improve.

There are certain clinical findings that occur from antithyroid drug toxicity and may develop at any time during the course of treatment irrespective of the state of clinical thyrotoxicosis. The most frequent are erythematous rash and urticaria, arthralgia, arthritis, neutropenia with secondary infection, a lupuslike syndrome, lymphadenopathy, fever, and hepatitis.

Evaluation and Management

In most patients with thyrotoxicosis from Graves' disease, the diagnosis is quite evident and needs only to be confirmed prior to therapy. In these cases only thyroid function tests (total and free T_4, T3, and TSH), basal complete blood count with peripheral smear for patients starting antithyroid drugs, and thyrotropin receptor antibody (TRAb) determination are needed. Though not essential, the TRAb determination confirms the diagnosis of the autoimmune hyperthyroidism of Graves' disease. The radioiodide uptake test is useful to assess the contribution of thyroid gland in the development of thyrotoxicosis, particularly for the diseases without hyperthyroidism (Table 16.4). The TRH test may assist in the differential diagnosis of the diseases with excessive TSH secretion and document the production of excessive thyroid hormone secretion in diseases with mild or atypical clinical presentations. When nodular thyroid disease presents with hyperthyroidism, the thyroid image tests with either ultrasound, radioiodide or technetium, and rarely with CT scan or magnetic resonance will define the nodular lesion and determine its functional status as to whether there is autonomous hypersecretion.

Therapy is directed toward the cause of thyrotoxicosis. In most diseases, the therapy is self-evident: surgical excision of the hyperfunctioning nodule or tumor, control of the transient thyrotoxic state with propranolol for inflammatory diseases, and removal of the thyroid hormone sources in foods and medications.

However, the treatment of Graves' disease cannot be directed as yet at the primary cause in the immune system, but at the source of excessive thyroid hormone secretion. Antithyroid drugs—either propylthiouracil (PTU) or methimazole (MTZ), radioiodine ablation, or subtotal thyroidectomy are the three forms of therapy currently available. For children, adolescents, young adults, and pregnant women the antithyroid drugs provide an effective control of the disease until remission without therapy occurs. If active disease persists on therapy or relapse develops after cessation of therapy, radioiodide ablation usually is selected as the next form of therapy. In adults beyond 35 years of age radioiodide therapy is the initial treatment of choice.

Subtotal thyroidectomy rarely is selected as therapy in adults, and has become much less popular as a second form of therapy in children and adolescents. Ablative therapy with radioiodide has proven safe and effective for the past four decades, and subtotal thyroidectomy offers no advantages to radioiodide therapy, but certain disadvantages such as general anesthesia with defined risks, hypoparathyroidism, damage to the recurrent laryngeal nerve, and the expenses and morbidity of major surgery. Hypothyroidism is expected to develop within weeks to years after both radioiodide treatment and subtotal thyroidectomy. Otherwise recurrence of hyperthyroidism very likely will occur.

Infiltrative ophthalmopathy usually improves as treatment for Graves' disease controls the hyperthyroidism. Persistence or progression of exophthalmos may cause very serious physical and psychological impairments. Abnormalities of vision may progress to blindness, and serious behavioral and personality changes may develop as the distressing physical appearance of the eyes progresses. Although there is minimal medical therapy available, there are recent surgical techniques that remarkably improve the visual acuity and appearance for patients with moderate to severe ophthalmic disease. Fortunately, infiltrative ophthalmopathy is rarely a progressive disease during adolescence and in the young adult. The ophthalmic symptoms associated with the thyrotoxic state improve and disappear with the development of euthyroidism.

References

Cranefield, P.F. (1962). The discovery of cretinism. *Bulletin History of Medicine, 36*, 489–511.

Curling, TB. (1850). Two cases of absence of the thyroid body. *Med Chir Trans, 33*, 303–306.

Fisher, D.A., Dussault, J., Foley, T.P. et al. (1979). Screening for congenital hypothyroidism: Results of screening one million North American infants. *Journal of Pediatrics, 94*, 700–705.

Foley, T.P. Jr. (1982). Acute, subacute and chronic thyroiditis. In S.A. Kaplan (Ed), *Clinical pediatric and adolescent endocrinology* (pp. 96–109). Philadelphia, PA: W.B. Saunders Co.

Foley T.P. Jr. (1985). Thyroid disease. In S.S. Gellis & B.M. Kagan (Eds.), *Current pediatric therapy* (12th ed.). Philadelphia, PA: W.B. Saunders Co.

Gaitan, E. (1975). Iodine deficiency and toxicity. In P.L. White & N. Selvey

(Eds.), *Proceedings Western Hemisphere Nutrition Congress-IV* (pp. 56–63). Acton, MA: Publishing Sciences Group Inc.

Graves, R.J. (1835). Clinical lectures. *London Medical Surgical Journal* (Part II), *7,* 516.

Ingbar, S.H. (1985). The thyroid gland. In J.D. Wilson & D.W. Foster (Eds.), *Williams textbook of endocrinology* (Chapter 21, pp. 682–815). Philadelphia, PA: W.B. Saunders Co.

Inoue, M., Taketani, N., Sato, T., & Nakajima, H. (1975). High incidence of chronic lymphocytic thyroiditis in apparently healthy school children: Epidemiological and clinical study. *Endocrinology Japan, 22,* 483–88.

Ord, W.M. (1878). On myxoedema, a term proposed to be applied to an essential condition in the "cretinoid" affection occasionally observed in middle-aged women. *Med Chir Trans, 61,* 57.

Parry, C.H. (1825). *Collections from the unpublished papers of the late Caleb Hillier Parry* (Vol. 2, p. 111). London.

Rallison, M., Dobyns, B., Keating, F., Rall, J., & Tyler, F. (1975). Occurrence and natural history of chronic lymphocytic thyroiditis in children. *Journal of Pediatrics, 86,* 675.

Robbins, J., & Edelhoch, H. (1986). Thyroid hormone transport proteins: Their nature, biosynthesis, and metabolism. In S.A. Ingbar & L.E. Braverman (Eds.), *Werner's the thyroid,* (Chapter 6, pp. 116–127), Philadelphia, PA: J.B. Lippincott Co.

Stanbury, J.B., & Hetzel, B.S. (1980). *Endemic goiter and endemic cretinism. Iodine nutrition in health and disease* (p. 164). New York: John Wiley and Sons.

Trowbridge, F.L., Matovinovic, J., McLaren, G.D., & Nichaman, M.Z. (1975). Iodine and goiter in children. *Pediatrics, 56,* 82.

Tunbridge, W.M.G., Evered, D.E., Hall, R. et al. (1977). The spectrum of thyroid disease in a community: The Wickham survey. *Clinical Endocrinology, 7,* 481–493.

17
Congenital Hypothyroidism: Intellectual and Neuropsychological Functioning

JOANNE F. ROVET

Screening newborns for congenital hypothyroidism (CH) has been effective in preventing the mental retardation that typically occurred when diagnosis was based on clinical symptomatology. This chapter provides an overview of the disease and its manifestations as well as a comparison of outcome in children with CH diagnosed clinically with those now being found via newborn screening. A detailed description of the author's own longitudinal study of screened CH children is also presented with special emphasis on the factors predicting better outcome.

Pathophysiology of Thyroid Metabolism and Congenital Hypothyroidism

Congenital hypothyroidism (CH), or cretinism, is a relatively common disease of newborns. It is caused by a deficiency in the production of thyroid hormone, a hormone that is needed for normal growth and development, including normal brain development. Recent estimates indicate the incidence of CH is about 1 in every 3,000 to 4,000 newborns (Ferreira, 1986), varying for different racial and ethnocultural groups (Addison et al., 1989; Brown et al., 1981; Rovet et al., 1986; Therrell et al., 1982; Frasier et al., 1982); twice as many females are affected as males. If CH is untreated, permanent and irrepairable brain damage will occur. Until recently, CH was one of the commonest contributors of mental retardation in children (Chrome & Stern, 1972).

Congenital hypothyroidism is caused by a defect in the thyroid gland (primary hypothyroidism), the pituitary gland (secondary hypothyroidism), the hypothalamus (tertiary hypothyroidism), or from a number of environmental factors such as the lack of iodine or exposure to goitrogens. Iodine insufficiency producing endemic cretinism occurs primarily among populations in developing nations (Delange, 1982; Ermans et al., 1980) and where salt is not standardly iodized (Klett et al., 1983)

Primary CH, which is far more common than secondary or tertiary forms, occurs (a) when the gland is totally missing, known as athyrosis, the cause of which is still not known (Dussault, 1987); (b) when the gland fails to descend to its final location during embryogenesis with a remnant in the throat or base of the tongue, known as an ectopic or lingual gland; (c) from a structural malformation of a properly descended gland, a hypoplastic gland; or (d) from an enzyme defect causing goiter, dyshormonogenesis (Degroot et al., 1984; Walfish, 1981). Price et al. (1981) estimated the relative frequencies of these different disorders as athyrosis 25%, ectopias 29%, goiters 29%, and hypoplasias, 17%. Similar figures have been provided for Japan (Irie et al., 1987), but other regions typically showing a much lower rate for goiters (e.g., Dussault et al., 1980) could reflect differences in ethnic and racial compositions, as well as different screening procedures. It has been suggested that ectopic and dyshormonogenetic etiologies represent less severe forms of the disorder as they provide some minimal functional thyroid activity in the fetal and neonatal periods (Klein, 1980). Because thyroid hormone does not permeate the placental barrier, maternal thyroxine is not believed to provide any supplementary contribution in utero (Burrow et al., 1983; Fisher & Klein, 1981). This means that a child without a thyroid gland lacks any alternative source of thyroxine until replacement therapy can be provided, unlike children with the other hypothyroid conditions, who may have partial thyroid function.[1]

The thyroid gland is located at the base of the neck. It matures from a thickening of the epithelium in the pharyngeal floor. At about the second month postconception, the immature thyroid leaves its point of origin as a small bilobate dimple to transcend to its final position, all the while undergoing histologic transformation. It becomes functional by about 20 weeks gestation at which point it is capable of secreting two hormones, triiodothyronine or T3 and tetraiodothyronine or T4 (also known as thyroxine). T3 is the active hormone, while T4 is a "prohormone," becoming activated only upon losing one of its iodine molecules at the target tissue site (Fisher & Klein, 1981; Walfish, 1981). Synthesis and secretion of these hormones is regulated by negative feedback with the hypothalamus, which

[1] The issue regarding the lack of maternal-fetal hormone transfer is controversial, being challenged recently (Morreale de Escobar et al, 1989) by evidence showing an association between maternal hypothyroidism during pregnancy and subsequent learning disabilities and behavior disorders (Man & Serunian, 1976; Morreale de Escobar et al., 1985; Thyroid Foundation of America Newsletter, 1986; Interagency Committee on Learning Disabilities, 1987; Konstantareas & Homatidis, 1988). As these disorders appear to be due to early and mid-trimester brain defects, it suggests that the fetus must be receiving hormone from the mother before its own thyroid system becomes functional (Morreale de Escobar, et al, 1989). Because most hypothyroid children have euthyroid mothers, any impairments in intelligence or behavior they demonstrate subsequently must be different from those of children with hypothyroid mothers reflecting late gestational or early postnatal defects when their own systems are supposed to take over but cannot.

produces thyrotropic releasing hormone (TRH) and the pituitary gland, which produces thyroid stimulating hormone (TSH).

Many tissues depend on thyroid hormone for normal growth, differentiation, and function. These include the skeleton, muscle, adipose tissue, lung, epithelium, kidney, gonads, and the central nervous system (Legrand, 1986). According to Fisher (1981) "thyroid hormones, while important at any age to maintain body metabolic homeostasis, are critical during childhood to control the manifold maturational events and the transformation of the newborn to the mature adult" (p. 67).

Most thyroid hormone receptors develop after the hormonal system has become functional. In the human infant, this occurs during the latter half of gestation (Pekonen & Pekonen, 1984) but may extend into the postnatal period (Fisher & Klein, 1981), with the timing varying for different tissues including different brain tissues (Morreale de Escobar et al., 1983). In rat brain, an increase in receptor concentration reaches a maximal level, signifying maximal TH sensitivity, at about 6 days after birth, and then decreases with age (Perez-Castillo et al., 1985). Because the rat is born less mature than the human infant (Fisher, 1987), this period of TH sensitivity corresponds to prenatal life in the human. This implies that major effects of thyroid hormone deficiency might actually be taking place prenatally, far before treatment can be given, regardless of diagnostic procedure.

Thyroid Hormone Deficiency and Early Brain Development

Detailed studies of animals (the rat in particular) have repeatedly demonstrated that altered thyroid states have a profound influence on early CNS development (Hamburgh et al., 1971; Myant, 1971). Thyroid hormones are necessary for normal rates of germinal cell proliferation, cessation of cell division, formation of precursor cells, axonal and dendritic growth, neuronal migration, and the formation of the correct number and type of synaptic relationships (Lauder & Krebs, 1986; Potter et al., 1982). Thyroid hormone is thought to act as a "biological time clock" coordinating the events involved in the ultimate differentiation of specific brain sites (Balazs et al., 1971) and the appropriate neuronal circuitry. Recent studies indicate that TH stimulates neurite outgrowth by changing the concentration and/or activity of certain proteins (i.e., tubulin and microtubule-associated proteins), which co-plymerize to form the microtubules required for neurite outgrowth (Fellous et al., 1979; Nunez, 1984). When TH is absent or insufficient (hypothyroidism), neuronal differentiation is curtailed (Nicholson & Altman, 1972; Pelton & Bass, 1973); when in excess (hyperthyroidism), development is accelerated but defects also occur because the cellular proliferation sequence terminates too quickly (Nicholson & Altman, 1972; Schapiro, 1966). This has implications for appropriate management to prevent the complications associated with overtreatment (Weichsel, 1978).

Thyroid hormone also plays a key role in synaptogenesis (Lauder & Krebs, 1986; Legrand, 1986). Hypothyroidism retards the developmental increase in density of synapses by reducing synaptosomal protein. Although this can be restored by relacement therapy, *it must begin within a critical time to have any effect* (Perez-Castillo et al., 1985) TH deficiency also affects glial cell production (Cheek, 1975; Holt et al., 1975; Legrand, 1986) and myelin formation (Malone et al., 1975).

Because each brain site has a unique developmental schedule (Dobbing & Sands, 1973), each tissue is affected differently depending on its period of critical sensitivity to TH hormone (Morreale de Escobar et al., 1983). Plioplys, Gravel, and Hawkes (1986), using a selective suppression of neurofilament antigen technique, found effects for axonal systems of the cerebellum (basket cells), cerebral cortex, and corpus callosum, which attained their mature pattern relatively late, but not for the earlier maturing olfactory system. For the human infant, these findings imply that the intellectual functions most affected by neonatal hypthyroidism will be those that develop late during the period prior to replacement therapy (including before birth), or after treatment, but before euthyroidism is achieved. The cerebellum, which contributes to skilled motor performance (Holmes, 1979) and the manipulaton of ideas (Leiner et al., 1986) is one CNS site that undergoes its major development, and presumably its period of greatest thyroid sensitivity, during early neonatal life (Legrand, 1986; Potter et al., 1982). The cortex (Horn, 1955), hippocampus (Lauder & Krebs, 1986), forebrain (Morreale de Escobar et al., 1983), visual cortex (Potter et al., 1982), auditory system (Hebert et al., 1986), have similarly been shown to undergo major development during postnatal life. The susceptibility of these various sites to TH deficiency can cause selective deficits in neuromotor (Murphy et al., 1986), language (Hulse et al., 1982; Rickards et al., 1987) and memory areas (Rovet et al., 1987); these in turn can subsequently produce academic dysfunction (Barclay, 1981).

Treatment and Diagnosis of Congenital Hypothyroidism

Congentital hypothyroidism is readily treated by the daily oral ingestion of L-thyroxine (Abbasi & Aldige, 1977; Redmond, 1982; Rezvani & DiGeorge, 1977). Although recommendations for appropriate starting and maintenance doses have been set for infants and young children (American Academy Pediatrics Committee on Genetics, 1987; Mitchell, 1987; Guyda, 1983), their adequacy has never been systematically evaluated. Indeed we have recently reported that between 1 and 4 months of age infants with higher levels of circulating thyroxine regardless of their starting dosage were subsequently more difficult children tempermentally, demonstrating heightened arousal and increased sensitivity to stress and/or stimulation (Rovet et al., 1989). Using an individualized titration treatment strategy, Thompson, McCrosson, Penfold, Woodroffe, Rose, and Robertson (1986) reported that requirements differed depending on etiology and whether or not the child was breast fed.[2]

Numerous studies have shown that the timing of hormonal replacement is critical for better outcome. In the past, 75% of hypothyroid children treated past the third month of life were retarded (Klein, 1980) whereas 75% of those treated earlier were normal (Hulse, 1984). However due to the late appearance of the outward signs and symptoms of the disease, most infants were unfortunately diagnosed past the critical period and so suffered permanent and irrepairable brain damage (Andersen, 1975). Furthermore, even though those treated earlier were not retarded, the majority still developed learning difficulties (MacFaul et al., 1978; Vandershueran-Lodeweyckx et al, 1980), behavioral problems (MacFaul et al., 1978; Money, 1956) and unspecified neurological disorders (Andersen, 1975; Maenpaa, 1972; Wolter et al., 1979).

Over the past decade pediatric medicine has witnessed a major breakthrough in allowing for the earlier diagnosis of CH. With the advent of radioimmunoassay techniques to measure TH in small blood samples (Chopra, 1972; Larsen & Broskin, 1975) and to rapidly distribute blood samples spotted onto filter paper cards (see following), screening programs have now been instituted world-wide for detecting hypothyrodism at birth. These now exist in every province in Canada (Ferreira, 1987) and every state in the United States (Therrell, 1987). Such programs operate efficiently and inexpensively (Layde et al., 1979; Laberge, 1983), often in conjunction with screening for other metabolic disorders, such as PKU and galactosemia. Guidelines have been established (American Academy Pediatrics, 1987) with few differences between the programs (Delange, 1979). According to Burrow and Dussault (1980) this represents one of the few areas of modern medicine where advance in medical progress has clearly "had a major impact on society" (p. ix).

The typical procedure in thyroid screening involves spotting a few drops of infant blood onto a filter paper card several days after birth. The card is then mailed to a central laboratory facility, usually state or regionally operated, where the blood spot is initially analyzed for T4 (Dussault et al., 1975; Foley, & Murphy, 1980), TSH (Delange, 1980; Walfish et al., 1983) or both (LaFranchi et al., 1979). Any sample with an abnormal reading is then reassayed for the other hormone. Assaying for TSH has a smaller false-negative rate but misses detecting the rarer secondary or tertiary

---

[2] The question of the influence of breast feeding concerns the fact that breast milk contains small quantities of thyroxine. It has not been adequately resolved whether this may be protective for the hypothyroid infant (Rovet, 1989).

types of hypothyroidism (Mitchell, 1987); assaying for both hormones, though most accurate, is costly (LaFranchi et al., 1985).[3,4] It is now widely accepted that there are no major advantages in screening for T4 or TSH (Dussault & Morrissette, 1983). Every case positively identified by newborn screening is subsequently recalled for confirmatory diagnosis, at which point treatment usually begins. This generally takes place within the infants' first month of life, although there are large variations from program to program. Cases that are not clear-cut, with values just above or below arbitrary cut-offs, may wait until diagnosis is completely confirmed before the child goes on treatment. In most programs children are usually taken off therapy for a short time once the cirtical period has been passed to ensure the child was indeed hypothyroid at birth (Davy et al., 1985; Mitchell, 1987), and did not have transient hypothyroidism (Delange et al, 1987).

Even though screening has drastically reduced the delay time in diagnosing CH, there is still a period of at least 2 to 3 weeks before treatment can begin and a further delay of about 1 to 2 months until euthyroidism is finally achieved (Mitchell, 1987). As discussed previously, studies with animals indicate that CNS damage may still occur from thyroid hormone deficiency in utero or during the early neonatal period, that is, during the time before hormone levels return to normal. These studies have shown that the type of impairment appears to be selective and depends on when specific cerebral sites are undergoing their maximal differentiation relative to the timing of the TH deficiency. This implies that for the children identified by screening, even though they may be spared from the more adverse consequences of hypothyroidism, they could still be at risk for selective neurocognitive impairments, the nature and severity of which will reflect different disease- and treatment-related factors and could later contribute to academic dysfunction. This chapter attempts to review the evidence on outcome in children diagnosed by screening. Its goals will be to identify those disease- and treatment-related factors that contribute to subsequent impairment. Using our own prediction studies, we will describe the factors that have the largest bearing on different specific components of intellectual functioning. We will also provide our preliminary results on school performance in these children.

---

[3] Because some children are missed due to an abnormally slow rise in TSH (Grant et al., 1986), several programs prefer a second voluntary blood sample 1 to 4 weeks after birth (Levine & Therrell, 1986).

[4] With newborn screening, it is also extremely important to maintain as low a rate of false positives as possible in order to prevent the inordinate and long-lasting emotional strain on a family associated with a false-positive diagnosis (Bodegard et al., 1983; Fyro & Bodegard, 1987; Thieffrey et al, 1983; Zetterstrom, 1988).

Outcome Following Clinical Diagnosis

Mental Development

The first record of brain damage due to hypothyroidism dates back to Paraclesus in the sixteenth century (cited by Rosman, 1976). However, his observations remained unchallenged for another 400 years, when Osler (1897) reported that mental retardation could be caused by a deficiency in thyroid hormone. Studies to directly assess the degree of impairment associated with neonatal hypothyroidism date back to the early 1930s when Gessell, Amatruda, and Culotta (1936), Lewis (1937), and Browne, Bronstein, and Kraines (1939), reported less severe retardation in cretins if treatment was adequate and given early. Interest in this disease then appears to have waned for about 20 years until Smith, Blizzard, and Wilkins (1957), and Collipp et al., (1965) reported on the relationships between disease severity, the age at onset of therapy, and ultimate mental prognosis in greater detail. However, it was not until the later 1970s and early 1980s when the specifics of early thyroid deficiency were more precisely determined, in some cases to establish normative data for properly evaluating the efficacy of the to-be-implemented newborn screening programs (Hulse, 1984; Alm et al., 1981).

Retrospective studies (Raiti & Newns, 1971; Maenpaa, 1972; Wolter et al., 1979; Macfaul, Dorner, Brett, & Grant, 1978; Hulse, 1984; Grant & Hulse, 1980; Frost & Parkin, 1986) showed that treatment before three months of age was critical to prevent mental retardation. Maenpaa (1972) found that 81% of his patients diagnosed before 3 months of age were not retarded in contrast to 47% of those diagnosed later. Hulse (1983) similarly reported that among children diagnosed before six weeks of age, none were retarded in contrast to 75% of those diagnosed past 6 months of age. Frost and Parkin (1986), assessing children from a defined geographic region in Northern England, reported an average IQ of 82 with lower IQ being associated with lower social class and later diagnosis. Wolter et al. (1979) established that IQs will be in the normal range provided that treatment is commenced before three months of age, even when the hypothyroidism appeared to be of prenatal onset. Finally Klein (1979) summarized outcome from 651 cretins described in the literature and found a mean IQ of 76 with only a handful in the 110 to 120 range. Treatment within two months was found to be critical especially for children with little functional thyroid activity at birth.

A number of studies have also observed specific deficits. For example, MacFaul et al. (1978) studying 30 patients reported that although their IQs were in the normal range, they were still significantly lower than those of a control sample of British school children. Performance IQs were significantly lower than verbal IQs. Moreover, even though these children were not retarded, 77% still showed at least one sign of impaired brain function, signifying hypothyroidism during prenatal life. The severity of these deficits, however, reflected the effects of postnatal hypothyroidism. Furthermore, because IQs did not differ for children with early symptoms treated between 11 and 16 weeks, from those treated after 16 weeks or those whose symptoms appeared later, it was suggested that the vulnerability of the brain to the adverse effects of hypothyroidism decreases with age in infancy. Alternatively, Virtanen et al. (1983) noted that although neurological damage appeared to originate prenatally, more serious damage appeared to occur if hypothyroidism was not treated until after three months of age, even when IQs were normal. These results suggest that brain damage due to CH precedes birth and progresses slowly until treatment is started.

Academic Achievement

Not surprisingly, hypothyroidism has been associated with an increased demand for special resources in education, even if the children were not retarded. Grant and Hulse (1980), reporting on 112 school-age hypothyroid children in the UK, found that of those with IQs in the normal range (75% of the sample), 65% were attending special schools or receiving remedial education. Von Hanack and Allis and Zabranksy et al. (cited in Klein, 1979) reported that even among hypothyroid children with IQs over 90, very few were able to attend regular schools because of varying problems with speech, perception, concentration, etc. Hulse (1984) reported that 28% were in special schools and that an additional 41% were in remedial classes. Thirty percent were unable to read (vs. 3.8% of controls), 48% had writing difficulties (vs. 17% of controls), and 49% had spelling difficulties (vs. 22% of controls). Vandershueren-Lodeweyckx, Malveau, Craen, Ernould, and Wolter (1980) found a mathematics learning disability was more prevalent than dyslexia in Belgian hypothyroid children.

Neurological Impairment

Neurological impairments are common in children with neonatal thyroid hormone deficiency. Smith et al. (1957) reported an increased incidence of spasticity, shuffling gait, incoordination, jerky movements, tremor, and abnormal reflexes, especially among those with the lowest IQs. Several studies have reported that cerebellar ataxia (Hagsberg & Westphal, 1970) or cerebellar disorder (Hanefield et al., 1974)) are frequent in these children regardless of IQ. Subsequently MacFaul et al. (1978) reported that while frank cerebellar disorders were not uncommon, clumsiness and minor motor disorders were likely to occur in about 50% of these patients. Similarly Wolter et al. (1979) reported that true cerebellar ataxia occurred only if the hypothyroidism was prenatal in origin *and* treatment was delayed past 7 months of age. Hulse (1984) reported that the majority of hypothyroid children were clumsy, even when treatment began early.

In addition to gross motor problems, Frost and Parkin (1986) found disturbances in fine motor skills in more than 50% of their cases. Kirkland, Kirkland, Robertson, Librik, and Clayton (1972) observed abnormal ocular findings in 44% of the children. Inadequate speech development, poor articulation (Money, 1956), and mild hearing loss (Vandershueren-Lodeweyx et al., 1983) have also been described.

In a study of Swedish children with congenital hypothyroidism, Alm, Larsson and Zetterstrom. (1981) reported that although age of starting treatment was associated with intellectual outcome, it was not correlated with neurological abnormalities, signifying that these could have been prenatal in origin. MacFaul et al. (1978) similarly reported a high incidence of minor abnormalities that were present early in life. Finally in a detailed collaborative study from five Belgian pediatric centers, Vandershueren et al. (1980) reported that neuropsychological defects suggestive of minimal brain dysfunction (e.g., short attention span, poor fine motor coordination, impaired spatial orientation) appeared to be prenatal in origin; the severity and number of sequelae depended on the length of delay in thyroid replacement therapy. These results suggest, therefore, that minor impairments will occur from hypothyroidism in utero, but their severity will depend on the duration of hypothyroidism in postnatal life.

Behavior Problems

In addition to mental and neurological impairment associated with neonatal hypothyroidism, hypothyroid children also tend to have an increased number of personality and behavioral disorders. Money (1956) reported that inertia, indifference, and stubbornness were characteristic of these children while Klein, Melzer, and Kenny (1972) found they were more likely to be high-strung and overly sensitive. MacFaul et al. (1978) and Hulse (1983), using actual rating scales, reported a high incidence of behavioral problems, which appeared to be more neurotic than antisocial in type. Frost and Parkin (1986) found that 47% demonstrated behavior problems, 20% of which were severe, with the majority being depressive variety.

Summary

These results suggest that while mental retardation can be clearly prevented by early diagnosis and treatment, minor neurological and neurocognitive and behavioral abnormalities occur in spite of early treatment. This appears to occur in about 50% of cases, especially those with hypothyroidism beginning in utero. The severity of such impairment appears, however, to depend on the length of delay before commencing treatment.

The implication of these findings for prognosis following neonatal diagnosis via newborn screening is that not all children will be entirely "corrected" by this diagnostic procedure (Hulse, 1984; Barnes, 1985).

Outcome Following Newborn Screening

With the advent in the early 1970s of radioimmunoassay techniques to measure thyroid hormones in blood, a number of groups subsequently implemented experimental programs to identify infants who were born with congenital hypothyroidism (Dussault et al., 1975; Foley & Murphey, 1980; New England Congenital Hypothyroidism Collaborative, 1981; Walfish, 1984). This has not been a simple task as criteria had to be established on when to screen (birth, 3 days, 5 days), which hormone or hormones to screen for, what cut-offs to use to minimize the number of false positive and false negative cases. Criteria also had to be set for (a) confirmatory diagnostic procedures when in most cases neither signs nor symptoms of the disease were evident; (b) when and if to establish etiology and by what means (Dussault, 1988; Illig, 1983; Muir et al., 1988); (c) whether skeletal maturity should be determined; (d) which thyroid function tests should be performed, and how frequently; (e) and clinical criteria (Letarte & LaFranchi, 1983). Optimal treatment and management procedures had to be established, as well as strategies for altering dose levels to prevent tissue thyrotoxicosis (Jennings et al., 1984) or undertreatment (Focarile et al., 1984). The interaction of dose with factors such as etiology, severity, and breastfeeding also had to be determined. Finally, methods for follow-up had to be implemented, although this has not been a consistent, systematic, or routine component of the majority of screening programs. Because of preliminary evidence that children with transient hypothyroidism were now being identified (Bainbridge et al., 1987; Delange et al., 1978), procedures also had to be developed to ensure that some children were not being unnecessarily or wrongly treated (Mitchell, 1987).[5]

The success of the earliest preliminary screening projects has led to the implementation of regionalized programs throughout the developed world as well as some regions in the third world (e.g., Brazil, India, Cuba). The most common screening procedures assay for low levels of T4, which is the method of choice in most centers in North America (except for Ontario and Illinois), or for high TSH levels, which is used primarily in Europe and

---

[5] In Toronto, all children are routinely taken off medication for a short period of time at age three; hormonal measurements taken at the end of this interval establish whether any children have normal thyroid function. Children so identified are removed immediately from the program. Only two cases of transient hypothyroidism have so far been detected in over 100 studied (Davy et al., 1985).

Japan, These procedures have been shown to be cost-effective in eliminating the burden of cretinism-associated mental retardation from society (Laberge, 1983; Layde et al., 1979). To date, four international meetings have been held on newborn screening, as well as three international symposia on congenital hypothyroidism itself. A national US organization has also been meeting on a regular basis.

Although the results of prospective studies on the children positively identified through screening are indeed encouraging, one must take into account a number of methodological factors before final conclusions can be drawn. This review will examine the published research on this topic with special emphasis on factors such as type of screening method, age at initiation of therapy, sampling, adequacy of controls, procedures used to evaluate outcome and their reliability and validity criteria, and the ages at which the assessments occurred. Other considerations will include "blind" assessment and degree of familiarity of the assessor with the study objectives and patient characteristics; the familiarity of the target child, or the control, with the assessor; and the location and frequency of assessment. Because most studies have been conducted over a relatively long period of time and are still ongoing, a number of other factors must also be accounted for, such as improved diagnostic and treatment procedures, secular changes in incidence of disease, changes in dose level and regimen, and intervening factors (e.g., remediation such as speech therapy, infant stimulation or special eduction; family crises; new siblings, etc.).

Method

The literature was reviewed in December 1988 for prospective studies that examined intellectual outcome in children with CH who were detected by newborn screening. Only articles written in English were selected. Sources of reference included medical journals, published abstracts, conference abstracts, and the proceedings of national and international symposia or workshops on this topic.

Results

Table 17.1 summarizes the studies that were found. These are listed alphabetically by country, and in some cases, within country by region or province. Because many of the studies are longitudinal in the sense that the same or an expanding cohort is being followed over an extended period of time, some of the investigators have published the same findings in more than one form. Accordingly, only the most recent publication of each team is listed, except if an earlier report contained unique information.

Twenty-two prospective studies from 12 countries were identified. There were 3 from Australia (Victoria State, South Australia, nation-

TABLE 17.1. Result of prospective studies on children with CM diagnosed by newborn screening.

Country	Authors	Year begun	Screening method	Tmt Age[g]	Dose	Controls	Assessment ages	Tests	# CH	IQ results	Additional	Comments
1. AUSTRALIA												
(a) Australia-Wide	Rickards et al. (1987)	1980	T$_4$	7-27		Yes	6 m, 1, 2 yr; 3.5 yr; 5 yr; 8 yr	Bayley or Griffiths; Stanford Binet; WPPSI; WISC-R	212	11.8% delayed/below average; 65.6% average; 65.6% average; 22.7% advanced/above average		—Distributions differ by state; —Retarded child rare
(b) Victoria State	Rickards et al. (1987)	1977	T$_4$	11-18		Yes	6 m; 1 yr; 2 yr	Bayley	68; 65; 62	MDI = 103.4, PDI = 96.3; MDI = 106.3, PDI = 84.8; MDI = 95.2[b], PDI = 88.9		Severe mental impairment eradicated but information on minor & specific deficiencies emerging—influenced by thyroidal and extrathyroidal factors.
Victoria State	Rickards et al. (1989)	1977	T$_4$	11-18		Yes —Random —Comparable ethnic & social background	5	WPPSI	45	VIQ = 95.9[c]; PIQ = 105[b]		
South Australia/ Adelaide	Thompson et al. (1986)	1977	T$_4$	<28	50[d]	No	<4 yr; 4-6 yr	Griffiths; WPPSI	26; 10	102.4; 112.2		
2. CANADA												
(a) Ontario	Rovet et al. (1986)	1976	TSH	14.3					101	Averaged IQ = 109.7[a]	Difference significant for PIQ	—Majority at lower limits of normal in school; —14% attentional problems

Study	Year	Hormone	Value	Range	Controls	Age	Test	N	Score	Findings
Rovet et al. (1987b)	1976	TSH	16.6	8.0[e]	79 siblings, 30 others	1,2,3 yr; 18m; 4-6 yr; 5 yr	Griffiths; Bayley; McCarthy; WISC-R		112.6; 105.6; 104.6[b]; 103.6[a]	—CH <sibs in language at 3, 5 yr —CH poorer on perceptual memory and neuromotor at 6 yr / —½ SD lower than controls after age 5. Subtle deficits in selective ability areas associated with fetal hypothyroidism & duratic
(b) Quebec										
Glorieux et al. (1983)	1974	T4	27	75-100[d]	Yes, matched	12 m; 18 m; 3 yr	Griffiths; Griffiths; Griffiths	45	115; 104[b]; 103[b]	—18 m Lower on Hearing & Speech, Performance —3 yr Lower on Practical Reasoning —GQ α T4, Bone Age / First manifestation of minimal brain dysfunction
Glorieux et al. (1985)	1974	T4	27	75-100[d]	Yes, matched	5 yr	Griffiths	105	102[a]	Lower on performance, Reasoning
Glorieux et al. (1987)	1974	T4	27	75-100[d]	Sibs	7 yr; 9 yr	WISC-R; WISC-R	43; 19	97[a]; 105	—High risk group (no ossification center at knee & T4 < 2.0) significantly lower than C or low risk —Low risk and C, no diff / Skeletal maturity and T4 level predict later outcome
3. FRANCE										
(a) Lyon										
David et al. (1983)	1976	TSH	34 (16-70)	10[e]	Yes	6 m	Brunet-Lezine	32; 52	1/32 retarded	—2/32 delayed motor / 8/32 delayed speech 10/32 pronunciation problems
(b) Paris										
Toublanc et al. (1989)	1976	TSH	29 ± 12	5-7[e]	Sibs & others	6 m, 1,2; 4 yr; 7 yr	Brunet-Lezine, Terman-Merrill; WISC-R	20	85-145; 95-130	—No difference controls, VIQ vs PIQ —Normal achievement —Bone age predicts outcome / Factors influencing outcome: 1. SES 2. hormones during first years 3. psychological disturbances during infancy

TABLE 17.1. *Continued*

Country	Authors	Year begun	Screening method	Tmt Age[g]	Dose	Controls	Assessment ages	Tests	# CH	IQ results	Additional	Comments
(c) Toulouse/ Rangeuil	Rochiccioli et al. (1983)	1977	TSH	29	20[d]		6 m, 1,3 4, 5 yr	Brunet-Lezine, Griffiths >30 m	35	6 m 98 1 yr 96 3 yr 99 4 yr 97 5 yr 97	—1 yr—transitory decrease in posture, —2,3 yr—transitory decrease in language, soft signs in some children, —DQ, posture language correl. severity hypo, —3 yr, A < E in posture, lang., sociabil.	studies stress psychomotor & neurological aspects
	Roge et al. (1987)	1977	TSH	29		Yes N = 59	1-8 yr	—1-5, Griffiths, —7 yr Lincoln Oseretsky, —6-8 yr WISC-R neurological		1 yr 96 2 yr 97 3 yr 100[a] 4 yr 99 5 yr 98	—Coding poor in 50%, —Eye-hand coordination difficulties	
									21	6 yr 103	VIQ = 101 PIQ = 105	
									12	7 yr 97	VIQ = 97 PIQ = 98	
									10	8 yr 94	VIQ = 97 PIQ = 93, —No correlation with etiology, —Correlation with BA, T$_4$, T$_3$ (1 yr)	
(d) Lille	Farriaux et al. (1988)		T$_4$	25	6[e]		1-2 y	Brunet-Lezine	101	1 yr 105.8	Correlation with BA, T$_4$ and BA	Initial T$_4$ and bone maturation most predictive
							5-7 y	WISC-R		5 yr 95 7 yr 96	—No effect etiol., tmt age, —VIQ < PIQ	

Location	Author	Hormone	N	Age	Screening	Assessment age	Test	Sample	IQ	Outcome
4. GREECE	Komianou et al. (1988)	TSH	28	8-10[e]	Yes—random	5-37 m	Grifiths	109, 19 cases longitudinally	105.6	—Stable IQs in CH assessed serially —E > A—not significant —Significant effects social class —BA correl outcome @ 17-19 m only
5. ITALY (a) Milan	Rondanini et al. (1989)		36		Prescreening treated @ 170 days.		Brunet-Lezine Stanford Binet Touwen & Prechtl Oseretsky		IQ innormal normal range	—Neurolog—delay in 55% vs 73% controls —1st year 0% CH vs 60% controls delayed in second year
(b) Naples	Tenore et al. (1987)				Clinical	2 y	Psychomotor	153	—Diagnosis<25 days, IQ = 100 —Diagnosis>41 days, IQ = 90 p < .001	—E < A —Psychomotor correl age at onset, severity, etiology, bone age.
(c) Pisa	Bargogna et al. (1989)	T₄				0-2	Neurol Uzgeris-Hun Brunet-Lezine	17	12 m—100 18 m—102 3 yr—99	—Normal neuropsychol development with mild & transient defects —Normal cognitive strategies —Slower walking —Mild hypotonia
						2-6	Terman Merrill			—Age 2—irritable & oppositional —Older—inhibited, lack of self-confidence Examined mother-child interaction

TABLE 17.1. Continued

Country	Authors	Year begun	Screening method	Tmt Age[g]	Dose	Controls	Assessment ages	Tests	# CH	IQ results	Additional	Comments
(d) Rome	Moschini et al. (1989)		T₄ and/or TSH	33	re Guyda		6-24 m, 3-5 yr, 6 yr	Brunet-Lezine, Stanford-Binet, WISC, neurolog., EEGs & ERPs	75	0.5 — 88, 1 — 88, 3 — 90, 4 — 97, 5 — 97, 7 — 100	—No effects etiology, —BA effects, —Abnormal EEGs 3/42, —VEP—64% increase in latency, —2/42 strabismus, —4/42 sensorineural hearing loss	—Pre- & perinatal risk factors, —MBD in small % age
6. JAPAN										Path contr.		
(a) Osaka	Maki et al. (1983)	1975	TSH	45		22 pathol. controls, 88 age-matched	6-12 m, 13-24, 25-36, >36	Tsumori & Inage, Suzuki, Binet, WPPSI, WISC-R	23	105 / 82, 101 / 82, 108 / 84, 102 / 87	—Unbalanced psycholog. profiles, —Low speech & sociability, —Problems in linguistic, coord.; movement imitative behavior	Some minimal brain dysfunction
(b) Tokyo	Irie et al. (1989)	1979	TSH	<28	5-10[e]		1-8 yr		247, 200, 123, 79, 35, 16, 3, 1, 37	1 — 105, 2 — 110, 3 — 106, 4 — 105, 5 — 106, 6 — 105, 7 — 109, 8 — 132	Overall mean IQ (N = 426), 107 ± 18	89.2%, IQ above 90
7. NETHERLANDS	Van Vliet et al. (1989)		TSH	14		Yes −34					Social, adaptive language below CA, not related to bone age	
8. NORWAY	Heyerdahl (1987)						2, 6	Bayley, WPPSI, Finnish neuro-develop	46	MDI = 98, VIQ = 85.6, PIQ = 94		Outcome correl. with # symptoms, T₄, psychosocial factors, not bone age or etiology

Country/Study	Reference	Year	Marker	Value	Sample/Controls	Age	Test	n	IQ	Findings	Comments
9. SPAIN (Catalonia)	Ibanez et al. (1989)	1983	TSH			6-12 m	neurol. DSST BAEP CAEP AEP			—Neurolog.: no anomalies, —25% hearing loss	Psychomotor & language deficits
						18-36 m	Gesell/Terman Merrill		IQ < 90 31% IQ normal 28%		
10. SWITZERLAND	Illig & Largo (1987)		TSH	10.4 (6-22)	125 healthy children	1 yr	Brunet Lezine	60	100	—CH<C for global & 1 subscale	Controls from longitudinal well-baby study
						4 yr	Snijders-Oonen	40	100	—CH<C for global & 2 subscales	
						7 yr	HAVIVA	20	99[a]	—CH<C for 1 subscale DQ significant lower for high risk risk group; low risk = control	
11. UK (a) NW Thames	Murphy et al. (1986)		T_4	27	3 & 5 y age-matched	1	Griffiths	58	104	VIQ = 101, PIQ = 112 —Significantly poorer on speeded pegboard at 5 —Absent gland &/or low bone age, behind but not significant	Mild clumsiness
						3	McCarthy	48	102		
						5	WPPSI, Bruininks	20	107		
(b)	Fuggle et al. (1989)		T_4	27	39—3 yr 52—5 yr	3	McCarthy		103	Signif lower scores for children with low bone age or T_4 Significant effect of T_4, T_4 correl PIQ, not VIQ No effect of age of onset	Initial severity associated with poorer outcome
						5	WPPSI		109		
12. US	New England CH Collaborative (NECHC, 1981)	1976	T_4	24	Siblings, False Positives	1-2	Bayley	67	105.3	No correlation between IQ, bone age, T_4, etiology, breast feeding	
				8.0[e]		3-4	Stanford-Binet				

TABLE 17.1. *Continued*

Country	Authors	Year begun	Screening method	Tmt Age[g]	Dose	Controls	Assessment ages	Tests	# CH	IQ results	Additional	Comments
	NECHC (1984)	1976	T$_4$	24	8.0[e]	Yes	3–4	Stanford Binet		3 4 5 Latest 105	102 No effect of etiology 106 106	Only predictor of outcome was adequacy of treatment
	NECHC (1985)	1976	T$_4$	24	8.0[e]	Yes	6	WISC-R WRAT Reitan	56	109	—No difference on WRAT —CH slower on finger tapping	CH less advanced in psychomotor develop
	NECHC (1987)	1976	T$_4$	24		44 Age-matched classmates 36 Siblings	3rd grade	PIAT	26	IQ = 109.2	PIAT: Math + 1.4 grades Reading + 1.4 grades Spelling + 1.4 grades	Blind testing Same % CH vs C needing help

[a,b,c] Significantly different than controls [a] $p < .01$, [b] $p < .01$, [c] $p < .001$.
[d] In μg/day.
[e] In μg/kg/day.
[g] In days.

wide), 3 from North America (New England states, Ontario, Quebec), 2 from Japan (Tokyo/Sapporo, Osaka), 1 from the UK, and the remainder from Europe, where perhaps some of the most extensive follow-up is being done. The European collaborative survey, which was conducted on 36 centers from 12 European countries (Delange et al., 1979), has provided additional findings on mental development in about 20% of the hypothyroid children positively identified by newborn screening in Europe (N = 790 children; 1560 assessments).

About half of the studies listed in Table 17.1 were based on T4 screening while the remainder were based on TSH. This factor was subsequently examined because it was thought it might contribute to sampling differences among the various studies. This is because TSH screening identifies children with TSH elevations regardless of their T4 level, which may be in the normal range; T4 screening by contrast will include only children with low T4 values. Because T4 at birth is one of the strongest contributors of later development (Glorieux et al., 1988), it was expected that the samples from studies based on TSH screening should be less severely affected initially and so do better on average subsequently than those involving T4 screening.

General Outcome

The results from Table 17.1 showed relatively good outcome for general IQ. Of the 22 studies examined, 10 reported no difference from controls or test norms (Australia-wide, South Australia, Lyon, Paris, Naples, Pisa, Osaka, Tokyo, UK, US). Because 6 of the 10 studies screened for TSH, there does not appear to be a difference due to screening method in subsequent outcome. Of the remaining 12 studies, 3 reported consistently lower but nonsignificant differences from controls (Greece, Milan, Switzerland), 7 reported significant differences that were not evident among CH infants but appeared when they were older (Australia-Victoria, Ontario, Quebec, Toulouse, Lille, Norway, Spain), and one study (Rome) reported differences up until age four only, at which point they disappeared. Finally, the European Collaborative survey (not reported in Table 17.1, Illig et al., 1985) found total group differences only at age three and for those children tested with the Bayley or Stanford-Binet, not the Griffiths or Terman-Merrill (differences were observed when specific subgroups were studied). Clearly then there does not seem to be any general consensus as to the specific impact of screening on general outcome.

The mean IQs or Developmental Quotients (DQ) range from a low of 88 (Rome) at age 6 to a high of 115 at 1 (Quebec) and 3 (Ontario) years (both using the Griffiths). Table 17.2 presents the mean IQ by age for the 22 studies and the European collaborative data. This was determined by averaging across the studies reporting age-specific data. Unfortunately because it did not account for the number of cases assessed at each age as

TABLE 17.2. Average IQ by age across the 22 studies.

Age	Number of studies	Average IQ
0.5 yrs	4	96.2
1.0 yrs	13	101.8
1.5 yrs	5	102.5
2.0 yrs	7	102.6
3.0 yrs	14	102.6
4.0 yrs	8	102.4
5.0 yrs	13	102.0
6.0 yrs	6	105.2
7.0 yrs	8	99.9
8.0 yrs	1	94.0
9.0 yrs	3	105.0

this was not always reported, this method gives more weight to the studies with smaller samples. However, as a supplementary analysis revealed that studies with smaller sample sizes did not differ in outcome from those with larger, we believe that this factor would not substantially affect the results in Table 17.2. The IQ results ranged from an average of 94 at age 8 to 105 at age 7. The overall IQ obtained by averaging across all studies and all ages is 102.0, which is clearly within the average range. There did not appear to be any relationship between the average IQ score and assessment age.

Table 17.3 presents the results according to test instrument. Scores on the Griffiths (1 to 6 years) tended to be somewhat higher than those obtained with Brunet-Lezine (0.5 to 2 years) or the Bayley (0.5 to 3 years). It is generally well-recognized that although the Griffiths may provide a detailed and thorough developmental assessment, this test tends to inflate scores given that it was standardized more than 25 years ago. Although it is now undergoing restandardization, the new norms are not yet available.

Selective Deficits

In spite of normal general IQs, a number of studies have reported that a significant proportion of the children appear to have subtle selective impairments, the exact nature and appearance of which vary from study to

TABLE 17.3. IQ averaged across tests.

Test	Number of studies	Average IQ
Bayley	7	99.5
Griffiths	14	104.0
Brunet-Lezine	7	96.9
Stanford-Binet	6	100.2
Wechsler	20	102.2

study. These are most evident in the studies that used additional measures beyond the standardized intelligence tests. Neuromotor impairments have ranged from abnormal EEGs and evoked potentials (Rome) plus neurological soft signs (Pisa, Toulouse), if neurologists were part of the team, to neuromotor deficits (Ontario, Japan), mild clumsiness (UK), and slower speeded motor performance (UK, US), when assessments were done by neuropsychologists. Both speech delays (Ontario) and a relative weakness in language skills (Australia, Quebec) have also been reported. This suggests that while the specific type of impairment depends on the focus of the study, a variety of selective impairments are still prevalent in these children.

Additional Factors

Outcome appears to be influenced by a number of disease-related factors, family circumstance variables, and perinatal risk factors. These have generally been examined individually; no study has as yet looked at the interaction of different variables with outcome.

The disease variables correlating with outcome include etiology and/or severity of illness at initial presentation (Quebec, Toulouse, Lille, Naples, Norway, UK), prenatal onset (Quebec, Ontario, Paris, Lille, Naples, Rome, UK), duration of hypothyroidism in infancy (Ontario, Naples), and treatment adequacy (US). The Quebec group has reported that a combination of low T4 and severely delayed skeletal maturity at diagnosis places the child at greatest risk for subsequent neurocognitive impairment (Glorieux et al., 1988). There does not appear to be a relationship between exact age at onset of replacement therapy and outcome but this may be limited by the small range of treatment ages within each of the different studies. The European Collaborative study (Illig et al., 1985) cited length of time children were deprived of thyroid hormone as the more important factor. We have found that *timing* of the hormonal deprivation relative to events in brain maturation is critical, not total duration (see subsequent section).

Four studies have addressed the issue of family background factors on outcome. Komianou et al. (Greece) reported significant effects of social class with the impact of CH being more severe among children from lower social classes. Illig and Largo (1987) noted that the effects of family SES were more pronounced among older than younger children. Heyerdahl (1987) in evaluating the impact of psychosocial variables, reported that family dysfunction also played a major role in outcome in these children. Finally Illig and Largo (1987) found that children with additional developmental and perinatal problems, as well as unfavorable environmental backgrounds, had the lowest scores (note that some studies have eliminated these children from their samples). However, because the majority of studies failed to examine the relationship on outcome of these "extrathyroidal factors'' (Rickards et al., 1987) along with thyroidal, it is difficult to specify their exact contributions.

Methodological Considerations

There are a number of methodological factors in addition to the specific disease effects that can also affect outcome. These include familiarity of the children with the tester, which may favorably bias the results, given that the children are apt to perform better if they know the tester. Such results would be expected from the studies involving repeated annual assessments. Only one study (US) has actually addressed this issue by using assessors who were blind as to disease status of both CH and control groups. Another factor concerns where and when the assessment is given—home, school, psychology office, medical office. It is expected that the children who were assessed as part of their medical follow-up may have done more poorly, due to increased anxiety associated with blood testing and physical examination.

One must also be aware that the studies describing outcome in older children are based on cohorts who were among the first to be identified by screening when procedures were not well set in place (New England Congenital Hypothyroidism Collaborative, 1982). Because these children may have received later and less adequate therapy, their outcome at school age may be poorer than that of the children who are now being diagnosed by screening. Therefore before final conclusions are drawn one should ideally await the results on these later cohorts of children when they reach school age (Mitchell, 1987).

One final issue concerns who is responsible for the follow-up. Studies that are part of the screening program may have different objectives than those which are independently funded research projects. Although this issue has never been properly evaluated, it is possible that because of different commitments, the two groups may differ in their biases, and orientations.

Summary and Conclusions

This analysis has shown that the children with CH diagnosed by newborn screening are clearly normal in their mental development, although their IQs tend to be slightly compromised on average relative to controls and standardization samples. The degree of compromise appears to depend on the sensitivity of the particular testing instrument used. Notwithstanding, these results signify a substantial and definite improvement in outcome from the children who were diagnosed clinically in the era prior to newborn thyroid screening. Although significant differences from normal controls were most evident among older children, IQs did not appear to decline with age. This is contrary to the report of the European Collaborative survey which has claimed an increasing impact of CH on mental development with age. Most evident from the studies reviewed in this analysis was the observation that older children demonstrated specific neurocognitive deficiencies, the exact nature of which varied from study to study and depended on the child's age and the important developmental issues at the time of testing. In conjunction with recent evidence from electrophysiological studies on these children, their deficits appear to be reflecting subtle persisting brain damage which was due to TH deficiency prior to treatment (Landenson et al., 1984; Richards et al., 1987), but only became manifested in performance once specific skills were required.

This review has also shown that variables reflecting disease duration and severity, as well as social and developmental problems, clearly influence outcome. However, because these were not specifically examined via multivariate methods, neither their relative contributions nor their interactions with disease variables can be determined. What is clearly needed is a multicenter collaborative study that will provide sufficiently large sample sizes of neonatally identified hypothyroid children, as well as adequate controls and psychometrically valid measurements. This will permit the detailed multivariate statistical analyses that are needed to draw the appropriate conclusions. It is important that such research also be designed to fit the principles of neurodevelopmental theory so that the most appropriate assessment procedures will be used at each age and the interpretation of results will be congruent with current knowledge about the developing brain. Because it is universally agreed that the children are not retarded, emphasis of future research should be directed more to describing the underlying deficits rather than to simply measuring IQ. Such research should also focus on how older children are functioning at school and whether they need special education or other services, as this will ultimately be the true test of the efficacy of newborn screening. The close screening network of the individual states in the United States as well as the large unilingual population base and good funding resources should make this an ideal area to now proceed. It is ironic that only one group in this country has as yet attempted a systematic evaluation of outcome in these children.

The Toronto Prospective Study

In 1981, we began an independently funded longitudinal prospective study of children with CH diagnosed by newborn screening who were being treated at the Hospital for Sick Children in Toronto, Canada. This was a research project that did not belong directly to any one screening program. All of the children studied had been positively identified through either a cord blood screening project at the Mount Sinai Hospital and serving the immediate Toronto area (Walfish et al., 1979) or the provincial screening

---

program for Ontario (Peter et al., 1987), or by both during the period that the two programs operated concurrently (Walfish et al., 1983). Both screening programs assayed for TSH. The goals of our study were to evaluate general and specific intellectual functioning in younger and older CH children, to determine the factors associated with poorer outcome, and to identify the children at greatest risk for subsequent impairment. A supplementary, but equally important goal was the more basic issue of studying the role of thyroid hormone on early brain development. As the bulk of our cohort has now become schoolage, our focus has more recently shifted towards defining the specific psychoeducational characteristics of these children. This section will describe our most recent findings.

Methods

Subjects

The CH sample consisted of 108 children (75 girls, 33 boys) who were born between 1976 and 1986. Of these 108 cases, 48 ranging from 1 to 6 years of age were available at the start of our study; the remainder were born after it began. Rates of refusal and attrition have been exceptionally low (5% and 1% per annum respectively). Two children were eliminated after they were found to have transient hypothyroidism at age three. Five more children assessed initially were also eliminated due to other illnesses or congenital abnormalities.

The sample included 30 children (28%) with athyrosis, 47 (44%) with ectopic glands, and 31 (29%) with dyshormonogenesis usually with goiter. Sixty-one percent, who were significantly retarded in their skeletal maturity at diagnosis (bone ages $\leq$ 36 weeks), presumably had hypothyroidism beginning at least one month prior to birth. The mean (SD) age at treatment onset was 16.7 (16.1) days; mean dosage was 7.9 (1.9) μg/kg thyroxine; mean time until euthyroidism was achieved, that is, the postnatal duration of hypothyroidism, which we defined as the age when the circulating T4 level reached 10.0 μg/dl, was 52.5 (49.4) days or 36 days from beginning treatment.

Controls included 110 nonaffected children. Most ($N = 78$) were siblings, 48 of whom were older and 30 were younger than the child with CH. (To adjust for this imbalance, every additional newborn sibling is also being added to the project when he or she becomes 12 months of age.) As most sibling controls have now been assessed at least twice and most CH repeatedly on an annual basis, there are presently 92 instances of a CH child and a control being tested at the same age. The 32 unrelated controls have included siblings of children participating in a different study and children of family friends.

To date, assessments have been conducted on between 70 and 100 CH children at each age between 1 and 6 years, 57 CH children at age 7, 40 at 8, and 25 at 9. Assessment of controls has included 13 one year olds, 11 one and one-half year olds, 22 two year olds, 17 three year olds, 30 four year olds, 32 five year olds, 21 six year olds, 26 seven year olds, 18 eight year olds, and 14 nine year olds. Given our recent shift in emphasis towards specifying psychoeducational characteristics, we are presently in the process of obtaining classmate participants who will serve to control for teacher and school influences on learning and ability.

Test Procedures

The study protocol has involved a detailed annual assessment of each CH child at about the time of his or her birthday. This initially took place when the child turned 12 months of age, or in the case of older children, when they reached their next birthday. Assessments have continued on an annual basis until each child has reached the fourth grade at school, or 9 or 10 years of age. Controls were assessed about the time of the CH child's initial assessment and as many times thereafter as parents would permit. These assessments were conducted primarily by two psychometrists, one of whom was totally blind to the study's objectives and the children's conditions or grouping. The other psychometrist, while knowing which child had hypothyroidism, was not aware of other factors such as etiology or disease status at diagnosis, which could presumably have affected the results. Most assessments were conducted at the Hospital for Sick Children, usually in conjunction with their annual endocrine clinic visit. A few were conducted in the children's homes where necessary.

The test battery has included a number of age-appropriate neuropsychological tasks tapping the following areas of mental and socioemotional functioning: general intelligence, language, neuromotor skills, perceptual abilities, memory, attention, social competence, temperament, behavior and school achievement. Intelligence was determined using the Griffiths scales for children from one to three years of age, the Bayley at 18 months, the McCarthy at 4 and 6 years, the WPPSI at age 5, and the WISC-R at ages 7 and 9 years. School achievement was assessed with the WRAT-R, the Keymath, and the Woodcock-Johnson Reading Mastery-Revised. Additional measures included the Reynell Language Scales before age 7 and the Carrow Elicited Language Test after 7; the Beery-Buktenica Test of Visuomotor Proficiency from age 2 on; the Bruinincks-Oseretsky Test of Neuromotor Development at 6 years; and the McCarthy Memory, Perceptual, and Motor scales at 3, 5, and 7 years. The present psychoeducational component also includes measures of attention, oral reading, oral and written language, and graphomotor skills.

Data Analyses

The data were analyzed in two ways: by matched comparisons with siblings and by comparing CH children with nonmatched controls of the same age. T-tests (dependent and independent, one-tailed) and Analyses of Variance were used for these comparisons.

Results

Neuropsychological Outcome

Few differences were evident between the CH and control children among the younger, preschool group. The analyses pairing CH children with their siblings tested at the same age indicated they did not differ in their performance on the Griffiths test. As Figure 17.1 shows, just as many CH scored higher than their siblings as lower. Contrasting CH children with controls similarly indicated there were no differences on any of the individual Griffiths scales at 1, 2, or 3 years of age (see Table 17.4). They also did not differ on the Bayley Mental Development Index (MDI) (CH = 107.0, Control = 112.7) or Psychomotor Development Index (PDI) (CH = 98.2, Control = 100.2). Results for the Reynell language scales, however, revealed delays in both receptive and expressive areas of language at 3 years of age. The mean z-scores for receptive skills were .301 for CH and 1.023

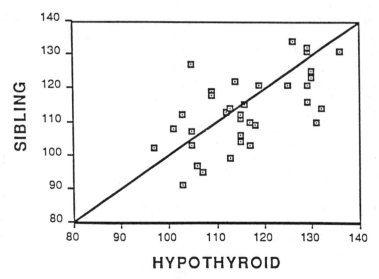

FIGURE 17.1. Hypothyroid versus sibling IQ in 1- to 3-year-old children in the Toronto Study.

TABLE 17.4. Griffiths test scores for preschool children.

	1 year		2 year		3 year	
	CH	C	CH	C	CH	C
GQ	112	110	113	114	115	116
Locomotor	116	121	117	119	120	118
Personal/social	109	107	118	117	119	116
Hearing/speech	113	106	107	109	115	120
Eye-hand	111	109	107	107	107	107
Perceptual	110	109	113	117	114	116
Reasoning			116	119	116	116

for controls ($t = -2.36, p < .01$; one-tailed) and for expressive skills were .433 for CH and .946 for controls ($t = -1.85, p < .05$; one-tailed).

The results on children after age three indicated poorer outcome for the CH than the controls in a number of different areas. The paired CH- sibling analyses revealed consistently lower IQs for CH than their siblings when they were tested at the same age (CH = 105.3; Sibling = 110.4; $t = -2.10, p < .05$), as shown in Figure 17.2; this difference was larger for the PIQ than the VIQ scale (6.0 vs. 4.0 points). Comparisons of older CH children with age-matched controls revealed that the CH scored significantly lower on the WPPSI Full Scale IQ (FSIQ) at age 5 (CH = 106.6; Control = 113.5; $t = -2.05$), the McCarthy General Cognitive Index (GCI) at age 6 (CH = 102.3; Control = 112.8; $t = -2.03, p < .05$) and the WISC-R at 9 years (CH = 98; Control = 115; $F = 12.8, p < .001$). Al-

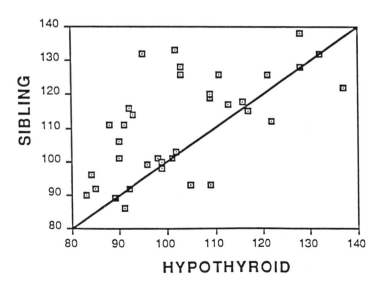

FIGURE 17.2. Hypothyroid versus sibling IQ in 5- to 9-year-old children in the Toronto Study.

though the CH children did not differ significantly from controls on the WISC-R IQ at age 7 years, the results were consistent (CH = 102.3; control = 106.6). (They were not administered an intelligence test when seen for their age 8 assessment.) In sum, it appears that the CH children are becoming increasingly discrepant from controls as they get older.

The CH group also demonstrated specific deficits relative to same age controls in receptive language at age 4 ($p < .001$), perceptual ($p < .001$) and neuromotor ($p < .05$) ability at age 6 and memory ($p < .01$) at age 8. As shown in Figure 17.3, the controls appear to score above average while the CH appear to score below.

Academic Achievement

There appears to be an increased incidence of learning problems associated with neonatal hypothyroidism (Figure 17.4), which seems to increase with age. Although the difficulties exhibited by CH children were not severe (e.g., there were few cases of frank dyslexia), they may be serious enough to be cause for concern for later "academic dysfunction" (Levine et al., 1981), that is, the child falls increasingly behind in his/her learning with age. Among CH children, 7% of 7 year olds, 10% of 8 year olds, and 16% of 9 year olds were in full-time special education classes as compared with a 1% rate for Ontario school children [a comparable rate of 0.8% has been reported for children in the United States between 6 and 11 years of age with learning disabilities (Office of Special Education, United States Department of Education)].

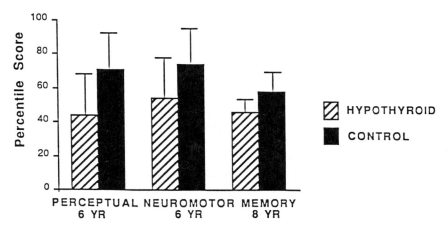

FIGURE 17.3. Specific neurocognitive deficits in hypothyroid children (Toronto Study).

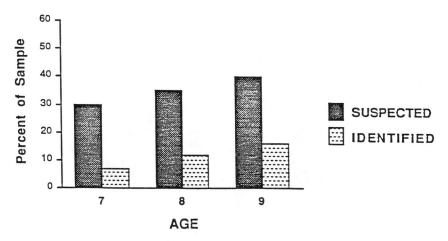

FIGURE 17.4. School learning problems in hypothyroid children (Toronto Study).

Also shown in Figure 17.4 are the number of children experiencing learning problems at school as determined by clinical judgment, parental interview, and a detailed analysis of achievement scores. This is clearly not just a "sensitivity" bias (i.e., people are more likely to call these children learning disabled because of their medical history), because it extends to achievement tests. We observed that by these criteria, 28% of the 7 year olds were suspected of having learning problems, 35% of 8 year olds, and 40% of 9 year olds. Unfortunately we do not as yet know how many of these children are actually receiving part-time remediation for their learning problems. Although we expect that this is a relatively small percentage, it should be substantially larger than the 3% value for children who qualify for special education on a withdrawal basis in the province of Ontario or than the 2.4% value for United States school-children (Office of Special Education, United States Department of Education).

In order to determine the kinds of school-related learning problems the children with CH are experiencing, we assigned the children to five groups on the basis of their WRAT-R scores using the Siegel and Heaven (1986) classification scheme. According to this scheme, children with a reading disability (RD) consist of those who scored below the twenty-fifth percentile in reading and above the thirty-fourth percentile in arithmetic and spelling; those with an arithmetic disability (AD) are those scoring below the twenty-fifth percentile in arithmetic and above the thirty-fourth in reading and spelling; those with a spelling disability are those scoring below the twenty-fifth percentile in spelling and above thirty-fourth in reading and arithmetic; those with a general disability (GD) are those scoring below the twenty-fifth percentile in all subject areas; and the no

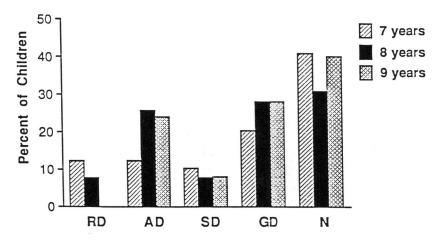

FIGURE 17.5. Specific learning disability subtypes in school-age hypothyroid children in the Toronto Study (RD = reading disability; AD = arithmetic disability; SD = spelling disability; GD = general disability; N = no problem).

problem group (N) are above the twenty-fifth percentile in all three subject areas. As shown in Figure 17.5, more than 60% of the CH children were in one of the four learning problem groups, the largest of which had a general disability.

Results for the Keymath test, which was given at age 9 only, indicated that 54% of the CH children were more than $\frac{1}{2}$ grade delayed in learning math concepts, 22% were behind by up to a $\frac{1}{2}$ grade while only 25% were at or above grade level. CH did not differ from controls in their reading or spelling scores at 7 and 8 years of age; however at age 9, CH scored significantly lower in passage comprehension (mean percentile scores: CH = 42.1, C = 63.6; $t = -2.64$, $p < .01$) and spelling (CH = 31.2, C = 62.8; $t = -2.08$, $p < .05$).

These results are preliminary and must obviously await more detailed testing, particularly on larger samples of children including patient cohorts who have been recently identified and treated by improved diagnostic and treatment procedures. Their results should also be compared with normal controls who have been carefully matched for social and educational background. Despite this however, present results do suggest that there are a greater number of CH children who experience difficulty at school than would be expected if they were totally unaffected. The majority of children in our sample appear to be in the bottom half of the class, and a substantial number needs or already receives additional remedial support. The clinical significance of these findings with respect to the longer term implications and their later success still needs to be established.

The Toronto Prospective Study (Continued)—Predicting Outcome

Etiology

Although comparisons between younger CH children and controls established no major differences in intellectual outcome overall when the preschool children with CH were compared among themselves, those with athyrosis (A) were significantly outperformed on a number of scales by the children with dyshormonogenesis (D) or ectopic glands (E). Figure 17.6 shows the results of comparisons between the three etiologies on the Griffiths Scales. It can be seen that at 12 months of age, A scored significantly lower than D or E on general quotient (GQ), locomotor, and eye-hand coordination scales ($p < .05$); at 2 years, A scored lower on GQ, personal social, and hearing and speech scales ($p < .05$); at 3 years, on GQ ($p < .001$), locomotor ($p < .001$), personal social ($p < .01$), and hearing and speech ($p < .001$). A also scored lower on the Reynell receptive language scale at 12 months, 3 and 4 years ($p < .05$) and the expressive scale at age 3 ($p < .01$); on the McCarthy GCI, verbal, quantitative, and memory scales at 4 years; WPPSI similarities and mazes, and McCarthy memory at 5 years; the Beery at 4, 5, and 7 years; the Bruinincks at 6 years; and WISC-R PIQ and WRAT-R spelling at 7 years. Clearly then, children born without a thyroid gland have a greater degree of impairment than children who have partial thyroid function at the time of diagnosis; this difference is evident from a very young age. Although there were fewer differences among the older children with different etiologies, this may be due to the small sample sizes of children at older ages rather than to a diminishing effect of this risk factor. It is important to await the results on the remaining cohort as it gets older. Nevertheless, one can conclude the athyroid condition compromises intellectual competence, especially in early, and potentially, in later childhood.

Disease Severity

This was examined in three ways: (a) by comparing children with and without bone-age retardation at diagnosis; (b) by determining the impact of their T4 levels at diagnosis; and (c) by grouping the children according to both parameters.

Bone-Age Retardation

The majority of children had knee radiographs taken at the time of confirmatory diagnosis. Of these, over 60% showed significant retardation in skeletal maturity suggestive of intrauterine hypothyroidism, the majority

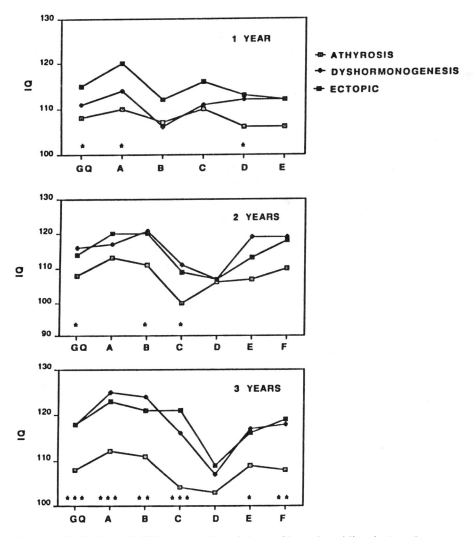

FIGURE 17.6. Mean Griffiths scores by etiology of hypothyroidism in 1- to 3-year-old children (A = locomotor skills; B = personal/social; C = hearing and speech; D = eye-hand coordination; E = perceptual performance; F = practical reasoning; *p < 0.05, **p < 0.01, ***p < 0.001).

of whom (but not all) were athyrotic. In comparing children with and without retarded bone ages, we found that the delayed group scored significantly lower in a number of areas of mental functioning (Rovet et al., 1987). This included most Griffiths scales, plus Reynell receptive and language scales, McCarthy verbal (4 years), quantitative (4, 6 years), memory (5 years), and motor (5 years) scales, Beery (5, 6, 7 years), WPPSI and WISC-R PIQ and spelling (8 years). These findings therefore suggest a substantial degree of impairment associated with hypothyroidism in utero.

T4 Levels

Children with T4 levels below 4.0 μU/dl at diagnosis subsequently scored lower than children with higher T4 levels in the following areas: Griffiths GQ, locomotor, hearing and speech, eye-hand and Reynell expressive and receptive scales at 12 months; Griffiths eye-hand and Reynell expressive at 3 years; McCarthy verbal at 4 years; McCarthy motor at 5 years; Beery, Bruinincks, and McCarthy verbal at 6 years. Therefore, severity of hypothyroidism at diagnosis is also an important determinant of outcome.

T4-Bone-Age Interactions

A series of bone-age X T4 ANOVAs were conducted on the outcome variables. The results revealed a greater number of significant main effects for boneage than T4, with few significant interactions (Rovet & Ehrlich, 1988). This was observed on most variables and at most ages. Hence bone-age delay appears to have a more longlasting and deleterious effect on outcome, regardless of the severity of hypothyroidism at diagnosis. Bone age is therefore a better indicator of severity and thus the impact of CH on the brain. The relationship among the various ''risk'' factors will be examined in a subsequent section.

In an additional series of analyses, the additive effects of both variables were examined by comparing three groups of children on the basis of the relative contributions of both variables: a high risk group had a low confirmatory T4 value ($<$4.0 μU/dl) and a delayed bone age (36 weeks or less); a moderate risk group had either a low T4 or a delayed bone age, not both; and a low risk group with both a normal T4 ($>$4.0) and a normal bone age. These data were analyzed by analyses of variance tests comparing all three groups. As shown in Table 17.5, children in the high risk group did significantly more poorly than those in the moderate or low risk groups in general cognitive, perceptual, language, and memory areas. Children in the low risk group generally, although not consistently, outperformed those at moderate risk.

These results therefore indicate that degree of fetal hypothyroidism is correlated with a number of neurocognitive deficits, which appear to affect a broad range of areas of intellectual function while low levels of thyroxine at diagnosis is associated primarily with subsequent deficits in verbal and motor skills. Therefore perceptual deficits are sensitive mostly to hypothyroidism in utero. Although bone-age retardation appears to have a more pronounced effect on subsequent outcome than T4 at diagnosis, clearly the child with either or both risk factors is more likely to have problems than the child with neither. As reported for the clinically identified cases,

TABLE 17.5. Performance as a function of initial severity of thyroid disease.

Age	Task	Risk group		Low	F	p-level
		High	Medium			
3 yrs	Beery	58.2	52.5	75.9	4.24	.02
4 yrs	McCarthy GCI	103.4	112.6	114.1	3.37	.04
	Verbal	49.6	56.9	56.8	4.49	.01
	Reynell ESS	−.14	.48	.62	3.27	.05
5 yrs	WPPSI PIQ	101.3	110.1	106.7	2.93	.06
	Beery	43.1	60.6	49.4	3.63	.03
6 yrs	Beery	33.0	45.2	63.3	4.77	.01
	Bruinincks	44.5	62.5	60.2	2.40	.10
7 yrs	WISC PIQ	99.6	104.2	113.2	3.44	.04
	Beery	40.3	59.2	77.3	6.31	.004
	Mcarthy Memory	45.5	52.9	47.7	4.55	.01
8 yrs	Woodcock Reading	48.9	69.5	56.5	3.32	.05

hypothyroidism in utero may contribute to subtle specific CNS damage, the severity of which reflects the degree and duration of hypothyroidism.

Age at Treatment Onset

The clinical studies showed that age at treatment onset was critical for later outcome. This variable has also been thoroughly investigated in the prospective studies of screening. Although it has been found to have little bearing on outcome, this does not necessarily mean that it has no effect. This is because the lack of a significant correlation may be an artifact of the limited variability given that the range of treatment ages is relatively limited among the different screening programs. In some, in fact, children tend to be standardly treated at the same postnatal age (e.g., day 21).

When we examined the effects of age of treatment onset using correlational techniques, few significant relationships were observed. The major exception was the consistent negative correlation between performance on the Bruinincks scale of neuromotor competence at age six and treatment age ($r = -0.30, p < 0.050$, signifying that poorer skills are associated with a longer delay in treatment. When the effects of treatment age were examined separately for different subgroups of hypothyroid children, however, those with ectopic glands were affected to the greatest degree by a delay in treatment onset. Children with ectopic glands, who were treated earlier, performed comparably to controls whereas those who were treated later performed more like the children who were lacking a thyroid gland and who did poorly regardless of when treatment was given (Rovet et al., 1986). This finding supports Delange et al's (1989) "vanishing thyroid hypothesis'' in which he has proposed that the thyroid glands of children with the ectopic etiological condition become increasingly less capable of functioning during the first month of life.

The results were additionally examined by grouping children according to week of starting therapy and then comparing groups using analysis of variance tests. The results revealed that despite the paucity of significant correlations with treatment age, a large number of significant effects were evident when the children were examined within weekly groups. Table 17.6 provides the results for tests yielding significant effects of week of treatment onset. Examination of the test measures reveals that it is primarily skills within the language domain that are affected. Children treated before the third week of life outperformed those treated past this age. These results therefore suggest the importance of commencing therapy before the third week of life to offset any subsequent language difficulties.

Dosage

The issue concerning what constitutes adequate therapy has never been fully studied (Weichsel, 1978). Current guidelines recommend starting dosages of about 50 μg/day (or 10 μg/kg) in order to achieve optimal levels of circulating thyroxine as quickly as possible and so minimize CNS

TABLE 17.6. Outcome by week of starting replacement therapy.

	Week				
	One	Two	Three	Four	p-level
12 Months					
Reynell Receptive	−.00	.05	−.52	—	.05
2 Years					
Beery	86.8	72.6	44.9	—	.05
3 Years					
Griffiths Hear/Speech	126	114	113	97	.01
Reynell Expressive	.93	.27	.68	−.30	.05
Reynell Receptive	1.27	.27	.28	−.50	.01
4 Years					
McCarthy GCI	117	110	108	88	.01
McCarthy Verbal	62	54	53	43	.01
McCarthy Perceptual	58	58	55	47	.05
Reynell Expressive	.26	.31	.45	−.93	.05
Reynell Receptive	.84	.30	.73	−1.17	.01
6 Years					
McCarthy GCI	104	103	105	88	.05
McCarthy Verbal	51	47	51	39	.05

(Note: Reynell scores are z-scores; Beery scores are percentiles; Griffith and McCarthy GCIs are standard scores [mean = 100; SD = 15]; and individual McCarthy scales are T-scores [mean = 50; SD = 10]).

damage from hypothyroidism (American Academy Pediatrics, 1987; Mitchell, 1987), but the long-term consequences of different dosages on outcome has never been systematically evaluated.

Because of a change in the treatment protocol at our hospital from 25 μg/day or less to 37.5 μg/day or higher midway through our study, we have been in an ideal position to evaluate the impact of this variable. We found an advantage of the higher dose level on fine motor skills at 1 and 2 years of age and perceptual performance abilities at 1 year. For one year olds, the mean fine motor scores were 108.2 for the low dose group and 113.8 for the high ($t = -2.37$, $p < .01$); their perceptual performance scores were 105.8 for low and 115.1 for high ($t = -2.84$, p $< .01$). Similarly at two years, the mean fine motor scores were 105.1 for the low dose group and 110.1 for high ($t = -1.94$, p $< .056$). The higher dosage group also obtained higher IQs at age 4 (112.6 vs. 105.1, $t = -2.20, p < .03$), age 6 (108.8 vs. 99.4, $t = -2.21$, $p < .03$) and age 7 (108.1 vs. 100.6, t $= -01.94$, p $< .059$).

More important than dosage, however, are the effects of actual circulating level of thyroxine (Rovet et al., 1989). We have reported that children with higher T4 levels in early infancy, which were in the upper end of the normal range, subsequently had more difficult temperaments due to heightened arousal levels and stronger stress reactions. In contrast, however, higher T4 levels in infancy were associated with better visuomotor skills at age three. Clearly, the long-term implications of these two contradictory findings must be determined.

Breast Feeding

Because breast milk is known to contain small quantities of thyroxine not found in commercial formula preparations (Bode et al., 1978; Hahn et al., 1983; Koldovsky & Thornburg, 1987; Mizuta et al., 1983; Sato & Suzuki, 1979; Varma et al., 1978), some believe that it can be protective for the hypothyroid infant in the period before treatment is given and/or normal hormone levels achieved (Sack et al., 1979; Bode et al., 1978). Research on hypothyroid infants has shown a benefit for children who were identified clinically (Tenore et al., 1977) but not for children found on screening (Letarte et al., 1980).

We examined this issue in children up to 7 years of age. The results revealed that breast fed children (BF) did have higher T4 levels than formula fed (FF) at 1 and 2 months of age ($p < .001$). BF also scored significantly higher on the Griffiths fine motor scale at 12 months of age ($p < .01$) but did not differ from FF past this age. Thus breast feeding confers a small advantage for infants but this does not appear to continue with age.

Family Background

We also examined for the effects of family background. The results revealed a normal distribution for parental IQ, education and SES and that these factors were significantly correlated with outcome in the CH children. However, because the correlations between parent IQ and child IQ were significantly larger for sibling controls (r (20) = .702 for the WISC-R) than for CH (r(72) = 0.354; t difference = 5.01, $p < .0001$), this suggests that hypothyroidism has affected intellectual outcome in the CH children.

Family functioning was assessed in the older children using the FACES III task. This is a well-validated parent-report questionnaire that accounts for the adequacy of family functioning along two orthogonal domains, cohesion and adaptability. Scoring of this scale allows for the assignment of families to 16 joint adaptability-cohesion classifications. Moreover these 16 groups then can be further classified as well-balanced or ideally functioning (4 classes), midrange or adequately functioning (8 classes), and extreme or dysfunctional (4 classes). For the 41 families administered this instrument to date, 35% are "balanced," 57% "midrange" and 8.0% "extreme." This represents a slight increase in the number of midrange families and a slight decrease in balanced, as compared with the standardization sample. There is no increase in the number of "extremes." These findings suggest that family stability may be mildly affected by having a child with CH. Correlations between family functioning variables and outcome are not yet available.

Summary

Clearly then, those factors associated with severity and duration of disease at diagnosis affect later outcome. Hypothyroidism, which is severe and begins during fetal life, appears to be the strongest predictor of subsequent impairment. Factors reflecting treatment adequacy and duration of disease appear to moderate these effects. The effects of breast feeding were minimal. The role of family factors on outcome must also be considered.

The Toronto Prospective Study (Continued)—Regression Studies

As an attempt to assess the relative contributions of the various disease-related factors, multiple regression analyses were conducted on the different outcome variables. In one series, we studied the effects of the following five predictor variables: bone age at diagnosis, confirmatory T4 level, age at treatment onset, dose level, and the postnatal duration of hypothyroidism using All Possible Subsets Regression Analyses (BMDP program

310   J.F. Rovet

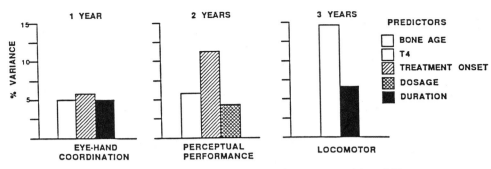

FIGURE 17.7. Significant predictors of early neurocognitive skills.

P9R). These were analyzed for main effects only and not for the interactions of different variable combinations. For the younger children, the results have revealed that even though they did not differ overall from controls as reported previously, their performance on most tests was still determined to a significant degree by factors associated with early thyroid disease. As shown in Figure 17.7, eye-hand coordination ability at 12 months of age reflected the involvement of treatment age, T4 level at diagnosis, and duration of hypothyroidism, accounting for 19.4% of the total variance; perceptual abilities at age two reflected significant contributions of treatment age, bone age, and dosage, $R^2 = 0.174$; locomotor skills at age three reflected the contributions of bone age delay and duration of hypothyroidism, $R^2 = 0.164$.

A definite pattern was evident in the results obtained with older children. As can be seen in Figure 17.8, levels of ability on tasks reflecting primarily verbal skills were associated with both prenatal (bone age) and

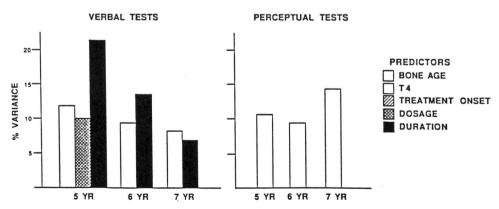

FIGURE 17.8. Significant predictors of later neurocognitive skills.

postnatal (duration of hypothyroidism) factors. By contrast performance in the perceptual area consistently reflected only the contribution of prenatal (bone age). We have interpreted these results as suggesting that the different cerebral substrates underlying specific mental skills appear to be developing at different rates in utero and early life and so have their greatest sensitivity to thyroid hormone deficiency at different times. Those substrates underlying perceptual skills appear to be developing primarily prenatally and in advance of those underlying verbal skills, which show substantial postnatal effects and may rely on other factors such as learning or early stimulation.

Summary

Specific abilities are affected by different combinations of thyroid-related factors. Outcome in the perceptual area appears to be influenced mostly by factors associated with intrauterine hypothyroidism whereas outcome in the verbal area appears to be influenced by both hypothyroidism in utero and during the first month or two of life.

Conclusions

A review of the literature and a detailed examination of our own prospective investigation have shown that neonatal screening definitely improves intellectual outcome in children with congenital hypothyroidism. Although the children with CH appeared to be functioning as adequately controls as infants, mild impairments in various aspects of mental function were evident in some of the children when they were older. Particularly affected appeared to be their skills in language, neuromotor, perceptual, and memory areas. Our studies showed that these deficits were associated with compromised performance at school and increased their need for special education. The children at greatest risk appeared to be those with evidence of severe hypothyroidism at birth, which began prenatally and extended over a lengthy period of time. Children with less severe forms of hypothyroidism (e.g., ectopic glands) and with adequate therapy appeared to be relatively unaffected. Treatment factors reflecting age at initiating therapy, dose level, and sources of supplementary thyroxine such as breast milk, have some moderating effects on outcome. Our regression studies examining the relative contributions of prenatal and neonatal hypothyroidism, treatment, and duration also showed that various selective abilities were differentially affected by different factors or factor combinations: verbal skills appeared to be most sensitive to factors associated with hypothyroidism in utero and early infancy, whereas perceptual skills were sensitive only to prenatal factors. This implies that some deficits may never be totally rectified by screening newborns. The longer term consequences of these deficits still need to be determined and alternate forms of screening, such as fetal screening (Yoshida et al., 1986), may still have to be considered. In the meantime, however, it is imperative to strive for the most effective therapy and postnatal health care so that given present circumstances, every child with hypothyroidism will be permitted to achieve the best possible outcome.

Acknowledgments. This work has been supported by research grants from two provincial agencies, the Ontario Ministry of Health and the Ontario Ministry of Community and Social Services and by a National Health Research & Development Program Scholarship from Health and Welfare Canada. I would like to acknowledge the contributions of my co-investigator and collaborator Dr. Robert Ehrlich for his direction and guidance; Dr. Paul Walfish for his seminal work in newborn thyroid screening in Toronto; Jeannine Pinsonneault for data management and analysis; Betty Adamo for compiling the bibliography; Dr. Neville Howard for first introducing me to this area of investigation; and Dr. Sheri Berenbaum and Dr. Elizabeth Hampson for their helpful comments. I am especially grateful to Donna Sorbara who has been responsible for organizing and conducting the study as well as maintaining the continued and devoted support of our families.

References

Abbasi, V., & Aldige, C. (1977). Evaluation of sodium L-thyroxine (TH) requirement in replacement therapy of hypothyroidism. *Journal of Pediatrics, 90,* 209–301.

Addison, G.M., Rosenthal, M., & Price, D.A. (1989). *Congenital hypothyroidism: Increased incidence in Asian families.* (Abstract) In F. Delange, D.A. Fisher & D. Glinoer (Eds.) *Research in Congenital Hypothyroidism* New York: Plenum

Alm, J., Larsson, A., & Zetterstrom, R. (1981). Congenital hypothyroidism in Sweden. *Acta Paediatrica Scandinavica, 70,* 907–912.

American Academy of Pediatrics Committee on Genetics. American Thyroid Association Committee on Neonatal Screening (1987). Newborn screening for congenital hypothyroidism: Recommended guidelines. *Pediatrics, 80,* 745–749.

Andersen, H.J. (1975). Hypothyroidism. In L.I. Gardner (Ed.), *Endocrinology and Genetic Disease of Childhood.* Philadelphia: W.B. Saunders Co.

Antonozzi, I., Monaco, F., & Dominici, R. (1982). Regional differences in central and southern Italy in congenital hypothyroidism. In H. Naruse & M. Irie (Eds.), *Neonatal screening* (pp. 119–120). Amsterdam: Excerpta Medica.

Bainbridge, R., Mughal, Z., Mimouni, F., & Tsang, R.C. (1987). Transient congenital hypoparathyroidism: How transient is it? *The Journal of Pediatrics, 111,* 866–868.

Balazs, R., Cocks, W.A., Eayrs, J.T., & Kovacs, S. (1971). Biochemical effects of thyroid hormones on the developing brain. In D.A. Hamburgh & E.J.W. Barrington (Eds.), *Hormones in development.* New York: Appleton-Century-Crofts.

Barclay, R.A. (1981). *Hyperactive children. A handbook for diagnosis and treatment.* New York: Guilford.

Bargagna, S., Cittadoni, L., Falciglia, G., Ferretti, G., Marcheschi, M., Chiovato, L., Giusti, F.L., Genzi, G.F. & Pinchera, A. (1989). *Neuropsychological development of congenital hypothyroid children.* (Abstract) In F.A. Delange, D.A. Fisher & D. Glinoer (Eds). *Research in Congenital Hypothyroidism.* New York: Plenum.

Barnes, N.D. (1985). Screening for congenital hypothyroidism: The first decade. *Archives of Disease in Childhood, 60,* 587–592.

Bode, H.H., Vanjonack, W.J., & Crawford, J.D. (1978). Mitigation of cretinism by breast-feeding. *Pediatrics, 62,* 13–16.

Bodegard, G., Fyro, K., & Larsson, A., (1983). Psychological reactions in 102 families with a newborn who has a falsely positive screening test for congenital hypothyroidism. *Acta Paediatrica Scandinavica, 304,* 1–21.

Brown, A.L., Fernhoff, P.M., Milner, J., McEwen, C., & Elsas, L.S. (1981). Racial differences in the incidence of congenital hypothyroidism. *The Journal of Pediatrics, 99,* 934–936.

Browne, A.W., Bronstein, I.P., & Kraines, F. (1939). Hypothyroidism and cretinism in childhood: VI. Influence of thyroid therapy on mental growth. *American Journal of Disease in Children, 57,* 517–523.

Burrow, B.N., & Dussault, J.H. (1980). *Neonatal thyroid screening.* New York: Raven Press.

Burrow, G.N., Bachrach, L.A., & Holland, H. (1983). Maternal-fetal thyroid function. In H. Naruse & M. Irie (Eds.), *Neonatal screening.* Amsterdam: Excerpta Medica.

Cheek, D.B. (1975). The fetus. In D.B. Cheek (Ed.), *Fetal and postnatal cellular growth. Hormones and Nutrition.* New York: Wiley & Sons.

Chopra, I.J. (1972). A radioimmunoassay for measurement of thyroxine in unextracted serum. *Journal of Clinical Endocrinological Metabolism, 34,* 938–949.

Collipp, P.J., Kaplan, S.A., Kogut, M.D., Tasem, W., Plachte, F., Schlamm, V., Boyle, D.C., Ling, S.M., & Koch, R. (1965). Mental retardation in congenital hypothyroidim: improvement with thyroid replacement therapy. *American Journal of Mental Deficiencies, 70,* 432–437.

Connelly, J. (1986). Congenital hypothyroidism. *Australian Pediatrics Journal, 22,* 165–166.

Crome, L, & Stern, J. (1972). *Pathology of mental retardation.* Edinburgh: Churchill Livingston.

David, M., Dorche, C., Coudier, J., Rolland, M.O. (1983). Follow up results of congenital hypothyroid infants found by neonatal screening. In H. Naruse & M. Irie (Eds.), *Neonatal Screening.* Amsterdam: Excerpta Medica.

Davy, T., Daneman, D., Walfish, P.G., & Ehrlich, R.M. (1985). Congenital hypothyroidism: The effect of stopping treatment at 3 years. *American Journal of Diseases in Childhood, 139,* 1028–1030.

DeGroot, L.J., Larsen, P.R., Refetoff, S., & Stanbury, J.B. (1984). Hypothyroidism in infants and children, and developmental abnormalities of the thyroid gland. In DeGroot (Ed.), *The thyroid and its diseases* (pp. 610–641).

Delange, F., DeVijlder, J., Morreale de Escobar, G., Ruchiccioli, P., Varrone, S. (1989). *Significance of early diagnostic data in congenital hypothyroidism.* In F.A. Delange, D.A. Fisher, D. Glinoer (Eds.), *Research in Congenital Hypothyroidism.* New York: Plenum.

Delange, F. (1980). Panel and general discussion: Optimal screening and recall procedures. In G.N. Burrow & J.H. Dussault, *Neonatal thyroid screening*. New York: Raven Press.

Delange, F., Beckers, C., Hofer, R., Konig, M.P., Monaco, F., & Varrone, S. (1979). Neonatal screening for congenital hypothyroidism in Europe. *Acta Endocrinological Supplement, 90,* 223–247.

Delange, F., Bourdoux, P., Ketelbant-Balasse, P., Van Humskerken, A., Glinoer, D., & Ermans, A.M. (1983). Transient primary hypothyroidism in the newborn. In J.H. Dussault & P. Walker (Eds.), *Basic and clinical endocrinology: Congenital hypothyroidism* (pp. 275–301). New York: Marcel Dekker, Inc.

Delange, F., Dodion, J., Wolter, R., Bourdoux, P., Dalhem, A., Glinoer, D., & Ermans, A. (1978). Transient hypothyroidism in the newborn infant. *The Journal of Pediatrics, 92,* 974–976.

Dobbing, J., & Sands, J. (1973). Quantitative growth and development of human brain. *Archives of Disease in Childhood, 48,* 757–767.

Dussault, J.H. (1987). *Thyroid binding immunoglobulins and congenital hypothyroidism*. Presented at the Lawson Wilkins Pediatric Endocrine Society Annual Meeting, Anaheim, California.

Dussault, J.H. (November, 1988). The Quebec network for genetic medicine—CHUL. *Congenital hypothyroidism*. Abstract No. 022. International Symposium on Newborn Screening, Sao Paulo, Brazil.

Dussault, J.H., Coulombe, P., LaBerge, C., Letarte, J., Guyda, H., & Khoury, J. (1975). Preliminary report on a mass screening program for neonatal hypothyroidism. *Journal of Pediatrics, 86,* 670–674.

Dussault, J.H., Mitchell, M.L., LaFranchi, S., & Murphey, W.H. (1980). gional screening for congenital hypothyroidism: Results of screening one million North American infants with filter paper spot T4-TSH. In G.N. Burrow (Ed.), *Neonatal thyroid screening* (pp. 155–165). New York: Raven Press.

Dussault, J.H., & Morissette, J. (1983). Higher sensitivity of primary thyrotropin in screening for congenital hypothyroidism: A myth? *Journal of Clinical Endocrinology and Metabolism, 56,* 849–852.

Ermans, A.M., Boudoux, P., Lagasse, R., Delange, F., & Thilly, C. (1980). Congenital hypothyroidism in developing countries. In G.N. Burrow & J.H. Dussault (Eds.), *Neonatal thyroid screening*. New York: Raven Press.

Farriaux, J.P., Dhondt, J.L., & Lebecq, M.R. (1988). *Intellectual outcome in hypothyroid children screened at birth*. Abstract No. 017. International Screening Symposium of Inborn Errors of Metabolism, Sa. Paulo, Brazil.

Fellous, A., Lennon, A.M., Francon, J., & Nunez, J. (1979). Thyroid hormones and neurotubule assembly in vitro during brain development. *European Journal of Biochemistry, 101,* 365–376.

Ferreira, P. (1986). *Neonatal screening in Canada 1985: A survey*. Unpublished manuscript.

Ferreira, P. (1987). Canadian neonatal screening practices. In B.L. Therrell (Ed.), *Advances in neonatal screening*. Amsterdam: Elsevier.

Fisher, D.A. (1975). Neonatal detection of hypothyroidism. *The Journal of Pediatrics, 86,* 822–824.

Fisher, D.A. (1987). Maturation of thyroid hormone actions. In B.L. Therrell Jr. (Ed.), *Advances in neonatal screening* (pp. 21–24). Amsterdam: Excerpta Medica.

Fisher, D.A., & Klein, A.H. (1981). Thyroid development and disorders of thyroid function in the newborn. *New England Journal of Medicine, 304*, 702–712.

Focarile, F., Rondanini, G.F., Bollati, A., Bartolucci, A., & Chiumello, G. (1984). Free thyroid hormones in evaluating persistently elevated thyrotropin levels in children with congenital hypothyroidism on replacement therapy. *Journal of Clinical Endocrinology and Metabolism, 59*, 1211–1214.

Foley, T.P., & Murphey, W.H. (1980). Hypothyroidism at birth, during the first week of life and approximately one month of age: Comparison to primary T4 with secondary TSH screening. In G.N. Burrow (Ed.), *Neonatal thyroid screening* (pp. 75–86). New York: Raven Press.

Frasier, S.D., Penny, R., & Snyder, R. (1982). Primary congenital hypothyroidism in Spanish-surnamed infants in Southern California. *Journal of Pediatrics, 101*, 315.

Frost, G.J., & Parkin, M.M. (1986). A comparison between the neurological and intellectual abnormalities in children and adults with congenital hypothyroidism. *European Journal of Pediatrics, 145*, 480–484.

Fuggle, P., Murphy, G., Thorley, G., & Grant, D.B. (1989). *Congenital hypothyroidism: Psychological outcome at three and five years.* (Abstract) In F. Delange, D.A. Fisher, D. Glinoer (Eds.) *Research in Congenital Hypothyroidism.* New York: Plenum.

Fyro, K., & Bodegard, G., (1987). Four-year follow-up of psychological reactions to false positive screening tests for congenital hypothyroidism. *Acta Paediatrica Scandinavica, 76*, 107–114.

Gesell, A., Amatruda, C.A., & Culotta, C.S. (1936). Effect of thyroid therapy on the mental and physical growth of cretinous infants. *American Journal of Disease in Childhood, 52*, 1117–1138.

Glorieux, J., Desjardins, M., Dussault, J.H., Letarte, J., Morissette, J., & Thibault, L. (1987). Follow-up congenitally hypothyroid children: Delineation of a group at higher risk. In B.L. Therrell Jr. (Ed.), *Advances in neonatal screening* (pp. 81–84). Amsterdam-New York-Oxford: Excerpta Medica.

Glorieux, J., Dussault, J.H., Letarte, J., Guyda, H., & Morissette, J. (1983). Preliminary results on the mental development of hypothyroid infants detected by the Quebec screening program. *The Journal of Pediatrics, 102*, 19–22.

Glorieux, J., Desjardins, M., Letarte, J., Morissette, J., & Dussault, J.H. (1988). Useful parameter to predict the eventual mental outcome of hypothyroid children. *Pediatric Research, 24*, 6–8.

Glorieux, J., Dussault, J.H., Morissette, J., Desfardins, M., Letarte, J., & Guyda, H. (1985). Follow-up at ages 5 and 7 years on mental development in children with hypothyroidism detected by Quebec screening program. *The Journal of Pediatrics, 107*, 913–915.

Grant, D.B., & Hulse, J.A. (1980). Screening for congenital hypothyroidism. *Archives of Disease in Childhood, 55*, 913–914.

Grant, G.A., Carson, D.J., McReid, M., & Hutchinson, J.M. (1986). Congenital hypothyroidism missed on screening. *Archives of Disease in Childhood, 61*, 189–197.

Guyda, H.J., (1983). Treatment of congenital hypothyroidism. In J.H. Dussault & P. Walker (Eds.), *Basic and clinical endocrinology: Congenital hypothyroidism* (pp. 385–396). New York: Marcel Dekker, Inc.

Hagberg, B., & Westphal, O. (1970). Ataxic syndrome in congenital hypothyroidism. *Acta Paediatrica Scandinavica, 59,* 323–327.

Hahn, H.B., Spiekerman, A.M., Otto, W.R., & Hossalla, D.E. (1983). Thyroid function tests in neonates fed human milk. *American Journal of Disease in Children, 137,* 220–222.

Hamburgh, M.N., Mendoza, L.A., Burkart, J.F., & Weil, F. (1971). Thyroid-dependent processes in the developing nervous system. In D.A. Hamburgh & E.J.W. Barrington (Eds.), *Hormones in development.* New York: Appleton-Century-Crofts.

Hanefield, F., Richter, I., Weber, B., & Zabransky, S. (1974). Neurological studies on children with hypothyroidism on long-term treatment. *Acta Pediatrica Scandinavica, 63,* 332–333.

Hebert, R., Laureau, E., Vanasse, M., Richard, J., Morrissette, J., Glorieux, J., Desjardins, M., Letarte, J., & Dussault, J. (1986). Auditory brainstem response audiometry in congenitally hypothyroid children under early replacement therapy. *Pediatric Research, 20,* 570–573.

Heyerdahl, S. (1987). Development of children with congenital hypothyroidism diagnosed by neonatal screening in Norway. In B.L. Therrell Jr. (Ed.), *Advances in neonatal screening* (pp. 81–84). Amsterdam: Excerpta Medica.

Holmes, G. (1979). The cerebellum of man. *Brain, 62,* 1–30.

Holt, A.B., Kerr, G.R., & Cheek, D.B. (1975). Prenatal hypothyroidism and brain composition. In D.B. Cheek (Ed.), *Fetal and post-natal cellular growth. Hormones and Nutrition.* New York: John Wiley & Sons.

Horn, G. (1955). Thyroid deficiency and inanition. The effects of replacement therapy on the development of the cerebral cortex of young albino rats. *Anatomical Research, 121,* 63.

Hulse, A.(1983). Congenital hypothyroidism and neurological development. *Journal of Childhood Psychology and Psychiatry, 24,* 629–635.

Hulse, J.A. (1984). Outcome for congenital hypothyroidism. *Archives of Disease in Childhood, 59,* 23–30.

Hulse, J.A., Grant, D.B., Clayton, B.E., Lilly, P., Jackson, D., Spracklan, A., Edwards, R.W.H., & Nurse, D. (1983). Population screening for congenital hypothyroidism. *British Medical Journal, 280,* 675–678.

Hulse, J.A., Grant, D.B., Jackson, D., & Clayton, B. (1982). Growth, development, and reassessment of hypothyroid infants diagnosed by screening. *British Medical Journal, 284,* 1435–1436.

Ibanez, L., Albisu, M., Potau, N., Vicens-Calvet, E., & the Catalan Collaborative Group (1989). *Follow-up of congenital hypothyroidism (CH) detected by neonatal screening in Catalonia.* (Abstract) In F. Delange, D.A. Fisher, D. Glinoer (Eds.) *Research in Congenital Hypothyroidism.* New York: Plenum.

Illig, R. (1983). Follow-up of thyroid function tests, skeletal maturation and scintigraphic findings with congenital hypothyroidism discovered by neonatal screening. In H. Naruse M., Irie (Eds.), *Neonatal screening.* Amsterdam: Excerpta Medica.

Illig, R., & Largo, R.H. (1987). Mental development in 60 children with congenital hypothyroidism prospective follow-up study at one, four and seven years of age. In B.L. Therrell Jr. (Ed.), *Advances in neonatal screening* (pp. 85–89). Amsterdam: Excerpta Medica.

Illig, R., Largo, R.H., Qin, Q., Torresani, T., Rochiccioli, P., & Larsson, A.

(1987). Mental development in congenital hypothyroidism after neonatal screening. *Archives of Disease in Childhood, 62,* 1050–1055.

Illig, R., Largo, R.H., & Rochiccioli, P. (1985). *European collaborative study on mental development in children with congenital hypothyroidism (CH) diagnosed by neonatal screening.* Poster No. 73. Second joint meeting: Lawson Wilkins Pediatric Endocrine Society and European Society for Paediatric Endocrinology, Baltimore.

Illig, R., Largo, R.H., Weber, M., Augsburger, T.H., Lipp, A., Wissler, D., Perrenoud, A.E., Torresani, T. (1986). Sixty children with congenital hypothyroidism detected by neonatal thyroid: Mental development at 1, 4, and 7 years: a longitudinal study. *Acta Endocrinologica, 279,* 346–353.

Interagency Committee on Learning Disabilities (1987). *Learning disabilities: A report to the U.S. congress.*

Irie, M., Nakajima, H., Inomata, H., Naruse, H., Suwa, S., & Takasugi, N. (1987). Screening of neonatal hypothyroidism in Japan. In B.L. Therrell Jr. (Ed.), *Advances in neonatal screening* (pp. 41–47). Amsterdam: Excerpta Medica.

Jennings, P.E., O'Malley, B.P., Griffin, K.E., Northover, B., & Rosenthal, F.D. (1984). Relevance of increased serum thyroxine concentrations associated with normal serum triiodothyronine values in hypothyroid patients receiving thyroxine: A case for "tissue thyrotoxicosis." *British Medical Journal, 289,* 1645–1647.

Kirkland, R.T., Kirkland, J.L., Robertson, M.C., Librik, L., & Clayton, G.W. (1972). Strabismus and congenital hypothyroidism. *The Journal of Pediatrics, 80,* 648–650.

Klein, A., Meltzer, S., & Kenny, F.M. (1972). Improved prognosis in congenital hypothyroidism treated before age three months. *Journal of Pediatrics, 81,* 912–915.

Klein, R. (1986). Screening for congenital hypothyroidism. *The Lancet,* August, 403.

Klein, R.Z. (1979). Neonatal screening for hypothyroidism. *Advances in Pediatrics, 26,* 417–440.

Klett, M., Schonberg, D., Bohnert, R. & Wille, L. (1983). *Influence of various clinical conditions on thyroid function in the newborn.* In H. Naruse & M. Irie (Eds.) *Neonatal Screening* Amsterdam: Excerpta Medica.

Koldovsky, O., & Thornburg, W. (1987). Hormones in milk. *Journal of Pediatric Gastro Nutrition, 6,* 172–196.

Komianou, F., Makaronis, G., Lambadaridis, J., Sarafidou, R., Vrachni, F., Mengreli, C., & Pantelakis, S. (1988). Psychomotor development in congenital hypothyroidism. *European Journal of Pediatrics, 147,* 275–278.

Konstantarcas, M., & Homatidis, S. (November, 1988). *Evidence of pre, peri- and neonatal complications in the births of autistic, mentally retarded and normal children.* Presented at Challenge and Change in Childhood Psychopathology Conference, Toronto.

Laberge, C. (1983). Cost benefit evaluation of neonatal thyroid screening programs. In G.N. Burrow & J.H. Dussault, *Neonatal thyroid screening.* New York: Raven.

Ladenson, P.W., Stokes, J.W., & Ridgway, E.C. (1984). Reversible alteration of the visual evoked potential in hypothyroidism. *The American Journal of Medicine, 77,* 1010–1014.

LaFranchi, S.H., Hannna, C.E., Krainz, P.L., Skeels, M.R., Miyahira, R. S., & Sesser, D.E. (1985). Screening for congenital hypothyroidism with specimen collection at two time periods: Results of the northwest regional screening program. *Pediatrics, 76,* 734–740.

LaFranchi, S.H., Murphey, W.H., Foley, J.P., Larsen, P.R., & Buist, N.R. (1979). Neonatal hypothyroidism detected by the northwest regional screening program. *Pediatrics, 63,* 180–191.

Larsen, P.R., & Broskin, K. (1975). Thyroxine immunoassay using capillary blood samples collected on filter paper. In D.A. Fisher & G.N. Burrow (Eds.), *Perinatal thyroid physiology and disease.* New York: Raven Press.

Lauder, J.M., & Krebs, H. (1986). Do neurotransmitter, neurohumors, and hormones specify critical periods? In W.T. Greenough & J.M. Juraska (Eds.), *Developmental neuropsychobiology.* New York: Academic Press.

Layde, P.M., Von Allmen, S.D., & Oakley, G.P. (1979). Congenital hypothyroidism control programs. A cost-benefit analysis. *Journal of the American Association, 241,* 2290–2292.

Legrand, J. (1986). Thyroid hormone effects on growth and development. In G. Hennemann (Ed.), *Thyroid hormone metabolism.* New York: Marcel Dekker.

Leiner, H.C., Leiner, A.L., & Dow, R.S. (1986). Does the cerebellum contribute to mental skills? *Behavioural Neuroscience, 100,* 443–454.

Letarte, J., Guyda, J., Dussault, J.H., & Glorieux, J. (1980). Lack of protective effect of breast-feeding in congenital hypothyroidism: Report of 12 cases. *Pediatrics, 65,* 703–705.

Letarte, J. & La Franchi, S. (1983). Clinical features of congenital hypothyroidism. In J.H. Dussault & P. Walker (Eds.) *Basic and clinical endocrinology: Congenital hypothyroidism* (pp. 351–383). New York: Marcel Dekker.

Levine, G.D., & Therrell, B.L. (1986). Second testing for hypothyroidism. *Pediatrics, 78,* 375–376.

Levine, M.D., Oberklaid, F., & Meltzer, L. (1981). Developmental output failure: A study of low productivity in school-aged children. *Pediatrics, 67,* 18–25.

Lewis, A. (1937). A study of cretinism in London. *Lancet, 2,* 1501–1525.

MacFaul, R., Dorner, S., Brett, E.M., & Grant, D.B. (1978). Neurological abnormalities in patients treated for hypothroidism from early life. *Archives of Disease in Childhood, 53,* 611–619.

Maenpaa, J. (1972). Congenital hypothyroidism: Aetiological and clinical aspects. *Archives of Disease in Childhood, 47,* 914–923.

Maki, I., Nose, O., Harada, T., Kai, H., Tajiri, H., Ogawa, M., Abe, M., Miyai, K., Mizuta, M., Takesada, M., Yabuuchi, H. (1983) Follow up study of treated hypothyroid infants on psychological and neurological development. In H. Naruse & M. Irie (Eds.), *Neonatal Screening.* Amsterdam: Excerpta Medica.

Malone, M.J., Rosman, N.P., Szoke, M., & Davis, D. (1976). Myelination of brain in experimental hypothyroidism. *Journal of the Neurological Sciences, 26,* 1–11.

Man, E.B., & Serunian, S.A. (1976). Thyroid function in human pregnancy. *American Journal of Obstetrics and Gynaecology, 125,* 949–957.

Mitchell, M.L. (1986). Screening for congenital hypothyroidism: A decade later. *Infant Screening Newsletter,* Vol. 11.

Mitchell, M.L., Larsen, R., Levy, H.L., Bennett, A.J., & Madoff, M.A. (1978). Screening for congenital hypothyroidism. *Journal of the American Medical Association, 239,* 2348–2351.

Mizuta, H., Amino, N., Ichihara, K., Harade, T., Nose, O., Tanizawa, O., & Miyai, K. (1983). Thyroid hormones in human milk and their influence on thyroid function of breast-fed babies. *Pediatric Research, 14*, 468–471.

Money, J. (1956). Psychologic studies in hypothyroidism. *Archives of Neurology in Psychiatry, 76*, 296–309.

Morreale de Escobar, G., Escobar del Rey, F., Ruiz-Marcos, A. (1983). Thyroid hormone and the developing brain. In J.H. Dussault & P. Walker (Eds.), *Congenital hypothyroidism*. New York: Dekker.

Morreale de Escobar, G., Obregon M.J., Escobar del Rey F (1989). *Transfer of thyroid hormone from mother to fetus* In F. Delange, D.A. Fisher, D. Glineoer (Eds), *Research in Congenital Hypothyroidism*. New York: Plenum.

Morreale de Escobar, G., Pastor, R., Obregon, M.J., & Escobar del Rey, F. (1985). Effects of maternal hypothyroidism on the weight and thyroid hormone content of rat embryonic tissues, before and after onset of fetal thyroid function. *Endocrinology, 117*, 1890–1900.

Moschini, L., Costa, P., Marinelli, E., Maggioni, G., Sorcini Carta, M., Fazzini, D., Diodato, A. Sabini, G., Grandolfo, M.E., Carta, S., Porro, G., Paolella, A., Gordiale, S., & Brinciotti, M. (1986). Longitudinal assessment of children with congenital hypothyroidism detected by neonatal screeing. *Helv Paediatrica Acta, 41*, 415–24.

Moschini, L., Sorcini, C., Costa, P., Antonozzi, I., Paolella, A., Porro, G., & Carta, S. (1989). *Mental development in 75 children with congenital hypothyroidism detected by neonatal screening.* In F. Delange, D.A. Fisher, D. Glinoer (Eds.) *Research in Congenital Hypothyroidism*. New York: Plenum.

Muir, A., Daneman, D., Daneman, A., & Ehrlich, R. (1988). Thyroid scanning ultrasound, and serum thyroglobulin in determining the origin of congenital hypothyroidism. *American Journal of Diseases in Children, 142*, 214–216.

Murphy, G., Hulse, J.A., Jackson, D., Tyner, P., Glossop, J., Smith, I., & Grant, D. (1986). Early treated hypothyroidism: Development at 3 years. *Archives of Disease in Childhood, 61*, 761–765.

Myant, N.B. (1971). The role of thyroid hormone in the fetal and postnatal development of mammals. In D.A. Hamburgh & E.J.W. Barrington (Eds.), *Hormones in development*. New York: Appleton-Century-Crofts.

New England Congenital Hypothyroidism Collaborative (November, 1981). Effects of neonatal screening for hypothyroidism: Prevention of mental retardation by treatment before clinical manifestations. *Lancet, 2*, 1095–1098.

New England Congenital Hypothyroidism Collaborative (1982). Pitfalls in screening for neonatal hypothyroidism. *Pediatrics, 70*, 16–20.

New England Congenital Hypothyroidism Collaborative (1984). Characteristics of infantile hypothyroidism discovered on neonatal screening. *The Journal of Pediatrics, 104*, 539–544.

New England Congenital Hypothyroidism Collaborative (1985). Neonatal hypothyroidism screening: Status of patients at 6 years of age. *The Journal of Pediatrics,107,*915–918.

Nicholson, J.L., & Altman, J. (1972). Synoptogenesis in the rat cerebellum. Effects of early hypo and hyperthyroidism. *Science, 176*, 530–531.

Nunez, J. (1984). Thyroid hormones and microtubules during brain development. International Endocrine Society Meeting, Quebec City.

Osler, W. (1897). Sporadic cretinism in America. *Transactions of the Congress of American Physicians and Surgeons, 4*, 169.

Pekonen, B., & Pekonen, F. (1984). Ontogenesis of the nuclear 3,4,3'-triiodothyronine receptor in the human fetal brain. *Endocrinology, 114,* 677–679.

Pelton, W., & Bass, N.H. (1973). Adverse affects of excess thyroid hormone on the maturation of rat cerebrum. *Archives of Neurology, 29,* 145–150.

Perez-Castillo, A., Bernal, J., Ferreiro, B., & Pans, T. (1985). The early ontogenesis of thyroid hormone receptor in rat fetus. *Endocrinology, 117,* 2457–2461.

Peter, F., Wang, S.T., & Strunc, G. (1987). Screening for congenital hypothyroidism with a one-day TSH assay. Immunoradimetric (IRMA) and immunofluorometric (IFMA) tests compared. In B.L. Therrell Jr. (Ed.), *Advances in neonatal screening* (pp. 61–62). Amsterdam: Excerpta Medica.

Plioplys, A.V., Gravel, C., & Hawkes, R. (1986). Selective suppression of neurofilament antigen expression in the hypothyroid rat cerebral cortex. *Journal of the Neurological Sciences, 75,* 53–68.

Potter, B.J., Mano, M.T., Belling, G.B., McIntosch, G.H., Hua, C., Cragg, B.G., Marshall, J., Wellby, M.L., & Hetzel, B.S. (1982). Retarded fetal brain development resulting from severe dietary iodine deficiency in sheep. *Neuropathology and Applied Neurobiology, 8,* 303–313.

Price, D.A., Ehrlich, R.M., & Walfish, P.G. (1981). Congenital hypothyroidism: Clinical and laboratory characteristics in infants detected by neonatal screeening. *Archives of Disease in Childhood, 56,* 845–851.

Raiti, S., & Newns, G.H. (1971). Cretinism: Early diagnosis and its relation to mental prognosis. *Archives of Disease in Childhood, 46,* 692–694.

Redmond, G.P. (1982). Therapy of congenital hypothyroidism in the era of mass screening. *Seminars in Perinatology, 6,* 181–189.

Rezvani, I., & DiGeorge, A.M. (1977). Reassessment of the daily dose of oral thyroxine for replacement therapy in hypothyroid children. *Pediatrics, 90,* 291–297.

Richards, G.E., Karyl, A., Norcross, A., & Cavallo, A. (1987). Evoked potential in newborns with congenital hypothyroidism. Abstract No. 476, *Pediatric Research Program, 21,* 253A.

Rickards, A., Coakley, J., Francis, I., Armstrong, S., & Connelly, J., (1988). *Results of follow-up at 5 years in a group of hypothyroid Australian children detected by newborn screening.* (Abstract) In F. Delange, D.A. Fisher, D. Glinoer (Eds.) *Research in Congenital Hypothyroidism.* New York: Plenum.

Rickards, A., Connelly, J., Coakley, J., & Armstrong, S. (1987). Psychological follow-up in Australian hypothyroid children detected by newborn screening. In B.L. Therrell Jr. (Ed.), *Advances in neonatal screening* (pp. 77–80). Amsterdam: Excerpta Medica.

Rives, S., & Toublanc, J.E. (1987). Cognitive and affective outcomes of congenital hypothyroid follow-up. In B.L. Therrell Jr. (Ed.), *Advances in neonatal screening* (pp. 109–112). Amsterdam-New York-Oxford: Exerpta Medica.

Rochiccioli, P., Roge, B., Alexandre, F., & Dutau, G. (1983). Study of perinatal and environmental factors of neuropsychological development of hypothyroid infants detected by neonatal screening. In H. Naruse & M. Irie (Eds.), *Neonatal screening.* Amsterdam: Excerpta Medica.

Roge, B., Rochiccioli, P., Alexandre, F., & Moron, P. (1987). An eight-year study of the mental and psychomotor development of hypothyroid children detected by neonatal screening. In B.L. Therrell Jr. (Ed.), *Advances in neonatal screening* (pp. 91–94). Amsterdam: Excerpta Medica.

Rondanini, G.F., Bollati, A., Della Porta, V., Cerabolini, R., Lenti, C., Rovej, L.,

Manzoni, A., & Chumello, G. (1989) *Long term neurological prognosis in early treated congenital hypothyroid children.* (Abstract) In F. Delange, D.A. Fisher, D. Glinoer (Eds.) *Research in Congenital Hypothyroidism. New York: Plenum.*

Rosman, N.P. (1976). Neurological and muscular aspects of thyroid dysfunction in childhood. *Pediatric Clinics of North America, 23,* 575–594.

Rovet, J.F. (1989). *Does breast feeding protect the hypothyroid infant diagnosed by newborn screening?* Manuscript submitted for publication.

Rovet, J., & Ehrlich, R. (1988). *Thyroid screening follow-up study.* Final report to the Ontario Ministry of Health.

Rovet, J., Ehrlich, R., & Sorbara, D. (1987a). Intellectual outcome in children with fetal hypothyroidism. *Journal of Pediatrics, 110,* 700–704.

Rovet, J., Ehrlich, R., & Sorbara, D. (1987b). Longitudinal prospective investigations of hypothyroid children detected by neonatal thyroid screening in Ontario. In B.L. Therrell Jr. (Ed.), *Advances in neonatal screening* (pp. 99–103). Amsterdam: Excerpta Medica.

Rovet, J.F., Ehrlich, R.M., & Sorbara, D. (1989). Effect of thyroid hormone level on temperament in infants with congenital hypothyroidism detected by screening of neonates. *Journal of Pediatrics, 114,* 63–68.

Rovet, J.F., Sorbara, D.L., & Ehrlich, R.M. (1986). The intellectual and behavioural characteristics of children with congenital hypothyroidism identified by neonatal screening in Ontario. The Toronto prospective study. In *Genetic disease: Screening and management* (pp. 281–315). New York: Liss.

Sack, J., Frucht, H., Amadeo, O., Brish, M., & Lunenfeld, B. (1981). Breast milk thyroxine and not cow's milk may mitigate and delay the clinical picture of neonatal hypothyroidism. *Acta Paediatrica Scandinavica, 277,* 54–56.

Sato, T., & Suzuki, Y. (1979). Presence of tri-iodothyronine, no detectable thyroxine and reverse tri-iodothyronine in human milk. *Endocrinology Japan, 26,* 507–513.

Schapiro, S. (1966). Metabolic and maturational effects of thyroxine in the infant rat. *Endocrinology, 78,* 527–532.

Siegel, L.S., & Heaven, R. (1986). Defining and categorizing learning disabilities. In S. Ceci (Ed.), *Handbook of cognitive, social and neuropsycholgoical aspects of learning disabilities, 1,* 95–121. Hillsdale, NJ: Ehrlbaum.

Smith, D.W., Blizzard, R.M., & Wilkins, L. (1957). The mental prognosis in hypothyroidism of infancy and childhood, *Pediatrics, 19,* 1011–1022.

Tenore, A., Militerni, R., D'Argenzio, G., DiMaio, S., Lubrano, P., Sandomenico, M.L., Mariano, A., & Varrone, S. (1987). Screening for congenital hypothyroidism in the region of Campania (Italy): Preliminary data on neuropsychomotor follow-up. In B.L. Therrell Jr. (Ed.), *Advances in neonatal screening* (p. 107). Amsterdam: Excerpta Medica.

Tenore, A., Parks, J.S., & Bongiovanni, A.M. (1977). Relationship of breast feeding to congenital hypothyroidism. In G. Chiumello & Z. Laron (Eds.), *Recent progress in pediatric endocrinology.* New York: Academic Press.

Therrell, B.L. (1987). National screening status report. *Infant Screening, 10,* 5–6.

Therrell, B.L., Meyer, D., Brown, L.O. et al. (1982). Incidence of primary congenital hypothyroidism in Texas by race and sex. In R.H. Dobbins (Ed.), *Proceedings of the 1982 National Newborn Screening Symposium* (pp. 57–60). Chicago, Illinois Department of Health.

Thieffry, A.F., Dhondt, J.L., Farriaux, J.P., & Parquet, P. (1983) Psychological cost of neonatal screening to families. In H. Naruse & M. Irie (Eds.), *Neonatal screening*. Amsterdam: Excerpta Medica.

Thompson, G.N., McCrossin, R.B., Penfold, J.L., Woodroffe, P., Rose, W.A., & Robertson, E.F. (1986). Management and outcome of children with congenital hypothyroidism detected on neonatal screening in South Australia. *The Medical Journal of Australia, 145*, 18–22.

Thyroid Foundation of America (1986). Newsletter. Massachusetts General Hospital, Boston.

Toublanc, J.E., Rives, S., & Job, J.C. (1989). *Factors related to the intellectual development of children treated for congential hypothyroidsm.* (Abstract) In F. Delange, D.A. Fisher, D. Glinoer (Eds). *Research in Congenital Hypothyroidism* New York: Plenum.

United States Department of Education. Information provided by Office of Special Education, U.W. Department of Education.

Vanderschueren-Lodeweyckx, M., Debruyne, F., Dooms, L., Eggermont, E., & Eeckels, R. (1983). Sensorineural hearing loss in sporadic congenital hypothyroidism. *Archives of Disease in Children, 58*, 419–422.

Vanderschueren-Lodeweyckx, M., Malvaux, P., Craen, M., Ernould, C., & Wolter, R. (1980). Neuropsychological study of treated children with thyroid dysgenesis. In G.N. Burrow & J.H. Dussault (Eds.), *Neonatal thyroid screening*. New York: Raven Press.

VanVliet G., Barboni T., Klees M., Cantraine F., Wolter F. Treatment strategy and long term follow up of congenital hypothyroidism. In F. Delange, D.A. Fisher, D. Glinoer (Eds.) *Research in Congenital Hypothyroidism*. New York: Plenum.

Virtanen, M., Maenpaa, J., Santavuori, P., Hirovonen, E., & Perheentupa, J. (1983). Congenital hypthyroidism: Age at start of treatment versus outcome. *Acta Paediatrica Scandinavica, 72*, 197–201.

Walfish, P.G. (1981). Thyroid physiology and pathology. In R. Collu et al. (Eds.), *Pediatric endocrinology*. New York: Raven Press.

Walfish, P.G. (Feb. 1984). The best way to screen for neonatal hypothyroidism. *Diagnostic Medicine, 7*:57–75.

Walfish, P.G., Gera, E., & Ehrlich, R.M. (1983). Primary TSH screening for neonatal hypothyroidism. Results of simultaneous cord and neonatal heel blood testing within the same infant population. In H. Naruse & M. Irie (Eds.), *Neonatal screening*. Amsterdam: Excerpta Medica.

Walfish, P.G., Ginsberg, J., Rosenberg, R.A., & Howard, N.J. (1979). Results of a regional cord blood screening program for detecting neonatal hypothyroidism. *Archives of Disease in Childhood, 54*, 1–7.

Weichsel, M.E. (1978). Thyroid hormone replacemnt therapy in perinatal period: Neurological consideration. *Pediatrics, 92*, 1035–1038.

Wolter, R., Noel, P., deCock, P., Craen, M., Enould, C.H., Malvaux, P., Verstraetan, F., Simons, J., Mertans, S., & Van Broeck, N. (1979). Neuropsycholgoical study in treated thyroid dysgenesis. *Acta Paediatrica Scandinavica Supplement, 277*, 41–46.

Yoshida, K., Sakurada, T., Takahashi, T., Furuhashi, N., Kiase, K., & Yoshinaga, K. (1986). Measurement of TSH in human aminiotic fluid: Diagnosis of fetal thyroid abnormality in utero. *Clinics of Endocrinology, 25*, 313–318.

Zetterstrom, R. (1988). *Psychological consequences of false-positive results.* Abstract No. 14. International Screening Symposium of Inborn Errors of Metabolism, San Paulo, Brazil.

18
Hyperthyroidism: Cognitive and Emotional Factors

JEAN E. WALLACE AND
DUNCAN J. MACCRIMMON

The Presentation of Clinical Hyperthyroidism

Historical Development

In describing the evolution of thought regarding hyperthyroidism, review usually begins with reference to Parry's 1825 account of symptoms that appeared in a girl shortly after her frightening ride in a wheelchair (cf Bauer et al., 1987). When Robert Graves described the syndrome in his clinical lectures of 1834–1835, the title, "Newly Observed Affections of the Thyroid Gland in Females–Its Connexion with Palpitation—with Fits of Hysteria" (Graves, 1940), indicated that enlargement of the thyroid gland was viewed as a disorder of emotional origin, even though he suggested that the cardiac symptoms could reflect an organic disorder of the heart. As the symptoms of Graves' disease became known and well documented, they continued to be associated with hysteria; Charcot considered hyperthyroidism a neurosis (cf Hoffenberg, 1974). Even as diagnostic procedures identified the physiological abnormalities of thyrotoxicosis, interest in the role of emotional disorder or environmental stressors in the development of the disease persisted and was particularly active during the 1930s when psychoanalytic concepts were applied. There is a 150-year history of conceptualizing hyperthyroidism as a disorder in which adaptive-defensive style and personal or environmental stressors have been viewed as predisposing and precipitating factors.

The clinical features of hyperthyroidism are now well established. In addition to physical signs such as gland enlargement, classical physiological symptoms include shakiness, weight loss, fatigue, palpitations, shortness of breath, muscular weakness, and sweating. A variety of cognitive and affective phenomena are also known to be associated with hyperthyroidism (MacCrimmon et al., 1979) and include nervousness, jumpiness, restlessness, tension, irritability and anxiety. The behavioral state has been called a "tense dysphoria."

Only in thyroid *hyper*function has emotional disturbance predating and contributing to development of the illness been emphasized. *Hypo-*

function, on the other hand, has been viewed as a deficiency state, with mental or emotional symptoms described as somatopsychic, postdating and arising entirely from the effects of gland dysfunction (Bauer et al., 1987). Critical thought about the reasons for this discrepancy has been sparse. This may be because hypothyroidism presents the physical symptoms of a profound deficiency state; hyperthyroidism at times presents a variety of behavioral symptoms such as active expression of emotional distress and instability in word and behavior, suggesting mental illness.

The clinical literature contains many case studies describing almost pure psychiatric presentation of occult hyperthyroidism. Contributing factors such as the joint occurrence of hyperthyroidism and pre-existing psychiatric disorder, or false-positive psychiatric diagnosis, cannot be disentangled in this literature. A number of these reports describe major depressive syndromes. A profoundly depressive picture is also found in apathetic hyperthyroidism and appears with greater frequency in elderly hyperthyroids. In these cases, neuromuscular, cognitive, and motivational deficits may predominate, as in hypothyroidism, more readily indicating a neurological condition. Parry's 1825 account unfortunately is usually cited without comment about the condition that caused the girl to be in a wheelchair, or whether it could have reflected acute hyperthyroidism in its neurological presentation (Swanson et al., 1981).

The Psychosomatic Hypothesis and Studies of Stress

In the past sixty years, descriptions of the many neurological and psychiatric presentations of hyperthyroidism have been published; the 1930s produced reports of emotional instability and environmental stressors in several series of patients; these appeared again in the 1950s and were thoroughly and critically reviewed by Gibson (1962). Population studies of environmental stress and hyperthyroidism first appeared in the 1940s. Published reports reflect the following research strategies for examining psychosomatic issues in this disease:

1. Description of predisposing adaptive-defensive style, and associated core conflicts, in persons who currently have the disorder, treated or untreated (Lidz, 1949; Ham et al., 1951; Mandelbrote & Wittkower, 1955; Alexander et al., 1961);
2. Description of stressful personal events occurring before the first onset of symptoms in a hyperthyroid sample (Wilson et al., 1962; Forteza, 1973). This approach also describes stressful personal events preceding recurrence of symptoms (Morillo & Gardner, 1980);
3. Study of the effect of immediate personally stressful events upon indices of thyroid activity in persons with gland hyperfunction (Alexander

et al., 1961; Flagg et al., 1965) or in normal persons (Dongier et al., 1956; Semple et al., 1988);

4. Study of the incidence of hyperthyroidism where it has been possible to identify and compare environmentally stressful and relatively non-stressful periods (see Ashkar, 1972; Hadden, 1974) in Denmark, Belgium, and Northern Ireland;

5. Study of the incidence of hyperthyroidism in two populations judged to be comparable, except for exposure to recent emotional trauma in one. This strategy has focused on recent immigrants: various ethnic groups to Canada (Spaulding, 1963) and Cubans to Miami (Ashkar, 1972).

Where sample selection strategies have been adequate and stated hypotheses have been objectively testable, results have not supported the hypothesis that personality features predispose a person to hyperthyroidism (Hermann & Quarton, 1965; Paykel, 1966). Gibson (1962) concluded that there were few positive findings regarding the importance of emotional factors in the etiology of thyrotoxicosis. Subsequent studies show that there is better evidence for familial-genetic predisposition (Zonana & Rimoin, 1975). The causal significance of personally stressful events preceding onset of symptoms can be evaluated only when one knows the incidence of similarly defined events in the general population during a similar span of time. For example, Forteza (1973) identified psychologically stressful events during the year preceding onset of illness in 78% of 72 hyperthyroid adults under the age of 50. There is evidence suggesting that stressful events occur in the general population with a similar frequency. Sampling reveals at least one significant loss, moderately threatening stressful, or undesirable uncontrollable event during the past 12 months ranging from 55% (Goldberg & Comstock, 1976) to 79% (Norman et al., 1985) in normal groups. Each study reported that the median frequency of such events was 2 in 12 months.

There is also evidence of a relationship between stressors and abnormally increased thyroid gland activity in vulnerable (currently and formerly hyperthyroid) individuals, where abnormal increases have been experimentally produced; these have not been producible in normals (Alexander et al., 1961; Semple et al., 1988). Nor has it been possible to produce increased TSH levels in phobic persons by engendering intense phobic anxiety during implosion psychotherapy (Nesse et al., 1982). Volpe (1977) sagely observed that the relation of environmental stress to development of thyroid gland hyperfunction should be examined within a population genetically at risk for this disorder.

Studies of incidence of hyperthyroidism during stressful and nonstressful intervals has been flawed by the presence of confounding variables such as diet or referral pattern; a recent, carefully controlled study produced negative findings (Hadden, 1974). Using the fact of recent immigration to define stress is problematic unless the base-rate incidence of the disorder in the population emigrated from is taken into consideration.

The symptoms of thyrotoxicosis will lead the clinician to take interest in emotional issues because they often include intense emotionality. Although psychosomatic formulations (and particularly psychodynamic ones) are likely to polarize opinion, both adherents and skeptics should take care to consider separately evidence of emotional distress that represents symptoms of the acute disorder and evidence that stressors preceded its onset.

The Relationship of Hyperthyroidism to Psychiatric Illness

In order to explore the role of hyperthyroidism in psychiatric disturbance, population survey is a worthwhile strategy and has been applied to hyperthyroid patients attending endocrine clinics. When relatively objective operational criteria for symptom report and diagnosis are used, the result is a symptom pattern that reflects the somatic and organic consequences of the hyperthyroid state rather than enduring psychiatric disorder (Wallace et al., 1980). This trend is typified by the work of Kathol and Delahunt (1986) who studied 33 newly diagnosed untreated hyperthyroid cases. Patients were interviewed using a structured questionnaire designed to elicit features of clinical hyperthyroidism, DSM-III major depression, organic affective syndrome, and anxiety disorder, and also depression and anxiety disorder using Feighner's criteria. When symptoms were taken at face value the incidence of depressive syndromes defined by either set of criteria was found to be greater than normal. When adjustment was made to de-emphasize somatic features, the incidence of diagnosable depression fell to a level more consistent with that found in the general population. It must be remembered that classification systems like DSM-III were developed to categorize psychiatric symptoms not attributable to medical illnesses, and that using such an approach in conditions like hyperthyroidism can lead to false-positive psychiatric diagnoses. Pursuing this idea, Kathol et al. (1986) followed 29 acute hyperthyroid patients until euthyroid status was achieved. Although when first seen, 9 had features of major depressive disorder and 23 showed a generalized anxiety syndrome, all the depressed patients and 21 of the 23 with anxiety showed complete resolution of these symptoms following successful antithyroid therapy only. This finding provides support of an earlier follow-up study indicating that psychiatric symptoms associated with hyperthyroidism will resolve when the thyroid disorder is successfully treated (MacCrimmon et al., 1979).

A strategy complementary to these approaches involves evaluating the incidence/prevalence of hyperthyroidism in psychiatric patient groups. The community study of Tunbridge et al. (1977) provides relevant benchmark abnormality rates of 1.9% to 2.7% for females and 0.16% to 0.23% for males in 2,779 randomly selected subjects. Acutely ill patients in general hospital psychiatry departments comprise one such group. Reported incidence in six such studies ranges from a high of 2 out of 50 cases (Weinberg & Katzell, 1977) to a low of 1 in 480 (Cohen & Swigar, 1979). Lower incidence seems to be associated with more recent studies and with university-affiliated hospital settings where comprehensive medical pre-screening is more likely to be routine. When state mental hospital samples are studied, the results appear to be similar. For example, rates of 8 in 1206 (McLarty et al., 1978) and 2 in 872 (Clower, 1984) have been reported.

Another strategy has been to examine the impact of hyperthyroidism on the course of established psychiatric illness. Checkley (1978) found that thyrotoxicosis had little effect on recurrences of manic depressive episodes; White and Barraclough (1988) concluded that once the thyroid disorder was treated, the associated mental illness followed its expected course and prognosis. Specific psychiatric diagnoses have also been used to select clinical populations. In this regard panic disorder, depression, alcohol abuse, anorexia nervosa, unipolar and bipolar affective disorder, and psychogeriatric conditions have been studied. Most studies report an incidence of hyperthyroidism in psychiatric populations that is not materially different from the general population base rate. They also indicate that the psychiatric and/or cognitive symptomatology present in acutely hyperthyroid samples will disappear when the thyroid disorder is treated.

In concentrating on the psychiatric antecedents and manifestations of hyperthyroidism, investigators generally have not, until rather recently, emphasized the acute effects and psychological concomitants of excess thyroid hormone. However, more objective screening tests of emotional functioning in acutely hyperthyroid patients have suggested that the disorder may be understood better by considering the influence of thyroid hormone on particular aspects of behavior.

The Effects of Thyroid Hormone on Brain/Behavior

From the early 1960s, the literature reflects three approaches to measuring the effect on brain function of increased thyroid hormone levels. These research strategies have investigated the effects of (a) normal-range variation, (b) experimentally administered excess hormone, and (c) endogenous excess hormone on dependent measures such as neurotransmitter systems, brain electrophysiology, mood/anxiety, and cognitive function.

Studies of Brain Physiology

Whybrow and Prange (1981) synthesized a wide range of clinical and laboratory observations and formulated the hypothesis that thyroid hormones, even in normal-range variation, promote increased beta adrenergic receptor activity within the central nervous system. Subsequent studies indicate complex effects when exogenous hormone is administered to experimental animals. Cortical areas may show increased beta receptor activity while some subcortical structures such as thalamus, striatum, and parts of the limbic system show decreases; cerebellum and brain stem appear to be unaffected (Perumal et al., 1984). However, Schmidt et al. (1985) reported that using lower doses of thyroid hormone or different genetic strains of rats produced an opposite pattern of beta receptor alteration. It thus appears that thyroid hormone, at the very least, alters central beta receptor activity and that these are region-specific effects. Such changes undoubtedly play a role in the genesis of the behavioral features of acute hyperthyroidism.

Tucker et al. (1984) proposed that degrees of brain arousal might covary with thyroid hormone levels. In a sample of normal university students, complex relations between triiodothyronine (T3) level, cortical area, EEG delta power, and cognitive activity were found. During cognitive activity, delta power was the most consistent predictor of T3 level; less left-hemisphere power and greater right occipital power predicted higher T3. No analysis of the relationship between thyroid hormone levels and EEG resting brain activity was reported. Normal-range thyroid hormone levels and EEG correlates do not appear to have been examined by others to date, but experimental manipulation of hormone levels in normals has been reported to produce EEG changes.

Examining the effect of exogenous T3 in normals, Wilson et al. (1964) and Kopell et al. (1970) reported EEG evidence of altered arousal patterns. Wilson et al. found reduced duration of arousal following photic stimulation. Kopell et al., using a visual average evoked potentials paradigm, reported an abnormal, augmented response to irrelevant stimuli, suggesting that excessive T3 compromised selective attention. They noted that hyperthyroid patients often seem to feel overwhelmed by unwanted sensory input.

Clinical EEG studies of hyperthyroid patients have been summarized by Markand (1984). Typically these report patterns characteristic of diffuse encephalopathies in quality if not always clearly abnormal in degree. EEGs following establishment of euthyroid status show a trend toward gradual disappearance of these features over sometimes lengthy periods of time. Zeitlhofer et al. (1984), using computer-analyzed quantitative EEG measurements, showed significant differences between hyperthyroid patients and matched controls. After three to six months of successful therapy, some, but not all, differences disappeared. The fact that EEG-expressed subtle organic effects of thyroid toxicity have persisted for long periods of time suggests that some reported follow-up evaluations were made before recovery was complete. But on the basis of continued EEG abnormalities observed two to three years after treatment began in 65% of one hyperthyroid sample, Siersbaek-Nielsen et al. (1972) suggested that brain dysfunction can persist indefinitely and that an episode of hyperthyroidism may cause irreversible damage to the brain in spite of apparently fully successful antithyroid therapy.

Objective Evaluation of Mood and Anxiety

There are relatively few reports of the influence of excess thyroid hormone on objectively measured emotional state. Wilson et al. (1964) noted that all their normal subjects showed marked changes in feeling tone after taking exogenous T3. Clyde Mood Scale results showed significant increases in depression and jitteriness and a decrease in friendliness (Wilson et al., 1962).

Beginning in the early 1960s standardized questionnaires and/or objective performance measures were used to document symptoms of acute hyperthyroidism. Investigators reported the extent to which thyrotoxic individuals resembled persons with psychiatric or organic diagnoses, with emphasis on symptom presentation rather than features judged to have predated the acute illness. A summary of these studies appears in Table 18.1.

Robbins and Vinson (1960) studied the symptom endorsement of pre-treatment thyrotoxic adults using a measure of somatizing or neurotic tendencies. Before treatment, hyperthyroids were significantly more endorsing of somatic symptoms than were normal controls but were not different from psychiatric patients classified as somatizers, nor from patients with brain dysfunction. The test reflected significant improvement after euthyroid status had been restored, showing significant decrease in symptom endorsement compared to patients with psychiatric or brain dysfunction diagnoses. However, their report of somatic complaints was still significantly greater than in normal controls. On the other hand, Paykel (1966) found no difference between formerly thyrotoxic women and matched medical controls using similar measures. Greer et al. (1973) reported that acutely thyrotoxic patients and patients suffering from anxiety state were not distinguishable using a self-report measure of anxiety.

Wilson et al. (1962) noted anxiety, irritability, change in libido, sleep disturbance, and either depressed or elated mood in their sample of acutely hyperthyroid patients. They then used the Clyde Mood Scale to compare a second sample of treated formerly hyperthyroid patients to a sample of normal controls, finding no differences. Referring to this and to their studies of the effects of exogenous T3 on mood, they proposed that a direct relationship exists between excessive levels of thyroid hormone and alterations in feeling tone. Using the same measures, Whybrow et al. (1969) found that thyrotoxic patients as a group scored one standard deviation above the test mean on scales reflecting unhappiness and dizziness. There was a significant posttreatment decrease in the group score for dizziness.

The Minnesota Multiphasic Personality Inventory (MMPI) has been used by several investigators of emotional status in hyperthyroidism.

TABLE 18.1. Evaluation of emotional status in hyperthyroidism.

Author	Sample size	Age range, mean, SD	Main measures	Controls	Pre-/post-treatment evaluation
Robbins & Vinson (1960)	7f 3m	— $\overline{X} = 42$ SD = 9	Maudsley Medical Questionnaire	Normal psychiatric neurologic	Pre-/post-interval not stated
Wilson et al. (1962)	16f	22–57 $\overline{X} = 37$ —	Clyde Mood Scale	Normal formerly thyrotoxic	Pre- only
Artunkal & Togrol (1964)	20f	— $\overline{X} = 36$	MMPI, Rorschach	Normal	Pre-/post-4–12 mo
Paykel et al. (1966)	35f	15–54 $\overline{X} = 34.2$ —	Eysenck Personality Inventory, Cornell Medical Index	Nontoxic goiter	Post- only 3–4 yr
Whybrow et al. (1969)	7f 3m	$\overline{X} = 45.4$ $\overline{X} = 42.8$	Clyde Mood Scale, MMPI, Brief Psychiatric Rating Scale	None	Pre-/post-4–15 mo
Greer et al. (1973)	14	$\overline{X} = 48$ SD = 16	IPAT Anxiety Scale	Psychiatric	Pre- only
MacCrimmon et al. (1979)	19f	25–55 $\overline{X} = 38.8$ SD = 10	MMPI, Psychiatric Status Schedule	Normal	Pre-/post-7–14 mo
Zeitlhofer et al. (1984)	30f	20–55 —	Clinical Self-Rating Scale, Ullrich Emotional Inventory	Normal	Pre-/post-3–6 mo
Trzepacz et al. (1988)	10f 3m	24–61 $\overline{X} = 38.9$ SD = 13.4	Beck Depr. Inventory, State-Trait Anxiety Scale, Symptom Checklist 90-R, S.A.D.S.	None	Pre- only

f: females.
m: males.

MMPI scales were initially validated using patients with the psychiatric diagnoses of (1) hypochondriasis, (2) depression, (3) hysteria, (4) psychopathic deviate, (6) paranoia, (7) psychasthenia, (8) schizophrenia, and (9) hypomania. MMPI interpreters now tend to refer to scale numbers rather than to these names because of the simplistic and unwarranted conclusions that these labels, some outdated, may encourage when taken singly and particularly when applied to medically ill people.

Whybrow et al. (1969) were among the first investigators to use the MMPI with pretreatment hyperthyroid patients. The group profile showed elevations of scales 1, 2, 3, and 8, similar to the profile pattern reported by Artunkal and Togrol (1964) where scales 2, 6, and 8 were significantly higher than those of matched controls. Whybrow et al. interpreted the hyperthyroid group profile to reflect a hysteroid elaboration of somatic complaints and distress that could be consequent to disturbance in cognitive function. They also rated, based on clinical interview, symptoms of somatic concern, anxiety, motor tension, and possible conceptual disorganization. Disturbance profound enough to constitute psychiatric illness was found in 70% of these thyrotoxic patients. On achievement of euthyroid status there was significant symptom reduction in each area. Posttreatment MMPIs reflected a significant decrease in all previously elevated scales except scale 3.

The Psychiatric Status Schedule (PSS) and MMPI were administered to acutely thyrotoxic women by MacCrimmon et al. (1979). Only the PSS depression-anxiety symptom scale was significantly elevated in pretreatment hyperthyroids when compared to controls. An elevated total score suggested that the hyperthyroid patients were generally more symptomatic, but the PSS no longer identified difference between patients and controls 21 days after patients began treatment. Acutely hyperthyroid patients generated a group MMPI profile with elevations significantly different from controls and from their own posttreatment profile on scales 1, 2, 3, 7, and 8, reflecting depressed mood, somatic sensitivity/preoccupation, anxiety/tension, and, possibly, a subjective sense of cognitive disorganization. The posttreatment MMPI profile was not statistically different from that of the controls but showed a persistent relative elevation on scale 3, suggesting a continued sense of somatic distress. An MMPI-derived Total Pathology Rating (TPR) correlated significantly with pretreatment serum T4 level. High initial T4 levels appeared to have a significant influence on the patients' posttreatment function in that TPR on follow-up assessment bore a significant relation to initial T4 level (Wallace et al., 1980).

More recently other standardized measures of emotional and psychiatric status have been applied to acutely hyperthyroid patients by Zeitlhofer et al. (1984) and Trzepacz et al. (1988). Zeitlhofer's Austrian study found that pretreatment thyrotoxic women endorsed more somatic complaints and greater pathology related to anxiety, depression, exhaustion, inhibition, and disturbance of general feeling. After satisfactory treatment the formerly hyperthyroid group produced scores no different from control group original scores except that they continued to endorse more somatic complaints. Trzepacz et al. used self-report and clinician-rated scales to characterize the emotional and psychiatric status of patients with untreated hyperthyroidism. Their procedures confirmed the predominance of potentially classifiable psychiatric disorder in a hyperthyroid sample, but the procedures were not repeated after treatment.

Summary

Each of these authors objectively measured symptoms of emotional distress and demonstrated the close correspondence between acute thyroid toxicity and symptoms of emotional illness, at times resembling psychiatric disorder but always including somatic concerns. Some have also found evidence of a return to normal-range values and/or the disappearance of differences from a control group after euthyroid status was again attained. Each has reported evidence that elevated thyroid hormone levels directly produced symptoms of psychopathology and could be regarded as the biological substrate for changes in mood and behavior, without any need to refer to pre-existing psychological adjustment, defensive patterns, or environmental stress.

Most of these findings verify resolution of emotional symptoms after adequate treatment and show that the psychological responses to illness which these acutely ill patients had in common also diminish. However, a careful review of those studies that followed patients until they were euthyroid also reveals recurring evidence that residual sensitivity to somatic symptoms characterizes many formerly hyperthyroid patients.

Robbins and Vinson (1960) were the first of several authors to document sensitivity to somatic dysfunction that continued into the euthyroid state, observed independently from any inferences about pre-illness characteristics. Artunkal and Togrol (1964), Whybrow et al. (1969), and Wallace et al. (1980) each reported persisting relative elevation on posttreatment group MMPI scale 3. It is not known whether the suggested persistence of somatic concerns is a result of having experienced hyperthyroidism or whether a pattern of somatic concern predated the illness. However, the relation between initial T4 and posttreatment MMPI TPR suggests an association between severity of thyrotoxicosis and persistence of somatic symptoms. There may be a question about whether all formerly thyrotoxic individuals regain pre-illness levels in their subjective sense of emotional and physical well-being.

Objective Evaluation of Cognitive Functions

Although it is now generally accepted that identifiable cognitive deficits will accompany other symptoms of clinical hyperthyroidism, there have been few studies that specify the deficits or the measures that are sensitive to them. There appear to have been no published studies of the effects on cognitive function of exogenous hormone in normals although Beierwaltes and Ruff (1958) indicated that changes in mental function had been examined in their study. There has been one study of the relation between normal hormone levels and some aspects of cognitive efficiency.

In a sample of normal university students, Tucker et al. (1984) studied the relationship between normal-range T3, T4, and TSH and a variety of cognitive functions, testing the hypothesis that thyroid function influences neuropsychological processes through normal neurophysiological mechanisms. Their tests included Digit Span, an analogous nonverbal test of tonal memory, and other experimental tasks. Higher T3 levels were associated with poorer performance on several tasks, particularly in males. They suggested that increased thyroid hormone levels led to increased CNS activation that may facilitate some functions and hinder others. Compared to subjects with lower T3 levels, those with higher levels tended to perform a test of verbal fluency better and tasks requiring nonverbal auditory and visual attention or immediate memory more poorly. Tucker et al. advance some arguments in favor of lateralized hormonal effects due to asymmetric anatomy and neurochemistry in the brain.

The majority of thyroid/cognition studies have described the cognitive correlates of thyroid toxicity without proposing specific mechanisms or lateralizing/localizing hypotheses. Hermann and Quarton (1965) asked hyperthyroid patients for specific examples of their complaints of irritability and concentration or memory difficulties. They concluded that each of these referred to increased distractibility and might reflect reduced ability to screen out or select sensory input. Complaints of anxiety seemed to refer to a "rootless sensory impression" rather than a sense of threat. The authors suggested that hyperthyroidism causes a lowered general sensory threshold and/or impaired sensory selection and recommended that objective measures of these descriptors be found. A summary of studies applying objective measures appears in Table 18.2.

The Money et al. 1966 study of intellectual function in posttreatment hyperthyroids remains the only report regarding school-aged children. Although an important contribution, the findings are limited by nonuniform data collection due to the mostly retrospective approach. Elapsed time between beginning of pharmacologic treatment varied greatly, from less than one up to 11 years. Most subjects were tested at an age appropriate to the WISC, but the WAIS and Wechsler Bellevue I had also been used.

Verbal exceeded Performance IQ in 18 of the 22 subjects; the means of 107.7 and 99.3 were significantly different. Similarly, the mean scores of Verbal Comprehension and Perceptual Organization factors (11.4 and 9.4) differed significantly. Unfortunately the factor score for Freedom from Distractibility, which subsequent research suggests would have been particularly relevant, could not be calculated because the full set of Wechsler subtests had not always been given. The superiority of Verbal over Perfor-

TABLE 18.2. Evaluation of cognitive status in hyperthyroidism.

Author	Sample size	Age range, mean, SD	Main measures	Controls	Pre-/post-treatment evaluation
Robbins & Vinson (1960)	7f 3m	— $\overline{X} = 42$ SD = 9	Stroop Color-Word	Normal psychiatric neurologic	Pre-/post-interval not stated
Artunkal & Togrol (1964)	20f	— $\overline{X} = 36$ —	Reaction time, Visual-motor perception/ coordination	Normal	Pre-/post-4–12 mo (N = 10)
Money et al. (1966)	17f 5m	6–20 $\overline{X} = 12.75$ —	Wechsler Intelligence Scales	None	Post- only 1 mo– 12 yr
Whybrow et al. (1969)	7f 3m	$\overline{X} = 45.4$ $\overline{X} = 42.8$	Trailmaking, Porteous Maze XIV	None	Pre-/post-4–15 mo
Wallace et al. (1980)	19f	25–55 $\overline{X} = 38.8$ SD = 10	Stroop Color-Word, Spokes, Paired Assoc. Learning	Normal	Pre-/post-6–12 mo
Alvarez et al. (1983)	21f 6m	— $\overline{X} = 35$ SD = 12	Tolouse-Pieron Concen-tration	normal (N = 20)	Pre- only
Zeitlhofer et al. (1984)	30f	20–55 —	Trailmaking, Reaction time, Visual perception	Normal	Pre-/post-3–6 mo (N = 23)
Perrild et al. (1986)	24f 2m	24–81 Mdn = 54 —	Tests of attention, learning/ memory, visual perception	Nontoxic goiter (N = 23)	Post- only 10 yr
Trzepacz et al. (1988)	10f 3m	24–61 $\overline{X} = 38.9$ SD = 13.4	Stroop Color-Word, Wechsler Memory, WAIS-R, Halstead-Reitan Battery	None	Pre- only

f: females.
m: males.

mance IQ in most subjects appeared largely attributable to slow perfor-mance on the timed nonverbal tests, but the authors also noted that some subjects showed "flighty attention" or muscular overactivity during test-ing. School records noted concentration difficulties, nervousness, and hyperkinetic activity in many, a decline in the quality of schoolwork coincident with onset of clinical symptoms, and improvement after treatment.

Money et al. noted that the interval between inception of treatment and time of testing appeared unrelated to the size of Verbal-Performance discrepancy. Referring to the persistence of a marked verbal-nonverbal discrepancy in the one serially assessed subject, they speculated that "perhaps there is an analogy . . . with the persistence of thyrotoxic exophthalmos, namely some sort of permanent residual" (Money et al., 1966, p. 280). Subsequent research may not yet have resolved this issue.

Robbins and Vinson (1960) had previously used the Stroop Color-Word test to measure cognitive deficit and found that pretreatment hyperthyroids made more errors, took longer, and were less able to screen out inappropriate visual stimuli than were matched normal controls. They made more errors than psychiatric patients and performed only slightly better than patients with brain dysfunction. After treatment their test performance was better than both patient groups and resembled normals. The authors hypothesized that distractibility or impaired selective attention was a significant cognitive effect of excess thyroid hormone.

Artunkal and Togrol (1964) studied acutely thyrotoxic Turkish women. Compared to controls, visual processing was slow and less accurate and motor coordination was poorer; Rorschach records resembled those of organically impaired individuals. Visual and motor measures repeated after treatment showed some, but not significant, improvement. The authors proposed that brain lesions accounted for their findings.

Whybrow et al. (1969) administered tests requiring visual tracking and planning. They reported impaired performance with reference to test norms. Posttreatment performance showed significant group improvement although the reported mean would still be considered to fall in the impaired range. Whybrow et al. asserted that there is a biological substrate to the cognitive and emotional symptoms of hyperthyroidism and on the basis of these findings argued that the central nervous system is extremely sensitive to the effects of thyroid hormone, contrary to earlier beliefs that the adult brain is relatively refractory to its effects.

In choosing tests, Wallace et al. (1980) emphasized the earlier-suggested cognitive deficits in selective attention, concentration or immediate memory, visual tracking, and visual-motor speed. Compared to controls, acutely thyrotoxic women showed impaired function in these areas and no identifiable disturbance in auditory attention or in simple motor speed. Posttreatment improvement in these functions caused differences between patient and control group performance to disappear. This study appears to be the only one that has controlled for test practice effects in evaluating cognitive factors.

Four subsequent reports regarding cognitive function have appeared. Alvarez et al. (1983) correlated thyroid indices with performance on a test of visual attention and motor efficiency. Thyroid indices did not correlate with impairment, but deficit in the patients' attention and efficiency was observed early in the task and became more pronounced the longer the test continued.

Zeitlhofer et al. (1984) reported that pretreatment hyperthyroids were motorically significantly faster than controls and no different in visual tracking. Impairment relative to controls appeared in tasks requiring visual reception and perception. Posttreatment comparisons, not controlled for practice effects, continued to show poorer visual reception/perception. Results are difficult to interpret because they were reported only in terms of significant group differences; no test data, group means, or standard deviations were provided.

Trzepacz et al. (1988) used a battery of neuropsychological measures in the study of untreated hyperthyroids. Unfortunately they did not correct for age or education in their use of test norms, nor report statistical analyses of test results. Consistent with earlier studies, they concluded that their hyperthyroid subjects as a group suffered from mild deficits in complex attention and immediate memory, and also in higher-level problem-solving abilities.

Perrild et al. (1986) conducted cognitive evaluations ten years after treatment in patients who earlier had been the subjects of EEG studies by this group of Danish investigators. They compared a group of formerly thyrotoxic patients to a group of patients previously treated for nontoxic goiter. They selected 15 appropriately referenced cognitive tests for general functions of attention, learning/memory, and abstraction, and for the specific functions of reading, writing, and visual perception. They classified each test performance using previously defined criteria for non-/slightly reduced, moderately reduced, or markedly/severely reduced cognitive functions. They then compared the frequency of these classifications for each test and for similarly rated evidence of EEG abnormality in the two groups. In the formerly thyrotoxic group, persisting EEG abnormalities were significantly more frequent, as was evidence of moderate or marked/severe cognitive impairment in 7 of the 11 tests of attention, memory, and abstraction. Several formerly thyrotoxic but only one former goiter patient showed impairments which the authors interpreted as evidence of localized cerebral dysfunction. Factors such as age or intercurrent illness were carefully considered in interpreting group differences.

In their thought-provoking discussion, the authors questioned earlier conclusions that there is complete recovery of cognitive normality in formerly hyperthyroid individuals. They suggested that thyrotoxicosis may be associated with a previously vulnerable brain and/or leave the brain with permanent and not necessarily subtle residual deficits, perhaps increasing its vulnerability to insult, after the episode of acute toxicity.

Summary

Excess thyroid hormone appears to compromise cognitive functions that emphasize input and information-processing efficiency rather than verbal retrieval and simple motor output. The compromised functions are those that are relatively more crucial when responding to novelty or coping with change. They require more cognitive flexibility and have been associated with the concept of "fluid" or adaptive, as opposed to crystallized, intelligence. It has been shown that such cognitive functions, believed to be mediated more by the right than the left hemisphere, are the ones that show greatest deficits in toxic and confusional states, and thus that toxicometabolic disorders produce cognitive deficits similar to those in patients with focal right-hemisphere disease (Lee & Hamsher, 1988). In their study of normal-range thyroid hormone levels, Tucker et al. (1984) advanced some arguments in favor of lateralized hormonal effects due to asymmetric anatomy and neurochemistry in the brain; Lee and Hamsher have pointed out that diffuse dysfunction can present a misleading "lateralizing" picture.

Studies assessing cognitive deficits, their pattern, and course over time sometimes suggest complete recovery of the central nervous system after hormone level has normalized, but at times the evidence is not entirely conclusive. Investigation of cognitive deficits in acute thyrotoxicosis, and the recovery from deficits after treatment, has been carried out infrequently and often with conceptual and methodological shortcomings. Standards for choosing assessment procedures, reporting and interpreting neuropsychological test findings, or controlling for confounding factors (Parsons & Prigatano, 1978) have often not been met. One must conclude that relatively little is known about the effects of thyroid hormone toxicity on the brain, particularly in the long term, and nothing is known about possible cognitive vulnerabilities that could be present prior to overt gland dysfunction and underlie subtle deficiencies in coping with stress.

These cognitive issues and current knowledge regarding emotional distress in clinical hyperthyroidism may need to be reexamined and reconceptualized, taking into consideration a persisting CNS or regulatory system deficit that may also underlie residual cognitive and/or emotional dysfunction. Recent research indicates that alterations in neuroendocrine regulation of the hypothalamic-pituitary-thyroid axis are associated with some psychiatric conditions, independent of excess thyroid hormone.

Thyroid Regulatory Mechanisms in Psychiatric Syndromes

Neuroendocrine Dynamics in Psychiatric Disorders

When considering this topic, a general model of neuroendocrine control mechanisms and their interactions should be borne in mind. The hypothalamus serves as a key link between the nervous and endocrine systems in reacting to the external environment, receiving converging integrating inputs from higher central nervous system levels and from the interior milieu via feedback loops. The hypothalamic influence is mediated through the release of specific neurohumeral substances, in this case thyrotropin releasing hormone (TRH), which travels via a venous portal system to the anterior pituitary gland. There the release of a pituitary trophic hormone (TSH) is induced; this enters general circulation to reach and stimulate the target organ, that is, to produce T3 and T4. Characteristic of neuroendocrine systems are multiple complex effects and feedback loops. Typical as well is the fact that hypothalamic trophic hormones such as TRH have multiple effects; for example, TRH may release other pituitary hormones such as prolactin and may even influence other brain regions. Moreover, peripheral hormones such as cortisol may inhibit TRH release. The homeostatic regulation of neuroendocrine systems is complex and may not yet be fully understood, but factors such as weight loss, ambient temperature, activity cycle, stress, anxiety, diet, and medications may influence neuroendocrine parameters.

With this knowledge of possibly confounding influences, it is intriguing to note a burgeoning literature reporting alterations in the dynamics of the hypothalamic-pituitary axis in euthyroid psychiatric patients (Loosen, 1985). The observation of an attenuated or blunted TSH response to exogenous TRH (the TRH Test) has been found regularly in approximately 25% of patients with major depressive disorder and to a lesser extent in mania, alcoholism, and borderline personality disorder. There is a degree of specificity in that schizophrenic patients infrequently show blunted TRH. Whether TRH blunting represents a trait or state feature is unclear. It may reflect response to antidepressant treatment in the short term, but it has also been associated with future suicidal behavior (Korner et al., 1987). Study of a wider range of neuroendocrine variables in a group of unipolar depressed patients suggests that blunted TSH response may be associated with a more general dysregulation of the hypothalamic endocrine axis (Roy et al., 1988). Replication of these findings in many recent studies indicates a degree of robustness and suggests that this area is worthy of continuing investigation.

Summary and Conclusions

Endocrine/neuroendocrine relationships have attracted behavioral science interest because this area has been thought to provide a "window" on the brain; that is, neuroendocrine dysfunction may reflect brain dysfunction. From the perspective provided by studies of hyperthyroidism, the following formulations and conclusions appear warranted. In terms of the specific etiology of hyperthyroidism, familial/genetic autoimmune mechanisms are of paramount importance. An etiological role for preexisting personality structure has not been substantiated, while the role of stress, if any, is likely to be that of a trigger factor or final precipitant befalling an already vulnerable individual. That excessive thyroid hormone affects central neurotransmission is a well-established fact in animal studies, and finds support in clinical EEG observations. These alterations affect emotional and cognitive processes, especially where the individual must accommodate and respond to new information. Higher levels of T4 for longer times prior to treatment may intensify these effects, which are now understood to be attributable to acute toxicity. After successful treatment, psychological difficulties usually resolve, but repeated hints of persisting somatic sensitivity have been found. Moreover, recent EEG and cognitive test evidence of prolonged and presumably permanent brain dysfunction has been presented.

Several hypotheses about residual, chronic effects may be offered. For example, it has been suggested by Perrild et al. (1986) that an episode of thyrotoxicosis accelerates brain "aging." Another is that, while the overly active gland has been suppressed by treatment, neuroregulatory dynamics are not necessarily returned to normal. Altered dynamics are found in some psychiatric conditions; these abnormalities may generate emotional or cognitive difficulties in their own right. Yet another mechanism of chronicity pertains to the fact that current treatments for hyperthyroidism do not address the underlying autoimmune disorder, which itself may directly cause CNS dysfunction. In this instance, residual features would not necessarily be of the same nature as acute deficits and might show greater individual variability, as in the CNS effects of other autoimmune disorders.

The investigator who is seeking a clearer view of the direct effect of excess thyroid hormone on the brain would be well-advised to consider the following issues regarding methodology. Posttreatment assessments of brain function should be timed from the reestablishment of a euthyroid state as was reported by Whybrow et al. (1969) rather than from the beginning of antihyperthyroid treatment. Outcome measures are very likely premature if they are applied before the patient has experienced a euthyroid condition for at least a year.

When procedures include the classification of levels of cognitive impairment or the interpretation of the pattern of test results to infer impairment of function, there should be explicit and neuropsychologically sound criteria for considering the effects of education and employment experiences on test performance. If test norms rather than the performance of well-matched controls are used to infer cognitive deficit, estimates of pre-illness intellectual level based upon demographic data and on the individual's best cognitive test performances should be made. When emotional and cognitive status are assessed, appropriate controls might best be sought from a group with a similar but nontoxic illness experience and/or from a genetically predisposed but never thyrotoxic group.

A variety of questions about brain neurophysiology in predisposed and formerly thyrotoxic persons has been raised by studies of brain function

340 J.E. Wallace and D.J. MacCrimmon

following thyroid gland hyperfunction. These questions merit further pursuit even though recent research emphasis has shifted from the hyperthyroid "window" on the brain to examination of the dynamics of hypothalamic-pituitary neuroregulatory mechanisms, in an effort to define associations between behavioral disorder and brain neurophysiology.

References

Alexander, F., Flagg, G., Foster, S., Clemens, T., & Blahd, W. (1961). Experimental studies of emotional stress. 1. Hyperthyroidism. *Psychosomatic Medicine, 23*, 104–114.

Alvarez, M., Gomez, A., Alavez, E., & Navarro, D. (1983). Attention disturbance in Graves' disease. *Psychoneuroendocrinology, 8*, 451–454.

Artunkal, S., & Togrol, B. (1964). Psychological studies in hyperthyroidism. In M.P. Cameron & M. O'Connor (Eds.), *Brain-Thyroid Relationships* (pp. 92–102). London: J.A. Churchill, Ltd.

Ashkar, F., Miller, R., Jacobi, G., & Naya, J. (1972). Increased incidence of hyperthyroidism in the Cuban population of greater Miami. *Journal of the Florida Medical Association, 59*, 42–43.

Bauer, M., Droba, M., & Whybrow, P. (1987). Disorders of the thyroid and parathyroid. In C. Nemeroff & P. Loosen (Eds.), *Handbook of Clinical Psychoneuroendocrinology* (pp. 41–70). New York: The Guilford Press.

Beierwaltes, W., & Ruff, G. (1958). Thyroxin and triiodothyronine in excessive dosage to euthyroid humans. *American Medical Association Archives of Internal Medicine, 101*, 569–576.

Checkley, S. (1978). Thyrotoxicosis and the course of manic-depressive illness. *British Journal of Psychiatry, 133*, 219–223.

Clower, C. (1984). Organic affective syndromes associated with thyroid dysfunction. *Psychiatric Medicine, 2*, 177–181.

Cohen, K., & Swigar, M. (1979). Thyroid function screening in psychiatric patients. *Journal of the American Medical Association, 242*, 254–257.

Dongier, M., Wittkower, E., Stephens-Newshan, L., & Hoffman, M. (1956). Psychophysiological studies in thyroid function. *Psychosomatic Medicine, 4*, 310–323.

Flagg, G., Clemens, T., Michael, E., Alexander, F., & Wark, J. (1965). A psychophysiological investigation of hyperthyroidism. *Psychosomatic Medicine, 27*, 497–507.

Forteza, M. (1973). Precipitating factors in hyperthyroidism. *Geriatrics, 28*, 123–126.

Gibson, J. (1962). Emotions and the thyroid gland: A critical appraisal. *Journal of Psychosomatic Research, 6*, 93–116.

Goldberg, E., & Comstock, G. (1976). Life events and subsequent illness. *American Journal of Epidemiology, 104*, 146–158.

Graves, R. (1940). Clinical lectures. *Medical Classics, 5*, 25-43.

Greer, S., Ramsay, I., & Bagley, C. (1973). Neurotic and thyrotoxic anxiety: Clinical, psychological, and physiological measurements. *British Journal of Psychiatry, 122*, 549–554.

Hadden, D., & McDevitt, D. (1974). Environmental stress and thyrotoxicosis: Absence of association. *Lancet, 2*, 577–578.

Ham, G., Alexander, F., & Carmichael, H. (1951). A psychosomatic theory of thyrotoxicosis. *Psychosomatic Medicine, 13*, 18–35.

Hermann, H., & Quarton, G. (1965). Psychological changes and psychogenesis in thyroid hormone disorders. *Journal of Clinical Endocrinology, 25*, 327–338.

Hoffenberg, R. (1974). Aetiology of hyperthyroidism—I. *British Medical Journal, 3*, 452–456.

Kathol, R., & Delahunt, J. (1986). The relationship of anxiety and depression to symptoms of hyperthyroidism using operational criteria. *General Hospital Psychiatry, 8*, 23–28.

Kopell, B., Wittner, W., Lunde, D., Warrick, G., & Edwards, D. (1970). Influence of triiodothyronine on selective attention in man as measured by the visual average evoked potential. *Psychosomatic Medicine, 32*, 495–502.

Korner, A., Kirkegaard, C., & Larsen, J. (1987). The thyrotropin response to thyrotropin-releasing hormone as a biological marker of suicidal risk in depressive patients. *Acta Psychiatrica Scandinavica, 76*, 355–358.

Lee, G., & Hamsher, K. (1988). Neuropsychological findings in toxicometabolic confusional states. *Journal of Clinical and Experimental Neuropsychology, 10*, 769–778.

Lidz, T. (1949). Emotional factors in the etiology of hyperthyroidism. *Psychosomatic Medicine, 11*, 2–8.

Loosen, P. (1985). The TRH-induced TSH response in psychiatric patients: A possible neuroendocrine marker. *Psychoneuroendocrinology, 10*, 237–260.

MacCrimmon, D., Wallace, J., Goldberg, W., & Streiner, D. (1979). Emotional disturbance and cognitive deficits in hyperthyroidism. *Psychosomatic Medicine, 41*, 331–340.

Mandelbrote, B., & Wittkower, E. (1955). Emotional factors in Graves' disease. *Psychosomatic Medicine, 17*, 109–123.

Markand, O. (1984). Electroencephalography in diffuse encephalopathies. *Journal of Clinical Neurophysiology, 1*, 357–407.

McLarty, D., Ratcliffe, W., Ratcliffe, J., Shimmins, J., & Goldberg, A. (1978). A study of thyroid function in psychiatric in-patients. *British Journal of Psychiatry, 133*, 211–218.

Money, J., Weinberg, R., & Lewis, V. (1966). Intelligence quotient and school performance in twenty-two children with a history of thyrotoxicosis. *Johns Hopkins Medical Journal, 118*, 275–281.

Morillo, E., & Gardner, L. (1980). Activation of latent Graves' disease in children. *Clinical Pediatrics, 19*, 160–163.

Nesse, R., Curtis, G., & Brown, G. (1982). Phobic anxiety does not affect plasma levels of thyroid stimulating hormone in man. *Psychoneuroendocrinology, 7*, 69–74.

Norman, G., McFarlane, A., & Streiner, D. (1985). Patterns of illness among individuals reporting high and low stress. *Canadian Journal of Psychiatry, 30*, 400–405.

Parsons, O., & Prigatano, G. (1978). Methodological considerations in clinical neuropsychological research. *Journal of Consulting and Clinical Psychology, 46*, 608–619.

Paykel, E. (1966). Abnormal personality and thyrotoxicosis: A follow-up study. *Journal of Psychosomatic Research, 10*, 143–150.

Perrild, H., Hansen, J., Arnung, K., Olsen, P., & Danielsen, U. (1986). Intellectual impairment after hyperthyroidism. *Acta Endocrinologica, 112*, 185–191.

Perumal, A., Halbreich, U., & Barkai, A. (1984). Modification of beta-adrenergic receptor binding in rat brain following thyroxine administration. *Neuroscience Letters, 48,* 217–221.

Robbins, L., & Vinson, D. (1960). Objective psychologic assessment of the thyrotoxic patient and the response to treatment: Preliminary report. *Journal of Clinical Endocrinology, 20,* 120–129.

Roy, A., Karoum, F., Linnoila, M., & Pickar, D. (1988). Thyrotropin releasing hormone test in unipolar depressed patients and controls: Relationship to clinical and biologic variables. *Acta Psychiatrica Scandinaviea, 77,* 151–159.

Schmidt, B., & Schultz, J. (1985). Chronic thyroxine treatment of rats downregulates the noradrenergic cyclic AMP generating system in cerebral cortex. *Journal of Pharmacology and Experimental Therapeutics, 233,* 466–472.

Semple, C., Gray, C., Borland, W., Espie, C., & Beastall, G. (1988). Endocrine effects of examination stress. *Clinical Science, 74,* 255–259.

Siersbaek-Nielsen, K., Hansen, J., Schioler, M., Kristensen, M., Stoier, M., & Olsen, P. (1972). Electroencephalographic changes during and after treatment of hyperthyroidism. *Acta Endocrinologica, 70,* 308–314.

Spaulding, W. (1963). The increased incidence of thyrotoxicosis in immigrants. *The Canadian Medical Association Journal, 88,* 287–289.

Swanson, J., Kelly, J., & McConahey, W. (1981). Neurologic aspects of thyroid dysfunction. *Mayo Clinic Proceedings, 56,* 504–512.

Trzepacz, P., McCue, M., Klein, I., Levey, G., & Greenhouse, J. (1988). A psychiatric and neuropsychological study of patients with untreated Graves' disease. *General Hospital Psychiatry, 10,* 49–55.

Tucker, D., Penland, J., Beckwith, B., & Sandstead, H. (1984). Thyroid function in normals: Influences on the electroencephalogram and cognitive performance. *Psychophysiology, 21,* 72–78.

Tunbridge, W., Evered, D., Hall, R., Appleton, D., Brewis, M., Clark, F., Grimley Evans, J., Young, E., Bird, T., & Smith, P. (1977). The spectrum of thyroid disease in a community: The Whickham Survey. *Clinical Endocrinology, 7,* 481–493.

Volpe, R. (1977). The role of autoimmunity in hypoendocrine and hyperendocrine function. *Annals of Internal Medicine, 87,* 86–99.

Wallace, J., MacCrimmon, D., & Goldberg, W. (1980). Acute hyperthyroidism: Cognitive and emotional correlates. *Journal of Abnormal Psychology, 89,* 519–527.

Weinberg, A., & Katzell, T. (1977). Thyroid and adrenal function among psychiatric patients (letter). *Lancet, 21,* 1104–1105.

White, A., & Barraclough, B. (1988). Thyroid disease and mental illness: A study of thyroid disease in psychiatric admissions. *Journal of Psychosomatic Research, 32,* 99–106.

Whybrow, P., & Prange, A. (1981). A hypothesis of thyroid-catecholamine-receptor interaction: Its relevance to affective illness. *Archives of General Psychiatry, 38,* 106–113.

Whybrow, P., Prange, A., & Treadway, C. (1969). Mental changes accompanying thyroid gland dysfunction. *Archives of General Psychiatry, 20,* 48–63.

Wilson, W., Johnson, J., & Feist, F. (1964). Thyroid hormone and brain function. II. Changes in photically elicited EEG responses following the administration of triiodothyronine to normal subjects. *Electroencephalography and Clinical Neurophysiology, 16,* 321–331.

Wilson, W., Johnson, J., & Smith, R. (1962). Affective change in thyrotoxicosis and experimental hypermetabolism. *Recent Advances in Biological Psychiatry, 4,* 234–243.

Zeitlhofer, J., Saletu, B., Stary, J., & Ahmadi, R. (1984). Cerebral function in hyperthyroid patients: Psychopathology, psychometric variables, central arousal and time perception before and after thyreostatic therapy. *Neuropsychobiology, 11,* 89–93.

Zonana, J., & Rimoin, D. (1975). Genetic disorders of the thyroid. *Medical Clinics of North America, 59,* 1263–1274.

Index

Subject Index